Smart Structures

NATO Science Series

A Series presenting the results of activities sponsored by the NATO Science Committee.
The Series is published by IOS Press and Kluwer Academic Publishers, in conjunction
with the NATO Scientific Affairs Division.

General Sub-Series

A. Life Sciences	IOS Press
B. Physics	Kluwer Academic Publishers
C. Mathematical and Physical Sciences	Kluwer Academic Publishers
D. Behavioural and Social Sciences	Kluwer Academic Publishers
E. Applied Sciences	Kluwer Academic Publishers
F. Computer and Systems Sciences	IOS Press

Partnership Sub-Series

1. Disarmament Technologies	Kluwer Academic Publishers
2. Environmental Security	Kluwer Academic Publishers
3. High Technology	Kluwer Academic Publishers
4. Science and Technology Policy	IOS Press
5. Computer Networking	IOS Press

The Partnership Sub-Series incorporates activities undertaken in collaboration with NATO's
Partners in the Euro-Atlantic Partnership Council – countries of the CIS and Central and Eastern
Europe – in Priority Areas of concern to those countries.

NATO-PCO-DATA BASE

The NATO Science Series continues the series of books published formerly in the NATO ASI
Series. An electronic index to the NATO ASI Series provides full bibliographical references (with
keywords and/or abstracts) to more than 50000 contributions from international scientists published
in all sections of the NATO ASI Series.
Access to the NATO-PCO-DATA BASE is possible via CD-ROM "NATO-PCO-DATA BASE" with
user-friendly retrieval software in English, French and German (© WTV GmbH and DATAWARE
Technologies Inc. 1989).

The CD-ROM of the NATO ASI Series can be ordered from: PCO, Overijse, Belgium.

3. High Technology – Vol. 65

Smart Structures
Requirements and Potential Applications in Mechanical and Civil Engineering

edited by

Jan Holnicki-Szulc

Institute of Fundamental Technological Research,
Polish Academy of Sciences,
Warsaw, Poland

and

José Rodellar

Department of Applied Mathematics III,
School of Civil Engineering,
Technical University of Catalunya,
Barcelona, Spain

Kluwer Academic Publishers

Dordrecht / Boston / London

Published in cooperation with NATO Scientific Affairs Division

Proceedings of the NATO Advanced Research Workshop on
Smart Structures - Requirements and Potential Applications
in Mechanical and Civil Engineering
Pultusk, Poland
June 16–19, 1998

A C.I.P. Catalogue record for this book is available from the Library of Congress.

ISBN 0-7923-5612-8 (HB)
ISBN 0-7923-5613-6 (PB)

Published by Kluwer Academic Publishers,
P.O. Box 17, 3300 AA Dordrecht, The Netherlands.

Sold and distributed in North, Central and South America
by Kluwer Academic Publishers,
101 Philip Drive, Norwell, MA 02061, U.S.A.

In all other countries, sold and distributed
by Kluwer Academic Publishers,
P.O. Box 322, 3300 AH Dordrecht, The Netherlands.

Printed on acid-free paper

TABLE OF CONTENTS

vi

PREFACE

Smart (intelligent) structures are being the subject of intense research and development in the last years. The conceptual impelling force for this research is the goal of designing and building structures able to monitor its state, detect health and failure problems, adapt its shape and properties to changing operation conditions and control its behaviour against environmental loads. Such a broad spectrum objective is requiring a multidisciplinary effort with contributions from material scientists, physicists, chemists, mathematicians and engineers of areas such as mechanical, civil, aeronautical, electrical, control, computers, etc.

The NATO Advanced Research Workshop *Smart Structures - Requirements and Potential Applications in Mechanical and Civil Engineering (SMART98)* was held in Pultusk (50 km from Warsaw, Poland) on 16-19 of June 1998. The workshop brought together more than 50 researchers from 14 countries representing several of the above disciplines. Living for three and a half days in a castle-style hotel close to River Narew, this group of researchers reported state of the art knowledge and shared new ideas, results and trends. This volume contains 43 contributions that were presented and discussed in the workshop, hoping that they will serve as a permanent reference for the community interested in smart structures. Although classification of the contributions is not always easy, we can identify the following subject groups:

- Basic research on smart materials and structures.
- Structural identification, monitoring and assessment.
- Control methods.
- Analysis and design of smart structures.

The first group concentrates mostly on composite materials with active phase (fibers from piezoelectrics or shape memory alloys and passive magnetic inclusions), reporting both theoretical and experimental results.

The second group of contributions are dominated by problems devoted to monitoring and assessment of bridges. However, issues related to supervision of historical buildings (research developed in European countries) are also discussed. On the other hand, some new methods (e.g. piezogenerated elastic waves) and their applications (e.g. magnetoelastic method) to stress fields identification and failure detection are presented.

The third group presents some methods useful for active control of structures and mechanical systems with a control theory background.

In the fourth group we can distinguish more practically oriented applications. They range from semi-active control of real, mechanical and civil engineering structures (e.g. vehicles and bridges) to fully active control applied to some special, flexible structures (e.g. tall masts, helicopter rotorcrafts, pantographs, long bridges with active cables). Some new technical applications of the concept of adaptive structure are also presented. In this concept no external energy source is needed for actuators. Controlled devices change only structural properties allowing desired redistribution and dissipation of externally imposed strain energy. The adaptive structure concept is more feasible in large structure applications and seems to be very promising in several civil engineering and mechanical problems.

We are deeply in debt with many people who have contributed to make the workshop a success: Dr. Przemyslaw Kolakowski and Prof. Grzegorz Kawiecki for their great job in the Organizing Committee; the young members of the Smart Structures Group of the Institute of Fundamental Technological Research, Dariusz Wiacek and Tomasz Bielecki, for helping with many details; Agata Holnicka and Magda Staniszewska for their continuous assistance during the workshop; and all the participants for the high level and rigor of their scientific contributions.

Finally, the financial and sponsoring support of the following institutions is highly appreciated:

- NATO Scientific Affairs Division.

- State Committee of Scientific Research, Warsaw, Poland.

- Stefan Batory Foundation, Warsaw, Poland.

- Poland Institute of Fundamental Technological Research, Warsaw, Poland.

- International Society of Structural and Multidisciplinary Optimization.

- European Panel of the Association for Control of Structures.

- Applied Research Group-Epsilon Ltd., Warsaw, Poland.

Jan Holnicki-Szulc and José Rodellar

ROBUST DECENTRALIZED H∞ CONTROL OF INPUT DELAYED INTERCONNECTED SYSTEMS

L. BAKULE AND J. BÖHM

Institute of Information Theory and Automation
Academy of Sciences of the Czech Republic
182 08 Prague, Czech Republic

1. Introduction

The paper presents a new approach to the solution of the problem of disturbances attenuation for interconnected dynamic systems by using decentralized control, such that the effect of the disturbances on the whole system is reduced to a required acceptable level. A class of uncertain, nominally linear interconnected continuoustime systems including both non-delayed and delayed inputs under unknown but norm-bounded parameter uncertainties described in the state space is considered. A decentralized, memoryless H_∞ control is designed by using the Riccati equation approach. The feedback control law guarantees the asymptotic stability of the overall closed-loop system and reduces the effect of the disturbance input on the controlled output to a prescribed level. The study of the case control design for a building structure under earthquakes excitation is presented. Though various solutions dealing with the problem of structural input delayed control appear in the literature [1], [8], [10], [13], this problem has not been solved yet [2]–[7], [11], [12], [15].

2. Problem statement

Consider the system to be controlled, including uncertainties and delay, in the form

$$\dot{x}_i(t) = (A_i + \Delta A_{ii}(t))x_i(t) + \sum_j \Delta A_{ij}(t)x_j(t) + B_i u_i(t)$$
$$+ B_{di} u_i(t - d(t)) + B_{wi} w_i, \tag{1}$$

where $x_i(t) = 0$ if $t < 0$, $i = 1, ..., N$. $x_i(t), u_i(t), w_i(t)$ is n_i-, m_i-, p_i- dimensional state, input, disturbance vector of the i–th subsystem, re-

1

J. Holnicki-Szulc and J. Rodellar (eds.), Smart Structures, 1–8.
© 1999 *Kluwer Academic Publishers. Printed in the Netherlands.*

spectively. N denotes the number of subsystems. $d(t)$ denotes a varying time delay. Suppose that all matrices in (1) satisfy the dimensionality requirements. A_i, B_i, B_{di}, B_{wi} are constant matrices. Uncertainty matrices are defined as follows:

$$\Delta A_{ii}(t) = D_{ii}F_{ii}(t)E_{ii}, \qquad \Delta A_{ij} = D_{ij}F_{ii}(t)E_{ij}, \qquad (2)$$

where $D_{ii}, D_{ij}, E_{ii}, E_{ij}$ are known real matrices of appropriate dimensions. These matrices specify the locations of uncertainties. $F_i(t) = (F_{i1} \ ... \ F_{iN})$ is an unknown matrix with Lebesgue measurable functions representing admissible uncertainties and satisfying the relation

$$F_i(t)^T F_i(t) \leq I. \qquad (3)$$

The time-variable delay is supposed to be identical for all subsystems. It is defined as

$$0 \leq d(t) \leq d^* < \infty, \qquad \dot{d}(t) \leq \overline{d} < 1, \qquad (4)$$

where $d^*, \overline{d}$ is the upper bound of the delay, and the derivative of delay $d(t)$, respectively.

We are interested in designing a state memoryless feedback controller

$$u_i(t) = -\frac{1}{\varepsilon_i} B_i P_i x_i(t) = -K_i x_i(t) \qquad (5)$$

for all i, such that the overall closed-loop system

$$\dot{x}(t) = (A - BK + \Delta A)x(t) - B_d Kx(t - d(t)) + B_w w \qquad (6)$$

satisfies, with the zero-initial condition for $x(t)$, the relation

$$\|x\|_2 \leq \gamma \|w\|_2 \qquad (7)$$

for all admissible uncertainties and all nonzero $w \in L_2[0, \infty]$. ε_i is a positive scalar, $\|.\|_2$ denotes the usual $L_2[0, \infty]$ norm. $x = (x_1^T, ..., x_N^T)^T$, $w = (w_1^T, ..., w_N^T)^T$ denotes the overall $n-$, and $p-$dimensional state, disturbance vector, respectively. The matrices in the overall system description given by (6) are defined as follows:

$$A = diag(A_1, ..., A_N), \qquad \Delta A(t) = \Delta[A_{ij}(t)], \qquad B = diag(B_1, ..., B_N),$$
$$B_d = diag(B_{d1}, ..., B_{dN}), \qquad B_w = diag(B_{w1}, ..., B_{wN}),$$
$$P = diag(P_1, ..., P_N), \qquad K = diag(K_1, ..., K_N). \qquad (8)$$

γ is a given scalar which defines the prescribed level of disturbance attenuation.

Our goal is to derive conditions for the solution of the state-space H_∞ decentralized stabilization problem for the system (1)-(4) by using state feedback (5) and to interpret the first order system result for structural systems.

3. Problem solution

Let us introduce first some necessary basic concepts. Consider the following system:

$$\dot{x} = (A + \Delta A)x + A_d x(t - d(t)) + B_w w, \tag{9}$$

where $A, \Delta A$ have the same meaning as the matrices in (6). A_d is a constant matrix.

Definition. Given a scalar $\gamma > 0$. The system (9) is said to be *quadratically stable with an H_∞- norm bound γ* (QSH) if it satisfies, for any admissible parameter uncertainty ΔA, the folllowing conditions:

1. The system is asymptotically stable.
2. Subject to the assumption of the zero-initial condition, the state x satisfies the inequality (7).

The main result is given in the form of the following theorem.

Theorem. Given a scalar $\gamma > 0$. Consider the system (1)-(4). Suppose that for some $\varepsilon_i > 0$ and $\eta_i > 0$, there exists a positive definite solution P_i of the following equation:

$$P_i A_i + A_i^T P - \frac{1}{\varepsilon_i} P_i B_i B_i^T P_i + \frac{1}{\varepsilon_i} P_i B_{di} B_{di}^T P_i + \gamma^2 P_i B_{wi} B_{wi}^T P_i$$
$$+ 2I_i + \eta_i P_i D_i P_i + \frac{1}{\eta_i} E_i = -\varepsilon_i Q_i, \quad i = 1, \ldots, N, \tag{10}$$

where $Q_i > 0$, $D_i = \sum_j D_{ij} D_{ij}^T$, $E_i = \sum_j E_{ij}^T E_{ij}$. Then the closed-loop system (1)–(5) is QSH.

Proof. Consider the asymptotic stability for zero-input disturbance. Choosing

$$L(t) = \sum_{i=1}^{N} L_i(t) = \sum_{i=1}^{N} \left(x_i^T P_i x_i + \frac{1}{\varepsilon_i} \int_{t-d}^{t} x_i^T(\tau) P_i B_i B_i^T P_i x(\tau) \, dx \right) \tag{11}$$

as a Lyapunov functional candidate for the system (1)-(5), then for positive constants c_1, c_2, the inequality

$$c_1 ||x(t)||^2 \le L(t) \le \sup_{-d^* \le d \le 0} c_2 ||x(t+d)||^2 \tag{12}$$

holds. The derivative of L along the solution of (6) is subject to some manipulations, which follow in a way of reasoning completely analogous to

4

the derivation of Proposition 1 in [6] and Theorem in [7]. It results in the inequality

$$\dot{L} = \sum_i \left[\begin{array}{c} x_i \\ \frac{1}{\varepsilon_i} B_i^T B_i x_{di} \end{array} \right]^T \left[\begin{array}{cc} S_i & -P_i B_{di} \\ -B_{di}^T P_i & -(1-\bar{d})\varepsilon_i I_{n_i} \end{array} \right] \left[\begin{array}{c} x_i \\ \frac{1}{\varepsilon_i} B_i^T B_i x_{di} \end{array} \right],$$
(13)

where $x_{di} = x_i(t - d(t))$ and

$$S_i = P_i A_i + A_i^T P - \frac{1}{\varepsilon_i} P_i B_i B_i^T P_i + \frac{1}{\varepsilon_i(1-\bar{d})} P_i B_{di} B_{di}^T P_i + P_i B_{wi} B_{wi}^T P_i$$
$$+ I_i + \eta_i P_i D_i P_i + \frac{1}{\eta_i} E_i, \quad i = 1, \dots, N.$$
(14)

By applying the same procedure as that used in [7], we conclude that the matrix in (13) is negative definite if $S_i < 0$. Thus, if there exists a positive definite solution of (10) for some ε_i, η_i such that

$$\dot{L} \leq -c_3 \| \left[\begin{array}{cc} x & \frac{1}{\varepsilon} BP x_d \end{array} \right]^T \| \leq -c_3 \|x\|^2 < 0,$$
(15)

where $\varepsilon = diag(\varepsilon_1, \dots, \varepsilon_N)$, $x_d = (x_{d1}^T, \dots, x_{dN}^T)^T$, then the closed–loop system (6) is asymptotically stable. Therefore, L is a Lyapunov functional for (1)-(5).

To prove the relation (7) for the system (6), denote the closed-loop transfer function for (6) by

$$H_{xw}(s) = (sI - A - \Delta A + \frac{1}{\varepsilon} BB^T P + \frac{1}{\varepsilon} B_d B_d^T P e^{-sd}) B_w.$$
(16)

Applying the same procedure as that used one in the derivation of Proposition 1 in [6], we obtain the relation

$$H_{xw}^*(j\omega) H_{xw}(j\omega) \leq \gamma^2 I_p,$$

which is in fact the relation (7). Q.E.D.

Let us consider with the interpretation of (1) for structural systems. We follow its derivation presented in [9].

Consider the structural dynamic system in the form

$$M\ddot{x}_m + (C + \Delta C)\dot{x}_m + (K + \Delta K)x_m = B_m u(t) + B_{dm} u(t - d(t)) + B_{wm} w = f(t),$$
(17)

where x_m is a generalized position vector, $f(t)$ is a force vector. $M = M^T$ is a mass matrix, $C = C^T \geq 0$ is a damping matrix, $K = K^T \geq 0$ is a stiffness matrix. $\Delta C(t) = \Delta C^T(t) \geq 0$, $\Delta K(t) = \Delta K^T(t) \geq 0$, for all $t \geq 0$, are the uncertainty matrices corresponding to C, K, respectively. B_m, B_{dm}, B_w are constant matrices. Suppose that

$$\Delta C(t) = D_c F_c(t) E_c, \quad \Delta K(t) = D_k F_k(t) E_k,$$
(18)

where $F_c(t)$, $F_k(t)$ satisfy the same inequality as $F_i(t)$ in (3). Denoting $x = (x_m^T, \dot{x}_m^T)^T$, we get

$$\dot{x} = \begin{bmatrix} 0 & I \\ -M^{-1}(K + \Delta K) & -M^{-1}(C + \Delta C) \end{bmatrix} x + \begin{bmatrix} 0 \\ M^{-1} \end{bmatrix} f(t). \quad (19)$$

The first order system description, when relating the matrices in (8) with (19), has the form

$$A = \begin{bmatrix} 0 & I \\ -M^{-1}K & -M^{-1}C \end{bmatrix}, \quad \Delta A = \begin{bmatrix} 0 & I \\ -M^{-1}\Delta K & -M^{-1}\Delta C \end{bmatrix},$$

$$B = \begin{bmatrix} 0 \\ M^{-1} \end{bmatrix} B_m, \quad B_d = \begin{bmatrix} 0 \\ M^{-1} \end{bmatrix} B_{dm}, \quad B_w = \begin{bmatrix} 0 \\ M^{-1} \end{bmatrix} B_{wm}. \quad (20)$$

The interconnected system decomposition corresponding to (1) has the form

$$M_i\ddot{x}_{mi} + (C_i + \Delta C_i)\dot{x}_{mi} + (K_i + \Delta K_i)x_m + \sum_{j=0, j\neq i}^{N} [M_{ij}\ddot{x}_{mj} + \Delta C_{ij}\dot{x}_{mj}$$

$$+ \Delta K_{ij}x_{mj}] = B_{mi}u_i(t) + B_{dmi}u_i(t - d(t)) + B_{wmi}w_i, \quad (21)$$

where $x_i = (x_{mi}^T, \dot{x}_{mi}^T)^T$. The relation between the matrices in (1) and (21) is straightforward. Only the corresponding indices must be supplied.

Note that the solution of (10) is independent of a particular value of time delay for the controller (5). A question arises how to understand this result because there exist generally a bound on time delay for systems with pure input delay feedback to satisfy the requirement on asymptotic stability. Suppose that the solution of (10) exists. Although the problem is continuous, the simulation of responses for a particular control design is made digitally, i.e. with a sampling determined by using a standard procedure. Then, the designer has in fact two possibilities how to proceed whith the bound on time delay. We can either to consider the bound on delay within the extent of maximally three sampling periods, as usually accepted for an effective disturbance rejection, or we may consider any finite delay without any requirement on disturbance rejection by employing only the property of guaranteed asymptotic stability. Therefore, this result demonstrates that any finite delay in the control loop does not affect the asymptotic stability of the closed–loop system when considering both the non-delayed and delayed inputs with a memoryless controller. Only the magnitude of delayed input matrices affects a solution. To compensate the effect of this term, an additional control effort entering via non-delayed term is neccesary.

The presented result has at least one important potential domain of application. It belongs to the design of reliable control systems [12], [14]. Consider the particular problem of multiple controller schemes proposed

6

by Šiljak [12]. To achieve the required reliability of controllers, using less reliable controllers an essential redundancy has been introduced by multiplicity of controllers in a parallel connection. Suppose a local controller consisting of several parallel actuators. Typically 2 or 3 parallel components are considered. An operation mode considers all parallel components with non-delayed action, while a failure mode includes some of parallel components operating with delay action. A reliability goal is the stabilization of the plant under both operation and failure modes. The system operating in a failure mode includes both nondelayed and delayed input terms simultaneously. The proposed method solves the problem in a failure mode. Note only that the standard approach supposes a total outage of some of parallel components. This case can be included as a delayed input with a large delay which in fact does not influence the distubance rejection.

4. Example

Consider a six-degree-of-freedom (DOF) building structure subject to a horizontal acceleration, as an external disturbance by Bakule *et al.* [3] in Example 3. This structure is decomposed into two disjoint subsystems (floors 1-3 and 4-6), with two actuators supplying active control forces, which are located at the 2nd and the 5th floors. Suppose that the actuator located at the 2nd floors consists of 3 parallely connected actuators. Consider a failure mode for one of parallel actuators at this floor. The experimentally identified damping and stiffness matrices in this structure have been reduced to tridiagonal matrices by neglecting off-tridiagonal terms. The open-loop system eigenvalues have not been changed by this neglection, but the simplified model does not satisfy standard modelling requirements, i.e. the sum of off-diagonal terms in C and K is not equal to the diagonal element with the opposite sign for the floors 2-5. To satisfy these requirements, the sum is considered to be a nominal value and the difference as uncertainties, i.e. $F(t)$ is considered as a constant matrix. The magnitude of uncertainties in the stiffness matrix grows when moving from the bottom to the upper floors. It confirms the physical expectations. The simulation study includes: 1) Decentralized control without delay (Fig. 1); 2) Decentralized control with the input delay in one parallel actuator component at the 2nd floor for $d = .5$ (Fig. 2). The responses of centralized controller without any delay is usually considered as a reference case for the comparison of responses. Fig. 9 in [3] can serve as such a reference.

5. Conclusion

A new extension of the problem of robust decentralized H_∞ control design with both non-delayed and delayed inputs has been derived and solved for

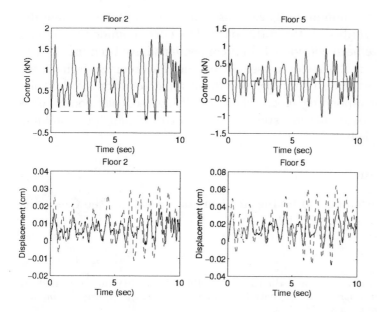

Figure 1. Figure 1. Decentralized control without delay,(——)with control, (- - -) without control.

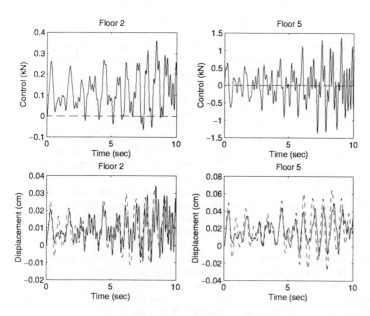

Figure 2. Figure 2. Decentralized control with delay, (——)with control, (- - -) without control.

8

a class of uncertain nominally linear interconnected continuous–time dynamic systems described in the state space. The norm bounded parameter uncertainties have been considered. Interconnections have been included as disturbance terms. The result presented by the Theorem is given as the condition of the solution of the derived Riccati-like equation. If the solution exists, a corresponding memoryless stabilizing feedback controller is designed. A case of simulation study of the control design for a building structure is supplied.

Acknowledgment

The authors were supported by the Academy of Sciences of the Czech Republic through Grant A2075802.

References

1. Agrawal, A.K., Fujino, Y. and Bhartia, B.K. (1993) Instability due to time delay and its compensation in active control of structures, *Earthquake Engineering and Structural Dynamics* **22**, 547–559.
2. Bakule, L. (1995a) Decentralized stabilization of uncertain delayed interconnected system, *7th IFAC-IFIP-IFORS Symposium on Large Scale Systems: Theory and Applications* **2**, Elsevier, 575–580.
3. Bakule, L. and Rodellar, J. (1995b) Decentralized control and overlapping of mechanical systems. Part I: System decomposition. Part II: Decentralized stabilization, *International Journal of Control* **61**, 559–587.
4. Bakule, L. and de la Sen, M. (1996) Decentralized stabilization of input delayed systems under uncertainties, *1st European Conference on Structural Systems* (A. Barrata, J. Rodellar, Eds.). World Scientific, Singapore, 40–47.
5. Cheng, Ch.-F. (1997) Disturbance attenuation for interconnected systems by decentralized control, *International Journal of Control* **66**, 213–224.
6. Choi, H.H. and Chung, M.J. (1995a) Memoryless H_∞ controller design for linear systems with delayed state and control, *Automatica* **31**, 917–919.
7. Choi, H.H. and Chung, M.J. (1995b) Memoryless stabilization of uncertain dynamic systems with time-varying delayed state and control, *Automatica* **31**, 1349–1351.
8. Chung, L.L., Lin, C.C. and Lu, K.H. (1995) Time-delay control of structures, *Earthquake Engineering and Structural Dynamics* **24**, 687–701.
9. Douglas, J. and Athans, M. (1994) Robust linear quadratic designs with real parameter uncertainty, *IEEE Transactions on Automatic Control* **39**, 107–111.
10. Lin, C.C. Sheu, J.F. and Chung, S.Y. (1996) Time-delay effect ant its solution for optimal output feedback control of structures, *Earthquake Engineering and Structural Dynamics* **25**, 547–559.
11. Petersen, I.R. and Hollot, C.V. (1986) Designing stabilizing controllers for uncertain systems using the Riccati eqation approach, *Automatica* **22**, 397–411.
12. Šiljak, D.D. (1991) *Decentralized Control of Complex Systems*, Acad. Press, N. York.
13. Udwadia, F. and Kumar, R. (1994) Time delay control of classically damped structural systems, *International Journal of Control* **60**, 687–713.
14. Veillette, R.J., Medanić, J.V. and Perkins, W.R. (1992) Design of reliable control systems, *IEEE Transactions on Automatic Control* **37**, 290–304.
15. Xie, L. and de Souza, C. (1992) Robust H_∞ control for linear systems with norm-bounded time-varying uncertainty, *IEEE Transactions on Automatic Control* **37**, 1188–1191.

SCALE INFLUENCE IN THE STATIC AND DYNAMIC BEHAVIOUR OF NO-TENSION SOLIDS

A. BARATTA
Dep. of "Scienza delle Costruzioni", University of Naples "FedericoII°"
Piazzale V. Tecchio, 80 - Fuorigrotta - 80125 - Naples - Italy

1. Introduction

In most cases, the structural analysis of masonry buildings is properly approached by the so called *no-tension model*. Starting from the early papers by Heyman [1], that had a deep impact in Italy where the problem of ancient architectural heritage is highly felt, studies on the statics and dynamics of structural systems made by materials exhibiting rather poor, *null in the limit,* tensile strength, have extensively developed (see, i.e. [2] through [8]). The further assumption, that considerably simplifies the solution of the problem, based on some agreement with actual observation, is that *compressed material exhibits indefinitely linear elastic behaviour*, apart from local problems that are not included in a first level approach.

This model is commonly referred to as *nt-elastic* (no-tension-elastic, briefly NRT). It seems rather reliable for safety assessments and succeeds in giving reason of the main features of the structural behaviour of masonry buildings (see i.e. [8]). It follows that this, relatively simple, mechanical model yields a firm reference point for technical well-founded assessment, as well as the current reinforced-concrete theory does.

After this statement is accepted, in the paper the same model is adopted to study the effect of the *scaling* of masonry buildings on their own safety and dynamic response, with the purpose to draw first conclusions about the correlation in the response of similar solids with different size, under the action of specified, possibly time-varying, loads; the treatment is intentionally confined to the case in which the original and the model solid are geometrically affine: structural systems are most often very complex, and more elaborated transformations may be rather hard to be realised. Although some preliminary result could be drawn from simple *dimensional* analysis, starting from the Buckingham's theorem (see i.e. [9]), a specialised *equational* approach is preferred, in that it yields much more details allowing to identify possible strategies to be investigated in order to improve the similitude between the model and the original.

The theory dealt with in the paper allows to manage two main problems.

The first one is the influence of the size of masonry fabrics on their static performance, and on their dynamic response, with possible application, among other, to evaluate size as a vulnerability factor in seismic risk assessment.

9

J. Holnicki-Szulc and J. Rodellar (eds.), Smart Structures, 9–18.
© 1999 *Kluwer Academic Publishers. Printed in the Netherlands.*

10

The second problem is concerned with shaking tests on small scale models of actual masonry buildings. The results in the paper yield a tool to cast the results obtained with reference to small-scale specimens onto an assessment rationale valid for actual buildings, provided that the latter can be well approached by the NRT assumption.

2. The Nt-Visco-Elastic Material

Correlation in the dynamic behaviour of natural and reduced-scale objects made by no-tension materials is investigated with reference to a rheological model in which dissipation occurs as the result of viscosity. It is assumed that the strain $\mathbf{e}$ is the sum of the *linearly elastic* strain ε_e and the inelastic component ε_f, the *fracture* strain. The stress σ is the resultant of the *elastic* and the *viscous* components, σ_c and σ_v, that are respectively proportional to the elastic strain ε_e and the elastic strain rate $\dot{\varepsilon}_e$. If $\mathbf{C}$ and $\mathbf{V}$ are the tensors of elastic and viscous constants, with $\mathbf{C} = \mathbf{D}^{-1}$

$$\varepsilon = \varepsilon_e + \varepsilon_f \quad ; \quad \sigma_c = \mathbf{C}\varepsilon_e \quad ; \quad \sigma_v = \mathbf{V}\dot{\varepsilon}_e \quad ; \quad \sigma = \sigma_c + \sigma_v \tag{1}$$

where, by the nt-assumption, σ_c and ε_f are defined in sign and the energy related to fractures is null

$$\varepsilon_f \geq 0 \quad ; \quad \sigma_c \leq 0 \tag{2}$$

$$\sigma_c \cdot \varepsilon_f = 0 \quad ; \quad \sigma_c \cdot \dot{\varepsilon}_f = 0 \quad ; \quad \sigma_v \cdot \varepsilon_f = 0 \quad ; \quad \sigma_v \cdot \dot{\varepsilon}_f = 0 \tag{3}$$

It follows trivially that the material obeys the Drucker's postulate, and therefore that all results of classical perfect elasto-plasticity with associate flow laws hold, including Limit Analysis. On the other side, dissipation is null, and the material is also (non-linearly) elastic. Hence all theorems of classical visco-elasticity hold true as well.

3. Geometrical Similarity of Scaled Bodies

Given a body $\mathbf{B} = \{x \in R^3 : f(x) \leq 0\}$ whose boundary is Ω (Fig. 1) with

$$\mathbf{x} \in \overset{o}{\mathbf{B}} \Leftrightarrow f(\mathbf{x}) < 0 \; ; \; \mathbf{x} \in \Omega \Leftrightarrow f(\mathbf{x}) = 0 \tag{4}$$

one says that the body $\mathbf{B}^\lambda$ *is scaled by* λ with respect to $\mathbf{B}$ ($\lambda > 0$) if

$$\mathbf{B}^\lambda = \left\{\mathbf{x}^\lambda \in R^3 : f\left(\mathbf{x}^\lambda / \lambda\right) \underset{DEF}{=} \hat{f}\left(\mathbf{x}^\lambda\right) \leq 0\right\} \tag{5}$$

It is easy to prove that the *correspondence between two points* P and P^λ is given by

$$P \in \mathbf{B} \to P^\lambda \in \mathbf{B}^\lambda \; ; \; P \equiv \mathbf{x} \; ; \; P^\lambda \equiv \mathbf{x}^\lambda = \lambda \mathbf{x} \tag{6}$$

with $\mathbf{x}$ the co-ordinate vector of P.

Eq. (6) *maintains directions* and *dimensional ratios*; in other words,

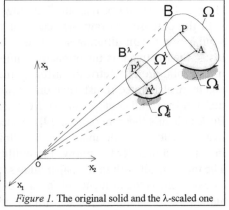

Figure 1. The original solid and the λ-scaled one

considering two points P and A and their correspondents P^λ and A^λ and two fibres AP, BQ one gets

$$A^{\overrightarrow{\lambda}P^\lambda} = \lambda \, \overrightarrow{AP} \;\; ; \;\; |AP|/|BQ| = |A^\lambda P^\lambda|/|B^\lambda Q^\lambda| \tag{7}$$

Moreover, for every point $R \in \Omega$ also $R^\lambda \in \Omega^\lambda$, and the outward normal to Ω in the point R is parallel to the normal to Ω^λ in R^λ.

4. Statics

Let consider the static equilibrium of the two bodies B and B^{λ}.

Let Ω_d be the constrained part of the boundary surface $\partial \mathsf{B}$, i. e. the part of Ω where displacements $\mathbf{u}_d(\mathbf{x})$ can be specified, $\Omega_p = \Omega - \Omega_d$ the part of Ω where surface tractions can be assigned. Let $\mathbf{u}$ and $\mathbf{r}$ be the displacement and reactions vector fields, respectively defined in B and on Ω_d, σ and ε the stress and strain tensor fields, ε_f the fractures' field, and $\mathbf{f}$ and $\mathbf{p}$ the body forces and surface tractions respectively defined in B and on Ω_p. Put

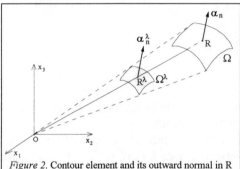

Figure 2. Contour element and its outward normal in R

$$\Omega_d^\lambda = \left\{ \mathbf{x}^\lambda \in \Omega^\lambda : \frac{\mathbf{x}^\lambda}{\lambda} \in \Omega \right\} \tag{8}$$

Assume the following hypotheses:
1) The constitutive law is the same both for the model B^λ and the original body B, i.e. eqs. (1) to (3) hold for both bodies. *In practice, the model is built up by the same material as the original..*
2) Body forces are the same both for the model B^λ and the original body B, i.e.

$$\mathbf{f}^\lambda\left(\mathbf{x}^\lambda\right) = \mathbf{f}\left(\mathbf{x}^\lambda / \lambda\right) = \mathbf{f}(\mathbf{x}) \tag{9}$$

In practice, the force on the unit volume is the same both in the model and in the original.
3) The intensity of surface tractions is reduced according to the factor λ, i.e.

$$\mathbf{p}^\lambda\left(\mathbf{x}^\lambda\right) = \lambda \mathbf{p}(\mathbf{x}^\lambda / \lambda) = \lambda \mathbf{p}(\mathbf{x}) \tag{10}$$

In practice, the tractions applied by contact on the unit surface on the boundary of the model result in forces equal to the original multiplied by λ.
4) Imposed displacements are reduced according to the factor λ^2, i.e.

$$\mathbf{u}_d^\lambda\left(\mathbf{x}^\lambda\right) = \lambda^2 \mathbf{u}_d\left(\mathbf{x}\right) \tag{11}$$

After these assumptions, it is possible to assess that the solution $[\mathbf{u}^\lambda(\mathbf{x}^\lambda), \varepsilon_f^\lambda(\mathbf{x}^\lambda)]$ of the equilibrium problem in the model is related to the solution $[\mathbf{u}(\mathbf{x}), \varepsilon_f(\mathbf{x})]$ in the original

12

body by the relationships

$$\mathbf{u}^\lambda\left(\mathbf{x}^\lambda\right) = \lambda^2 \mathbf{u}\left(\mathbf{x}^\lambda / \lambda\right) = \lambda^2 \mathbf{u}(\mathbf{x})$$

$$\boldsymbol{\varepsilon}_f^\lambda\left(\mathbf{x}^\lambda\right) = \lambda \boldsymbol{\varepsilon}_f\left(\mathbf{x}^\lambda / \lambda\right) = \lambda \boldsymbol{\varepsilon}_f(\mathbf{x})$$

and

$$\varepsilon^\lambda\left(\mathbf{x}^\lambda\right) = \lambda \varepsilon\left(\mathbf{x}^\lambda / \lambda\right) = \lambda \varepsilon(\mathbf{x})$$

$$\sigma^\lambda\left(\mathbf{x}^\lambda\right) = \lambda \sigma\left(\mathbf{x}^\lambda / \lambda\right) = \lambda \sigma(\mathbf{x}) \qquad (12)$$

$$\mathbf{r}^\lambda\left(\mathbf{x}^\lambda\right) = \lambda \mathbf{r}\left(\mathbf{x}^\lambda / \lambda\right) = \lambda \mathbf{r}(\mathbf{x})$$

In fact, considering that

$$\sigma_{ij,k}^\lambda\left(\mathbf{x}^\lambda\right) = \frac{\partial \sigma_{ij}^\lambda\left(\mathbf{x}^\lambda\right)}{\partial x_k^\lambda} = \lambda \frac{\partial \sigma_{ij}\left(\lambda \mathbf{x}/\lambda\right)}{\partial x_k^\lambda} = \lambda \sum_{r=1}^{3} \frac{\partial \sigma_{ij}(\mathbf{x})}{\partial x_r} \frac{\partial x_r}{\partial x_k^\lambda} \qquad (13)$$

one can agree that if [$\mathbf{u}(\mathbf{x})$, $\sigma(\mathbf{x})$, $\varepsilon(\mathbf{x})$, $\varepsilon_f(\mathbf{x})$, $\mathbf{r}(\mathbf{x})$] is the solution of the full-size problem, then [$\mathbf{u}^\lambda(\mathbf{x})$, $\sigma^\lambda(\mathbf{x})$, $\varepsilon^\lambda(\mathbf{x})$, $\varepsilon^\lambda_f(\mathbf{x})$, $\mathbf{r}^\lambda(\mathbf{x})$], as given in eqs. (12), is the solution of the reduced size problem.

In other words, *full similitude*, in the sense of eq.(12), is to be expected from static tests on no-tension models.

5. Scaling in Nt-Visco-Elastodynamics

5.1. FULL-SCALE BODY

Let consider the dynamic equilibrium problem of the full-scale body **B** made by nt-elastic material, with given boundary and initial conditions.

Let $\mathbf{u}(\mathbf{x},t)$, $\sigma(\mathbf{x},t)$, $\varepsilon(\mathbf{x},t)$, $\varepsilon_f(\mathbf{x},t)$ and $\mathbf{r}(\mathbf{x},t)$ be the response of the original body under permanent loads $\mathbf{f}_o$ ($\mathbf{x}$), $\mathbf{p}_o$ ($\mathbf{x}$) and time-varying actions $\mathbf{f}_t$ ($\mathbf{x},t$), $\mathbf{p}_t$ ($\mathbf{x},t$). It is assumed that permanent body forces $\mathbf{f}_o$ ($\mathbf{x}$) are gravitational in essence, so $\mathbf{f}_o$ ($\mathbf{x}$) = $\rho(\mathbf{x})$ **g**, $\rho(\mathbf{x})$ being the specific mass of the material and **g** being the vector of gravity acceleration. The response of the system obeys the following equations

$$\begin{cases} \mathbf{Div}\,\sigma\left(\mathbf{x},t\right) + \mathbf{f}_t\left(\mathbf{x},t\right) + \mathbf{f}_o\left(\mathbf{x}\right) - \rho\left(\mathbf{x}\right)\ddot{\mathbf{u}}\left(\mathbf{x},t\right) = 0 & in\ \overset{\circ}{\mathbf{B}} \\ \sigma\left(\mathbf{x},t\right)\alpha_n\left(\mathbf{x}\right) = \mathbf{p}_o\left(\mathbf{x}\right) + \mathbf{p}_t\left(\mathbf{x},t\right) & on\ \Omega_p \\ \sigma\left(\mathbf{x},t\right)\alpha_n\left(\mathbf{x}\right) = \mathbf{r}\left(\mathbf{x},t\right) & on\ \Omega_d \\ \varepsilon\left(\mathbf{x},t\right) = \nabla\mathbf{u}\left(\mathbf{x},t\right)\ ;\ \dot{\varepsilon}\left(\mathbf{x},t\right) = \nabla\dot{\mathbf{u}}\left(\mathbf{x},t\right) & in\ \mathbf{B} \end{cases} \qquad (14)$$

complying with

$$\sigma_c\left(\mathbf{x},t\right) = \mathbf{C}\left(\mathbf{x}\right)\left[\varepsilon\left(\mathbf{x},t\right) - \varepsilon_f\left(\mathbf{x},t\right)\right]\ ;\ \sigma_v\left(\mathbf{x},t\right) = \mathbf{V}\left(\mathbf{x}\right)\left[\dot{\varepsilon}\left(\mathbf{x},t\right) - \dot{\varepsilon}_f\left(\mathbf{x},t\right)\right]$$

$$\sigma\left(\mathbf{x},t\right) = \sigma_c\left(\mathbf{x},t\right) + \sigma_v\left(\mathbf{x},t\right)\ ;\ \sigma_c\left(\mathbf{x},t\right) \le 0\ ;\ \varepsilon_f\left(\mathbf{x},t\right) \ge 0 \qquad (15)$$

$$\sigma_c\left(\mathbf{x},t\right)\cdot\varepsilon_f\left(\mathbf{x},t\right) = \sigma_v\left(\mathbf{x},t\right)\cdot\varepsilon_f\left(\mathbf{x},t\right) = 0\ ;\ \sigma_c\left(\mathbf{x},t\right)\cdot\dot{\varepsilon}_f\left(\mathbf{x},t\right) = \sigma_v\left(\mathbf{x},t\right)\cdot\dot{\varepsilon}_f\left(\mathbf{x},t\right) = 0$$

and with initial and boundary conditions (being T the duration of the disturbance)

$$\mathbf{u}\left(\mathbf{x},0\right) = \mathbf{u}_o\left(\mathbf{x}\right)\ ;\ \dot{\mathbf{u}}\left(\mathbf{x},0\right) = \dot{\mathbf{u}}_o\left(\mathbf{x}\right)\quad \forall \mathbf{x} \in \mathbf{B}$$

$$\mathbf{u}\left(\mathbf{x},t\right) = \mathbf{u}_g\left(t\right)\ ;\ \ddot{\mathbf{u}}\left(\mathbf{x},t\right) = \ddot{\mathbf{u}}_g\left(t\right)\quad \forall \mathbf{x} \in \Omega_d\ ;\ \forall t \in \left(0,T\right) \qquad (16)$$

5.2. SCALED BODY

Let $\mathbf{u}^\lambda\left(\mathbf{x}^\lambda,t^\lambda\right)$, $\boldsymbol{\varepsilon}^\lambda\left(\mathbf{x}^\lambda,t^\lambda\right)$, $\boldsymbol{\sigma}^\lambda\left(\mathbf{x}^\lambda,t^\lambda\right)$, $\boldsymbol{\varepsilon}_f^\lambda\left(\mathbf{x}^\lambda,t^\lambda\right)$ and $\mathbf{r}^\lambda\left(\mathbf{x}^\lambda,t^\lambda\right)$ be the response of the scaled body under permanent loads $\mathbf{f}^\lambda{}_o$ (x), $\mathbf{p}^\lambda{}_o(x)$ and time-varying actions $\mathbf{f}^\lambda{}_t(\mathbf{x}^\lambda,\ t^\lambda)$, $\mathbf{p}^\lambda{}_t(\mathbf{x}^\lambda,t^\lambda)$. The motion of the body is ruled by the following equations

$$
\left\{
\begin{aligned}
& \mathbf{Div}^\lambda\,\boldsymbol{\sigma}^\lambda\left(\mathbf{x}^\lambda,t^\lambda\right)+\mathbf{f}_t^\lambda\left(\mathbf{x}^\lambda,t^\lambda\right)+\mathbf{f}_o^\lambda\left(\mathbf{x}^\lambda\right)-\rho^\lambda\left(\mathbf{x}^\lambda\right)\ddot{\mathbf{u}}^\lambda\left(\mathbf{x}^\lambda,t^\lambda\right)=0 && in\ \overset{\circ}{\mathbf{B}}{}^\lambda \\
& \boldsymbol{\sigma}^\lambda\left(\mathbf{x}^\lambda,t^\lambda\right)\boldsymbol{\alpha}_n^\lambda\left(\mathbf{x}^\lambda\right)=\mathbf{p}_o^\lambda\left(\mathbf{x}^\lambda\right)+\mathbf{p}_t^\lambda\left(\mathbf{x}^\lambda,t^\lambda\right) && on\ \Omega_p^\lambda \\
& \boldsymbol{\sigma}^\lambda\left(\mathbf{x}^\lambda,t^\lambda\right)\boldsymbol{\alpha}_n^\lambda\left(\mathbf{x}^\lambda\right)=\mathbf{r}^\lambda\left(\mathbf{x}^\lambda,t^\lambda\right) && on\ \Omega_d^\lambda \\
& \boldsymbol{\varepsilon}^\lambda\left(\mathbf{x}^\lambda,t^\lambda\right)=\nabla^\lambda\mathbf{u}^\lambda\left(\mathbf{x}^\lambda,t^\lambda\right)\ ;\quad \dot{\boldsymbol{\varepsilon}}^\lambda\left(\mathbf{x}^\lambda,t^\lambda\right)=\nabla^\lambda\dot{\mathbf{u}}^\lambda\left(\mathbf{x}^\lambda,t^\lambda\right) && in\ \mathbf{B}^\lambda
\end{aligned}
\right.
\tag{17}
$$

with the constitutive equations

$$
\boldsymbol{\sigma}_c^\lambda\left(\mathbf{x}^\lambda,t^\lambda\right)=\mathbf{C}^\lambda\left(\mathbf{x}^\lambda\right)\left[\boldsymbol{\varepsilon}^\lambda\left(\mathbf{x}^\lambda,t^\lambda\right)-\boldsymbol{\varepsilon}_f^\lambda\left(\mathbf{x}^\lambda,t^\lambda\right)\right]
$$

$$
\boldsymbol{\sigma}_v^\lambda\left(\mathbf{x}^\lambda,t^\lambda\right)=\mathbf{V}^\lambda\left(\mathbf{x}^\lambda\right)\left[\dot{\boldsymbol{\varepsilon}}^\lambda\left(\mathbf{x}^\lambda,t^\lambda\right)-\dot{\boldsymbol{\varepsilon}}_f^\lambda\left(\mathbf{x}^\lambda,t^\lambda\right)\right]\ ;\quad \boldsymbol{\varepsilon}_f^\lambda\left(\mathbf{x}^\lambda,t^\lambda\right)\geq 0
$$

$$
\boldsymbol{\sigma}^\lambda\left(\mathbf{x}^\lambda,t^\lambda\right)=\boldsymbol{\sigma}_c^\lambda\left(\mathbf{x}^\lambda,t^\lambda\right)+\boldsymbol{\sigma}_v^\lambda\left(\mathbf{x}^\lambda,t^\lambda\right)\ ;\quad \boldsymbol{\sigma}_c^\lambda\left(\mathbf{x}^\lambda,t^\lambda\right)\leq 0
\tag{18}
$$

$$
\boldsymbol{\sigma}_c^\lambda\left(\mathbf{x}^\lambda,t^\lambda\right)\cdot\boldsymbol{\varepsilon}_f^\lambda\left(\mathbf{x}^\lambda,t^\lambda\right)=\boldsymbol{\sigma}_v^\lambda\left(\mathbf{x}^\lambda,t^\lambda\right)\cdot\boldsymbol{\varepsilon}_f^\lambda\left(\mathbf{x}^\lambda,t^\lambda\right)=0
$$

$$
\boldsymbol{\sigma}_c^\lambda\left(\mathbf{x}^\lambda,t^\lambda\right)\cdot\dot{\boldsymbol{\varepsilon}}_f^\lambda\left(\mathbf{x}^\lambda,t^\lambda\right)=\boldsymbol{\sigma}_v^\lambda\left(\mathbf{x}^\lambda,t^\lambda\right)\cdot\dot{\boldsymbol{\varepsilon}}_f^\lambda\left(\mathbf{x}^\lambda,t^\lambda\right)=0
$$

and initial and boundary conditions

$$
\mathbf{u}^\lambda\left(\mathbf{x}^\lambda,0\right)=\mathbf{u}_0^\lambda\left(\mathbf{x}^\lambda\right)\ ;\quad \dot{\mathbf{u}}^\lambda\left(\mathbf{x}^\lambda,0\right)=\dot{\mathbf{u}}_0^\lambda\left(\mathbf{x}^\lambda\right)\qquad \forall\mathbf{x}^\lambda\in\mathbf{B}^\lambda
$$

$$
\mathbf{u}^\lambda\left(\mathbf{x}^\lambda,t^\lambda\right)=\mathbf{u}_g^\lambda\left(t^\lambda\right)\ ;\ \dot{\mathbf{u}}^\lambda\left(\mathbf{x}^\lambda,t^\lambda\right)=\dot{\mathbf{u}}_g^\lambda\left(t^\lambda\right)\ \forall\mathbf{x}^\lambda\in\Omega_d^\lambda,\ \forall t^\lambda\in\left(0,T^\lambda\right)
\tag{19}
$$

T^λ being the duration of the transient loads on the scaled structure.

5.3. CORRELATION BETWEEN THE MODEL AND THE ORIGINAL RESPONSE

It is assumed that the scaled body is made by the same material as the original body, with a chance to control the dissipation matrix $\mathbf{V}^\lambda$; i.e. it is assumed that $\mathbf{C}^\lambda(\mathbf{x}^\lambda) = \mathbf{C}(\mathbf{x}^\lambda/\lambda)$ while $\mathbf{V}^\lambda(\mathbf{x}^\lambda) = \beta\mathbf{V}(\mathbf{x}^\lambda/\lambda)$ with $\beta \geq 0$, whence

$$
\left\{
\begin{aligned}
& \boldsymbol{\varepsilon}=\boldsymbol{\varepsilon}_e+\boldsymbol{\varepsilon}_f\ ;\quad \boldsymbol{\sigma}_c=\mathbf{C}\boldsymbol{\varepsilon}_e\ ;\quad \boldsymbol{\sigma}_v=\mathbf{V}\dot{\boldsymbol{\varepsilon}}_e\ ;\quad \boldsymbol{\sigma}=\boldsymbol{\sigma}_c+\boldsymbol{\sigma}_v \\
& \boldsymbol{\varepsilon}^\lambda=\boldsymbol{\varepsilon}_e^\lambda+\boldsymbol{\varepsilon}_f^\lambda\ ;\quad \boldsymbol{\sigma}_c^\lambda=\mathbf{C}\boldsymbol{\varepsilon}_e^\lambda\ ;\quad \boldsymbol{\sigma}_v^\lambda=\beta\mathbf{V}\dot{\boldsymbol{\varepsilon}}_e^\lambda\ ;\quad \boldsymbol{\sigma}^\lambda=\boldsymbol{\sigma}_c^\lambda+\boldsymbol{\sigma}_v^\lambda \\
& \boldsymbol{\varepsilon}_f\geq 0\ ;\ \boldsymbol{\sigma}_c\leq 0\ ;\ \boldsymbol{\sigma}_c\cdot\boldsymbol{\varepsilon}_f=0\ ;\ \boldsymbol{\sigma}_c\cdot\dot{\boldsymbol{\varepsilon}}_f=0\ ;\ \boldsymbol{\sigma}_v\cdot\boldsymbol{\varepsilon}_f=0\ ;\ \boldsymbol{\sigma}_v\cdot\dot{\boldsymbol{\varepsilon}}_f=0 \\
& \boldsymbol{\varepsilon}_f^\lambda\geq 0\ ;\ \boldsymbol{\sigma}_c^\lambda\leq 0\ ;\ \boldsymbol{\sigma}_c^\lambda\cdot\boldsymbol{\varepsilon}_f^\lambda=0\ ;\ \boldsymbol{\sigma}_c^\lambda\cdot\dot{\boldsymbol{\varepsilon}}_f^\lambda=0\ ;\ \boldsymbol{\sigma}_v^\lambda\cdot\boldsymbol{\varepsilon}_f^\lambda=0\ ;\ \boldsymbol{\sigma}_v^\lambda\cdot\dot{\boldsymbol{\varepsilon}}_f^\lambda=0
\end{aligned}
\right.
\tag{20}
$$

By the same reason, $\rho^\lambda(\mathbf{x}^\lambda) = \rho(\mathbf{x}^\lambda/\lambda)$ and one is forced to put $\mathbf{f}^\lambda{}_o(\mathbf{x}^\lambda) = \mathbf{f}_o(\mathbf{x}^\lambda/\lambda) = \rho(\mathbf{x})\mathbf{g}$. Assume the following hypotheses as in Sec. 4, specialised to the dynamic case:

1) Permanent body forces are the same both for the model B^λ and the original B

$$\mathbf{f}_0^\lambda\left(\mathbf{x}^\lambda\right) = \mathbf{f}_0\left(\mathbf{x}^\lambda / \lambda\right) = \mathbf{f}_0(\mathbf{x}) \tag{21}$$

2) Time-varying body forces are the sum of a term similar to $\mathbf{f}_t$ $(\mathbf{x},t)$ apart from a contraction in the time scale plus a correction term proportional to the divergence of the viscous stress. With $t^\lambda = \lambda t$

$$\mathbf{f}_t^\lambda\left(\mathbf{x}^\lambda,t^\lambda\right) = \mathbf{f}_t\left(\mathbf{x}^\lambda / \lambda, t^\lambda / \lambda\right) + \frac{(\lambda - \beta)}{\beta} \mathbf{Div}^\lambda \sigma_v^\lambda\left(\mathbf{x}^\lambda,t^\lambda\right) = \mathbf{f}_t(\mathbf{x},t) + \frac{(\lambda - \beta)}{\lambda} \mathbf{Div}\,\sigma_v(\mathbf{x},t) \tag{22}$$

3) Permanent surface tractions are reduced by the scale factor λ

$$\mathbf{p}_o^\lambda\left(\mathbf{x}^\lambda\right) = \lambda \mathbf{p}_o\left(\mathbf{x}^\lambda / \lambda\right) = \lambda \mathbf{p}_o(\mathbf{x}) \tag{23}$$

4) Time-varying surface tractions are reduced by the scale factor λ, are contracted in the time scale and are corrected by a term proportional to the surface dissipative stress vector

$$\mathbf{p}_t^\lambda\left(\mathbf{x}^\lambda,t^\lambda\right) = \lambda \mathbf{p}_t\left(\mathbf{x}^\lambda / \lambda, t^\lambda / \lambda\right) + \frac{(\beta - \lambda)}{\beta} \sigma_v^\lambda\left(\mathbf{x}^\lambda,t^\lambda\right)\alpha_n^\lambda\left(\mathbf{x}^\lambda\right) = \lambda \mathbf{p}_t(\mathbf{x},t) + (\beta - \lambda)\sigma_v(\mathbf{x},t)\alpha_n(\mathbf{x}) \tag{24}$$

5) Imposed displacements are reduced according to the factor λ^2 and are contracted in the time scale

$$\mathbf{u}_d^\lambda\left(\mathbf{x}^\lambda,t^\lambda\right) = \lambda^2 \mathbf{u}_d\left(\mathbf{x},t\right) \tag{25}$$

After these assumptions, it is possible to assess that the solution $[\mathbf{u}^\lambda(\mathbf{x}^\lambda,t^\lambda),\varepsilon_f^\lambda(\mathbf{x}^\lambda,t^\lambda)]$ of the dynamic problem in the model is related to the solution $[\mathbf{u}(\mathbf{x},t),\varepsilon_f(\mathbf{x},t)]$ in the original body by the relationships

$$\mathbf{u}^\lambda\left(\mathbf{x}^\lambda,t^\lambda\right) = \lambda^2 \mathbf{u}\left(\mathbf{x}^\lambda / \lambda, t^\lambda / \lambda\right) = \lambda^2 \mathbf{u}(\mathbf{x},t) \;;\; \varepsilon_f^\lambda\left(\mathbf{x}^\lambda,t^\lambda\right) = \lambda \varepsilon_f\left(\mathbf{x}^\lambda / \lambda, t^\lambda / \lambda\right) = \lambda \varepsilon_f(\mathbf{x},t)$$

$$\varepsilon^\lambda\left(\mathbf{x}^\lambda,t^\lambda\right) = \lambda \varepsilon\left(\mathbf{x}^\lambda / \lambda, t^\lambda / \lambda\right) = \lambda \varepsilon(\mathbf{x},t) \;;\; \dot{\varepsilon}^\lambda\left(\mathbf{x}^\lambda,t^\lambda\right) = \dot{\varepsilon}(\mathbf{x},t) \;;\; \dot{\varepsilon}_f^\lambda\left(\mathbf{x}^\lambda,t^\lambda\right) = \dot{\varepsilon}_f(\mathbf{x},t)$$

$$\sigma_c^\lambda\left(\mathbf{x}^\lambda,t^\lambda\right) = \lambda \sigma_c(\mathbf{x},t) \;;\; \sigma_v^\lambda\left(\mathbf{x}^\lambda,t^\lambda\right) = \beta \sigma_v(\mathbf{x},t) \;;\; \sigma^\lambda\left(\mathbf{x}^\lambda,t^\lambda\right) = \lambda \sigma_c(\mathbf{x},t) + \beta \sigma_v\left(\mathbf{x},t^\lambda\right) \tag{26}$$

$$\mathbf{r}^\lambda\left(\mathbf{x}^\lambda,t^\lambda\right) = \lambda \mathbf{r}(\mathbf{x},t) + \frac{\beta - \lambda}{\beta} \sigma_v^\lambda\left(\mathbf{x}^\lambda,t^\lambda\right)\alpha_n^\lambda\left(\mathbf{x}^\lambda\right)$$

Finally, it can be assessed that *the response of the model* $[\sigma^\lambda(\mathbf{x}^\lambda,t^\lambda),\varepsilon^\lambda(\mathbf{x}^\lambda,t^\lambda),\varepsilon_f^\lambda(\mathbf{x}^\lambda,t^\lambda),\mathbf{u}^\lambda(\mathbf{x}^\lambda,t^\lambda),\mathbf{r}^\lambda(\mathbf{x}^\lambda,t^\lambda)]$ *is related to the response of the original* $[\sigma(\mathbf{x},t),\varepsilon(\mathbf{x},t),\varepsilon_f(\mathbf{x},t),\mathbf{u}(\mathbf{x},t),\mathbf{r}(\mathbf{x},t)]$ *by the relationships (26) provided that the applied loads and imposed displacements are adapted as specified by eqs. (21) to (25).*

6. The role of active control technology

After the previous discussion, the problem is: how to apply corrective entities whose expression is given by eqs. (21) to (25)? This is the point where the experience gained by active control technology may help in making effective the approach presented in the previous sections. Consider in fact that one has set up the small model of a building on a shaking table (Fig. 3), and that one aims at predicting the seismic response of the original full-scale body under null surface forces $\mathbf{p}_o(\mathbf{x})$ and $\mathbf{p}_t(\mathbf{x},t)$, and under body forces $\mathbf{f}_o(\mathbf{x})$ and $\mathbf{f}_t(\mathbf{x},t)$.

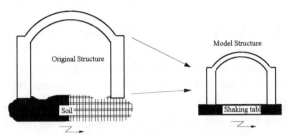

Figure 3. The original building under an earthquake and its scaled model on the shaking table

The time-varying body forces are dependent on some three-components ground accleration $\mathbf{a}(t)$, so that one can assume that $\mathbf{f}_t(\mathbf{x},t) = -\rho(\mathbf{x})\,\mathbf{a}(t)$, while the permanent body forces are given by the self-weight $\mathbf{f}_o(\mathbf{x}) = \rho(\mathbf{x})\mathbf{g}$, $\mathbf{g}$ being the gravity acceleration vector.

It is clear that, since $\mathbf{p}_o(\mathbf{x}) \equiv \mathbf{0}$ and the model is made by th same material as the full-scale original body, $\beta = 1$, $\rho^\lambda = \rho$ and eqs. (21) and (23) are ininfluent, while eq. (22) requires that the instantaneous body forces are given by

$$\mathbf{f}_t^\lambda\!\left(\mathbf{x}^\lambda,t^\lambda\right) = -\rho\!\left(\mathbf{x}^\lambda / \lambda\right)\mathbf{a}\!\left(t^\lambda / \lambda\right) + (\lambda - 1)\mathbf{Div}^\lambda\boldsymbol{\sigma}_v^\lambda\!\left(\mathbf{x}^\lambda,t^\lambda\right) \tag{27}$$

i.e. it is the sum of one term that reproduces the actual inertia forces at a contracted time-scale, and a second term that is the correction to be applied to compensate the dissipation effect. Eq. (24) similarly requires that additional surface forces are applied. Remembering that $\mathbf{p}_t(\mathbf{x},t) \equiv \mathbf{0}$, eq. (24) yields

$$\mathbf{p}_t^\lambda\!\left(\mathbf{x}^\lambda,t^\lambda\right) = (\lambda - 1)\boldsymbol{\sigma}_v^\lambda\!\left(\mathbf{x}^\lambda,t^\lambda\right)\boldsymbol{\alpha}_n^\lambda\!\left(\mathbf{x}^\lambda,t^\lambda\right) \tag{28}$$

The instantaneous resultant of the corrective forces is

$$\mathbf{R}\!\left(t^\lambda\right) = (\lambda - 1)\left[\int_{B^\lambda} \mathbf{Div}^\lambda\boldsymbol{\sigma}_v^\lambda\!\left(\mathbf{x}^\lambda,t^\lambda\right)dV^\lambda - \int_{\Omega_p^\lambda} \boldsymbol{\sigma}_v^\lambda\!\left(\mathbf{x}^\lambda,t^\lambda\right)\boldsymbol{\alpha}_n^\lambda\!\left(\mathbf{x}^\lambda\right)dA^\lambda \right] \tag{29}$$

Considering that, by the Theorem of Divergence

$$\int_{B^\lambda} \mathbf{Div}^\lambda\boldsymbol{\sigma}_v^\lambda\!\left(\mathbf{x}^\lambda,t^\lambda\right)dV^\lambda = \int_{\Omega^\lambda} \boldsymbol{\sigma}_v^\lambda\!\left(\mathbf{x}^\lambda,t^\lambda\right)\boldsymbol{\alpha}_n^\lambda\!\left(\mathbf{x}^\lambda\right)dA^\lambda \tag{30}$$

it is concluded that at any instant of the motion

$$\mathbf{R}\!\left(t^\lambda\right) = (\lambda - 1)\int_{\Omega_d^\lambda} \boldsymbol{\sigma}_v^\lambda\!\left(\mathbf{x}^\lambda,t^\lambda\right)\boldsymbol{\alpha}_n^\lambda\!\left(\mathbf{x}^\lambda\right)dA^\lambda \tag{31}$$

which means that, comparing with the last eq. (26), the resultant of the corrective forces at the connection between the table and the model equals the resultant of the active corrective forces, $\mathbf{R}(t^\lambda) = -\mathbf{R}_r(t^\lambda)$. The latter can then be applied, in the respect of the global equilibrium of the system, by suitably modifying on-line the movement of the shaking table. The equivalent base acceleration $\mathbf{a}^\lambda_e(t^\lambda)$ should then be found by equating the resultant of the body forces due to such acceleration to the resultant of the body and surface corrective forces plus the scaled accelerogram, whence

$$\mathbf{a}_e^\lambda\left(t^\lambda\right) = \mathbf{a}\left(t^\lambda/\lambda\right) + \chi^\lambda\left(t^\lambda\right) \tag{32}$$

with

$$\chi^\lambda\left(t^\lambda\right) = \frac{1}{M^\lambda}(\lambda-1)\left[\int_{\Omega_p}\sigma_v^\lambda\left(\mathbf{x}^\lambda,t^\lambda\right)\alpha_n^\lambda\left(\mathbf{x}^\lambda,t^\lambda\right)dA - \int_{B^\lambda}\mathbf{Div}^\lambda\sigma_v^\lambda\left(\mathbf{x}^\lambda,t^\lambda\right)dV\right] \tag{33}$$

and M^λ the global mass of the model on the table. The righthand member in eq. (33) depends on the response of the model. If the instantaneous (compressive) stress rates $\dot\sigma_c^\lambda$ at a number of suitably chosen points (e.g. Gauss-points) are monitored during the experiment, the correspondent elastic strain rate and the relevant viscous stress can be recorded

$$\dot\varepsilon_e^\lambda = \mathbf{C}^{-1}\dot\sigma_c^\lambda \ ; \ \sigma_v^\lambda = \beta\mathbf{V}\dot\varepsilon_e^\lambda = \mathbf{VC}^{-1}\dot\sigma_c^\lambda \tag{34}$$

whence the integrals at the second member of eq. (33) can be evaluated through any quadrature formula correspondent to the locations of the monitored points.

Since the shaking table possesses its own dynamics, eq. (32) dos not directly yield the input for the motion basement. The input must be first filtered through the inverse transfer functiion (TF) of the table-specimen composed system, and the *drive* to be supplied to the table is given by

$$\overline{\mathbf{a}}_e^\lambda\left(t^\lambda\right) = \overline{\mathbf{a}}\left(t^\lambda/\lambda\right) + \overline{\chi}^\lambda\left(t^\lambda\right) \tag{35}$$

with

$$\overline{\mathbf{a}}\left(t^\lambda/\lambda\right) = \mathrm{TF}^{-1}\left[\mathbf{a}\left(t^\lambda/\lambda\right)\right] \ ; \ \overline{\chi}^\lambda\left(t^\lambda\right) = \mathrm{TF}^{-1}\left[\overline{\chi}^\lambda\left(t^\lambda\right)\right] \tag{36}$$

Thus, the final strategy is to add to the control of the shake-table (LMS system) a feedback control (FCS) allowing to correct in real time the expected acceleration drive $\overline{\mathbf{a}}(t^\lambda/\lambda)$ by the term $\overline{\chi}^\lambda(t^\lambda)$, according to the block diagram in Fig.4, where the analogue machine performs all calculations necessary to evaluate eq. (33).

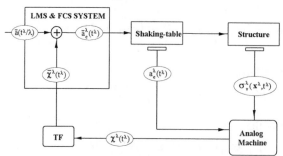

Figure 4. Block diagram of the control system

The control system aims at supplying the *drive* accelerogram $\overline{\mathbf{a}}_e(t^\lambda/\lambda)$ as given by eq (36) to the shaking table.

7. Conclusions

The results presented in the paper allow to deal with the two main problems announced in the Introduction.

The first problem consists in comparing two similar buildings, different only in regard to their size. Consider the larger one as the *primary*, and the smaller as the *secondary* building, i.e. the latter is scaled with $\lambda < 1$.

If the static performance is considered, full similarity holds according to eqs. (12). Clearly, stresses are reduced by the factor λ in the secondary building, so the latter would be safer than the primary building. Anyway, it is generally agreed that in masonry fabrics crushing is not the main source of static collapse (see e.g. [1]), and the increase in compressive stress makes not a practical difference between the two buildings. Indeed, the attenuation in the stress may be less than $(1-\lambda)$, in that in general surface tractions (the *accidental* loads) are not reduced as assumed in eq.(10). Most important is the fact that displacements are increased by a factor $1/\lambda^2$, which makes larger buildings prone to loss of stability by geometrical second-order effects much more than smaller buildings. So one can conclude that the smaller structure is intrinsically safer than the larger one, both because it suffers less intensive stress and because the size factor is a source of geometrical instability.

Consider now the performance of the two buildings acted on by the same dynamical disturbance $f_t(x,t)$; for instance an earthquake that strikes on a site where both buildings stand up. A first statement is that, since the statical equilibrium is less stable for the primary building, the latter is more prone to collapse after the dynamical force has exhausted its action and the building remains damaged by shaking. The second statement is concerned with the dynamic response, and it is somewhat complementary with the previous one. For simplicity, although it is to be intended that the two buildings are made by the same materials, assume nevertheless as a first approach that $\beta=\lambda$, i.e. that the secondary building dissipates less energy than the primary one. In this case the forcing function is the same [see eq.(22)] for both buildings, as it has been assumed in the case of an earthquake, and assuming that surface tractions are reduced in the smaller building as in eq.(24), the transient motion is scaled by the factor λ^2 in the secondary building. This implies that the change of configuration has much more influence on the dynamics of a large structure than in a small one, that second-order effects are much more pronounced, and finally that the secondary building is again safer than the primary one. Moreover, one has to consider that, since the same materials constitute both buildings, β is actually equal to unity, which means that the secondary building is damped more than would be required for similitude, and this is one additional factor of safety. In conclusion, the *seismic vulnerability* of a building is enhanced by its size.

The second problem is the experimentation on small-scale models of no-tension full-scale structures. The approach presented in the paper tries to avoid recourse to elaborated manipulations of the material constituting the model, that would be necessary in order to preserve some similitude ratio. It has been proved that, even if the model is built up by the same material as the original, static tests should be fully reliable, in that a definite correlation exists between the behaviour of the original solid and the relevant small-scale model, at least in the range of small displacements, i.e. to the extent that geometrical changes do not affect the structure's performance.

Dynamic tests present some difficulty, mainly arising from the role of viscous dissipation, if present. Proportionality in the elastic component of stress is crucial in

18

this context, because of the constraint on the sign of the stress/strain tensors, that is implied by the no-tension assumption. The problem has been approached by identifying additional forces to be applied to the model in way that similarity with the original's response is preserved. Such corrective terms are expressed in eqs. (22) and (24).

If the material does not exhibit viscous dissipation, the matrix $\mathbf{V} = 0$ and β can be given any value (e.g. $\beta = \lambda$), the corrective terms are null, and therefore the applied loads don't need any modification, apart from the contraction of the time-scale and the reduction of the intensity of time-varying surface tractions. The same applies if $\beta = \lambda$, i.e. if it is possible to attenuate the viscous properties in the model (or to enhance them in the original). In both cases, full similarity of the response in the model and in the original can be guaranteed, to the extent that the no-tension assumption can be approximated in the model and that the same assumption is reliable with respect to the original.

Apart from these two favourable cases, no definite relationship between the response of the model and of the original has been found. The research line that can be suggested after the developments in the paper, is to design some sophisticated experimental device allowing to actuate on-line changes in the time-history applied to the shaking table, in way to realise the corrective terms in eqs. (22) and (24). A simpler approach, that can be properly applied for survival capacity under not sharply defined dynamic actions, like seismic motion, is to search for a-priori elaboration of the expected input signal, in way to compensate for the error deriving from the loss in the viscous resistance. First attempts to deal with both these technological problems are presented in [10,11]

Acknowledgement
Paper supported by grants of the Italian National Council of Researches (C.N.R.)

References
1. Heyman, J. (1969) The Safety of Masonry Arches, *Int. J. of Mech. Sci.* **2**, 363-384
2. Franciosi, V. (1980) L' attrito nel Calcolo a Rottura delle Murature, *Giorn. del Genio Civile* **8**, 215-234
3. Baratta, A. and Toscano, R. (1982) Stati Tensionali in Pannelli di Materiale Non Reagente a Trazione, *Proc. VI Nat. Conf. AIMETA* **II**, 291-301
4. Di Pasquale S. (1982) Questioni di Meccanica dei Solidi Non Reagenti a Trazione, *Proc. VI Nat. Conf. AIMETA* **II**, 251-263
5. Franciosi, V. (1986) L' Arco Murario, *Restauro* **87/88**, 5-56
6. Del Piero, G. (1988) Constitutive Equations and Compatibility of the External Loads for Linearly Elastic Masonry-Like Materials, *Meccanica* **3**, 150-162
7. Baratta, A. (1991) Statics and Reliability of Masonry Structures, in F. Casciati & J.B. Roberst (eds), *Reliability Problems: General Principles and Applications in Mechanics of Solids and Structures*, CISM, Udine, 1991, pp. 205-235
8. Baratta, A. (1995) The No-Tension Approach for Structural Analysis of Masonry Buildings, *Proc. of the 4th Int. Masonry Conference* **2**, pp. 265-280, The British Masonry Society, London.
9. Szücs, E. (1980) *Similitude and Modelling*, Elsevier Sc. P.C., New York.
10. Baratta, A., Clemente, P., Rinaldis, D. and Garofalo, G. (1996) Applications of Active Control Technology to Shake-Table Tests, *Proc. of the Ist European Conference on Structural Control*, pp. 64-71, Barcelona.
11. Baratta, A., Rinaldis, D. and Zuccaro, G. (1997) Preliminary Design of Experimental Facilities for Dynamic Tests on Controlled Structural Specimens, *Proc. of the IInd Int. Workshop on Structural Control*, Hong Kong.

PASSIVE MAGNETIC DAMPING COMPOSITES

A. BAZ
Mechanical Engineering Department
University of Maryland
College Park, MD 20742, USA

Abstract

The theoretical and experimental performance characteristics of a new class of Passive Magnetic Damping Composites (PMDC) are presented in this paper. These composites rely in their operation on arrays of specially arranged permanent magnetic strips that are embedded inside a visco-elastic damping matrix. Proper interaction between the magnetic and visco-elastic forces can enhance the damping characteristics of the PMDC in such a way that makes them suitable for vibration and noise control applications. The completely passive nature of the PMDC renders them to operate effectively without the need for any electronic sensors, actuators, associated circuitry or external energy sources. In this manner, the PMDC treatments provide an efficient and reliable means for damping out undesirable structural vibration and associated noise radiation.

A Finite element model is developed to describe the characteristics of the passive magnetic composites. The model accounts for the coupling between the permanent magnets, the visco-elastic damping and the base elastic structure. Such coupling is determined for different lay-up configurations of the magnetic strips. The analysis is guided by the theory of magneto-elasticity which is developed for untreated and undamped base structures.

The predictions of the finite element model are validated experimentally for beams/PMDC systems subjected to various boundary and loading conditions. The static and dynamic performance of the beams/PMDC systems are compared with the performance of beams treated with viscoelastic layers which are reinforced with non-magnetic strips of similar mass and elastic properties as the magnetic strips used in the beams/PMDC systems. Such comparison emphasizes clearly the merits of the new class of PMDC.

1. Introduction

Surface treatments that are a hybrid between active and passive damping have been considered as viable alternatives to conventional passive damping treatments. Such hybrid treatments aim at using one active control mechanism or another to augment the

J. Holnicki-Szulc and J. Rodellar (eds.), Smart Structures, 19–25.

passive damping to compensate for its performance degradation with temperature or frequency. Among the commonly used hybrid treatments are the Passive Constrained Layer Damping with Shunted Networks (PCLD/SN) treatments [1], the Active Constrained Layer Damping (ACLD) treatments [2] and the Active Piezoelectric-Damping Composites (APDC) [3]. In the PCLD/SN treatments, a piezo-electric film is used to passively constrain the deformation of a visco-elastic layer that is bonded to a vibrating structure. The film is used also as a part of a shunting circuit which is actively tuned to improve the damping characteristics of the treatment over a wide operating range. Similar configuration is employed in the ACLD treatments. However, the piezo-film is actively strained in such a manner to enhance the shear deformation of the visco-elastic damping layer in response to the vibration of the base structure. In the APDC treatments, an array of piezo-ceramic rods embedded across the thickness of a visco-elastic polymeric matrix are electrically activated to control the compressional damping characteristics of the matrix which is directly bonded to the vibrating structure.

Therefore, in the three hybrid damping treatments described, one can identify three distinct damping augmentation mechanisms. In the PCLD/SN, the augmentation results from the energy dissipation in the shunted electric circuitry whereas in the ACLD and the APDC treatments, the augmentation is attributed to the enhanced shear and compressional deformations of the visco-elastic layers, respectively.

Although the PCLD/SN, ACLD and APDC treatments have proven to be very successful in damping out structural vibration, they require the use of piezoelectric films, amplifiers and control circuits. As simplicity, reliability, practicality and effectiveness are our ultimate goal in controlling the vibration and noise; the concept of the Passive Magnetic Damping Composites (PMDC) is introduced to eliminate the need for these films, associated circuitry as well as any external energy [4]. In this way, the PMDC provides a viable and efficient alternative to the ACLD.

It is therefore the purpose of this paper to give an overview of the development efforts, governing equations and performance characteristics of the Passive Magnetic Damping Composites (PMDC) treatments [5-6]. Emphasis is placed on demonstrating the effectiveness of the PMDC in controlling the vibration of flexible beams as compared to conventional Passive Constrained Layer Damping (PCLD).

Therefore, the present review paper is organized in five sections. In Section 1 a brief introduction is given. In Section 2, the concept of the PMDC treatment is presented. Finite element modeling of the dynamics of the PMDC treatment is given in Section 3. The vibration damping characteristics of beams fully treated with the PMDC treatment are presented in Section 4. The performance characteristics of these beams are compared, in Section 4, with those treated with conventional PCLD treatments. In Section 5, the conclusions of the present study are summarized.

2. Concept of PMDC treatments

The PMDC treatment relies in its operation on arrays of specially arranged permanent magnetic strips that are bonded to visco-elastic damping layers. The interaction between the magnetic strips can improve the damping characteristics of the PMDC by virtue of enhancing either the compression or the shear of the viscoelastic damping layers as shown in Figure (1).

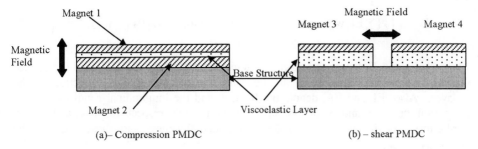

(a)– Compression PMDC   (b) – shear PMDC

Figure 1. Configurations of the PMDC treatment

In the compression PMDC configuration of Figure (1-a), the magnetic strips (1 and 2) are magnetized across their thickness. Hence, the interaction between the strips generates magnetic forces that are perpendicular to the beam longitudinal axis. These forces subject the viscoelastic layer to across the thickness loading which makes the treatment act as a Den Hartog dynamic damper. In the shear PMDC configuration of Figure (1-b), the magnetic strips (3 and 4) are magnetized along their length. Accordingly, the developed magnetic forces, which are parallel to the beam longitudinal axis, tend to shear the viscoelastic layer. In this configuration, the PMDC acts as conventional constrained layer damping treatment.

It is important to note that arranging the interacting magnets in attraction or repulsion affects considerably the damping characteristics of the PMDC as will be discussed in Section 4.

3. Finite element modeling of the PMDC

The finite element model of the PMDC is formulated using the Lagrangian dynamics approach. In that approach, the magnetic energy W, kinetic energy T and the potential energy U of the entire composite are utilized to derive the equations of motion of the PMDC as follows:

$$\frac{d}{dt}(\frac{\partial L}{\partial \{\dot{q}\}}) - \frac{\partial L}{\partial \{q\}} = \{F\} \qquad (1)$$

where t and L denoting time and the Lagrangian with $L = T - U + W$. Also, $\{q\}$ and $\{F\}$ denote the independent coordinate and external force vectors. In this class of magneto-elastic problems, the independent coordinate vector $\{q\} = \{\Delta \quad A\}^T$, where Δ and A are the nodal deflection and the magnetic potential vectors. Note that the magnetic potential A is defined as $B = \text{curl } A$ where B is the magnetic induction.

The expressions of W, T and U are given by:

$$W= \iint_V H.dB\,dV, \quad T = \frac{1}{2}\sum_{j=1}^{3}\int_{0}^{L_i} m_j(w_{t_j}^2 + u_{t_j}^2)\,dx,$$

and

$$U = \frac{1}{2}[\int_{0}^{L_i}\{\sum_{j=1}^{3} EI_j w_{xx_j}^2 + EA_j u_{x_j}^2 + F_m w_{x_1}^2 + Gh\gamma^2 + E_2(w_1 - w_3)^2\}dx] \quad (2)$$

where H, F_m and V denote the magnetic field intensity, magnetic force and the control volume. In equation (2), w_j and u_j are the transverse and longitudinal deflections of the jth layer. Also, EI_j and EA_j denote the flexural and the longitudinal rigidity of the jth layer. Subscripts t and x denote differentiation with respect to time and space. Furthermore, G, h and γ are the shear modulus, thickness and shear angle of the viscoelastic layer.

Equations (1) and (2) yield the following finite element equations:

$$\begin{bmatrix} [M_s] & 0 \\ 0 & 0 \end{bmatrix}\begin{Bmatrix} \ddot{\Delta} \\ \ddot{A} \end{Bmatrix} + \begin{bmatrix} [K_s] & [K_{sm}] \\ 0 & [K_m] \end{bmatrix}\begin{Bmatrix} \Delta \\ A \end{Bmatrix} = \begin{Bmatrix} F \\ M_g \end{Bmatrix} \quad (3)$$

where $[M_s]$, $[K_s]$ and $[K_m]$ are the mass, the structural stiffness and the magnetic stiffness matrices as determined in Ref.[5-6]. Also, $[K_{sm}]$ is the magneto-elastic coupling matrix and the $\{M_g\}$ is the vector of global magnetization.. In equation (3), the nodal deflection vector $\{\Delta\} = \{u_1, u_3, w, w_x\}^T$.

4. Performance of the PMDC

4.1. EXPERIMENTAL SET-UP

The test beams are clamped in a cantilever configuration and is positioned on a sliding table connected to an electromagnetic shaker (Model F4, Wilcoxon Research, Bethesda, MD) giving the table a swept sinusoidal excitation. The amplitude of vibration of the free end of the beam is monitored by a laser sensor (Model MQ-Aeromat Corp. Providence, NJ) mounted on the sliding table. The output signal of the laser sensor is fed to a spectrum analyzer (Model CF910, ONO sokki).

4.2. COMPRESSION CONFIGURATION OF PMDC

The performance of the MCLD treatment is determined experimentally using a cantilever aluminum beam which is 30 cm long, 0.508 mm thick and 25.4 mm wide. The visco-elastic material used in the core (Neoprene Potomac Rubber Inc., Upper Marlboro, MD) is 1.524 mm thick with Young's modulus E'=0.5e6 N/m^2, loss factor η=0.27 and density ρ=150 kg/m^3. The constraining layers are made of 1.524 mm thick magnetic strips (Master Magnetics, Castle Rock, CO) with Young's Modulus E=0.6e9 N/m^2, density ρ=3543 kg/m^3, residual magnetic induction B_r= 0.19 Tesla and coercive force H_c=151197.5 A/m.

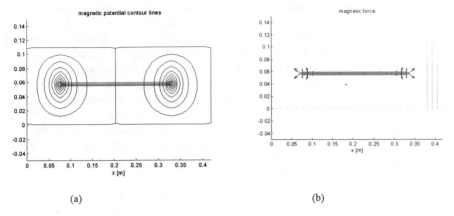

(a)   (b)

Figure 2. The magnetic potential and forces acting on the compression configuration of PMDC
(magnetic strips in attraction)

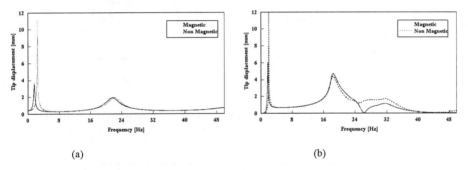

(a)   (b)

Figure 3. Frequency Response Function of the compression configuration of PMDC
[magnets in attraction, (a) – theoretical, (b) – experimental]

Figures (2-a) and (2-b) show the theoretical predictions of the magnetic potential and forces acting on the PMDC when the constraining layers are arranged in attraction. These predictions are used to determine the Frequency Response Function (FRF) of the PMDC with magnetic and non-magnetic constraining layers as shown in Figure (3-a). Figure (3-b) shows the corresponding experimental predictions. It is evident that there is a close agreement between theory and experiments.

4.3. SHEAR CONFIGURATION OF PMDC

The beam used, in this section, is a cantilever aluminum beam, 30 cm long, 0.05 cm thick and 2.5 cm wide. The visco-elastic layers are 0.3125 cm thick, while the aluminum constraining layers are 0.025 cm thick. The considered visco-elastic material has storage modulus G'=0.4E6 N/m^2, loss factor η=0.4 and density ρ=150 kg/m^3. The magnets used are made of neodymium blocks 0.375x0.8x2.5 cm, magnetized through the thickness, with residual induction B_r=1.08 Tesla. . Each magnet has been divided into 4 elements along the length and 2 elements along the thickness. The dimensions of the surrounding region have been adjusted in order to be able to assume a vanishing vector potential at its boundaries and in order to have the magnets placed in the center and preserve a symmetric configuration. Figures (4-a) and (4-b) show the contours of the magnetic induction and the magnetic forces acting on the magnets and the constraining layers. The corresponding Frequency Response Function (FRF) is shown in Figure (4) for magnets in attraction and repulsion. It is evident that magnets in attraction are more effective in attenuating structural vibration than magnets in repulsion. The displayed results are obtained by dividing the beam into 40 elements to accurately model the beam dynamics. The first three modes of the beam, with no magnetic forces, are 5.31, 74.29 and 133.81 Hz.

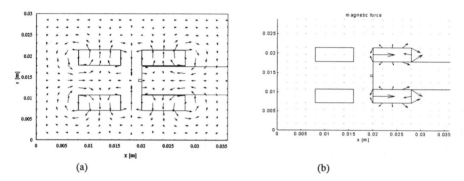

(a)   (b)

Figure (4). Magnetic induction and forces acting on the shear configuration of PMDC

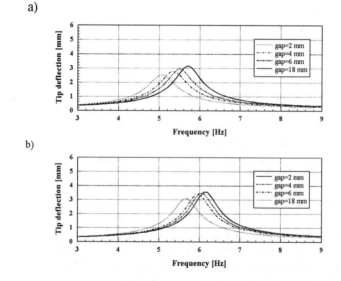

Figure 5. Frequency Response Function of the shear configuration of PMDC [magnets in:
(a) -attraction and (b) – repulsion]

5. Conclusions

This paper has presented an overview of our development studies of the new class of Passive Magnetic Damping Composites (PMDC) treatments. The proposed PMDC treatments rely in their operations on arrays of viscoelastic damping layers that are controlled completely passively by a specially arranged network of permanent magnets. The considered configurations of the PMDC enhance the shear or the compression damping characteristics of conventional Passive Constrained Layer Damping (PCLD) treatments without the need for any electronic sensors, control circuitry or external energy. This is in contrast to the Active Constrained Layer Damping (ACLD) treatments, the Passive Constrained Layer Damping with Shunted Networks (PCLD/SN) and the Active Piezoelectric Damping Composites (APDC) which require extensive electronics for their effective operation. Such excellent feature of the PMDC makes its operation simple, reliable and efficient as compared to other surface damping treatments.

Acknowledgments

This work is funded by The U.S. Army Research Office (Grant number DAAH-04-96-1-0317). Special thanks are due to Dr. Gary Anderson, the technical monitor, for his invaluable technical inputs.

References

1. H. Ghoneim, "Bending and Twisting Vibration Control of a Cantilever Plate via Electromechanica Surface Damping", 2nd Smart Structures and Materials Conf., Vol. SPIE-2445, 1995, pp. 28-39.
2. Baz, and J. Ro, "Vibration Control of Plates with Active Constrained Layer Damping", J. of Smar Materials and Structures, Vol. 5, 1996, pp. 272-280.
3. W. Shields, J. Ro and A. Baz, "Control of Sound Radiation from a Plate into an Acoustic Cavity Usin Active Piezoelectric Damping Composites", J. Of Smart Materials and Structures, Vol. 7, No. 1, 1998 pp. 1-11.
4. Baz, "Magnetic Constrained Layer Damping", U.S.Patent application, 1997.
5. M. Ruzzene, A. Baz and B. Piombo, "Finite Element Modeling of Magnetic Constrained Laye Damping", J. of Computers & Structures, 1998.
6. M. Ruzzene, J. Oh and A. Baz, "Magnetic Compressional Damping Treatments of Beams", J. o Computers & Structures, 1998.
7. W. Kamminga, "Finite element Solution for Devices with Permanent Magnets" Journal of Applie Physics 8 (1975), pp. 841-855.
8. P. P. Silvester and R. L. Ferrari, Finite Elements for Electrical Engineers, Cambridge University Press 1996.

DESK-TOP SELECTED LASER SINTERING OF STEREOMETRIC SHAPES

Z.M.BZYMEK, T.MANZUR, C.ROYCHOUDHURI, L.SHOW, L.SUN
AND S.THEIS
University of Connecticut
Storrs, CT, USA

1. Introduction

Solid Freeform Fabrication (SFF) in manufacturing is a rapidly developing technology. The University of Connecticut has a special research position in the development of both CO_2 and diode laser based technologies for SFF. Usually, the SFF technologies refer to the fabrication of physical parts directly from computer based solid models described by STL (STereo Lithography) or VRML (Virtual Reality Modeling Language) files generated by Computer-Aided Design (CAD) systems. Most SFF processes produce parts by building each consecutive layer using a row by row pattern, though it is possible to build the part using other patterns. The SFF technology represents a larger than average challenge to designers who, besides making decisions concerning optimum shape and functionality of the entire part, have to take into consideration several other manufacturing factors. These factors can be divided into two groups. The first group depends on design for manufacturing, Computer-Aided Design model generation, and part description files, as well as model slicing, precision of part design, and shape of the part. The second group is made up of factors that depend on fabrication technology i.e. laser path files, and the orientation of residual stress phenomena.

On the basis of experience gained in previously developed SFF part manufacturing the authors describe the concepts and some conclusions from the research in progress. Developing of most suitable CAD methods for precise parts design, generation of the part description file, optimization of the part building structure for maximum part stiffness, highest possible precision and minimum fabrication time.

Developing principles for the optimal part rendering method for SFF manufacturing from the point of view of design with the goal of minimizing thermal expansion, shrinkage, possibility of shape changes, precision limits and minimizing the necessity of additional processing of sintered parts by conventional methods.

The above goals, if achieved, will form a foundation for Computer-Aided Sintering (CAS). This will create a path for further development of design for a new method of manufacturing, in which the material is delivered in powder form and leaves the factory in the form of solid stock.

J. Holnicki-Szulc and J. Rodellar (eds.), Smart Structures, 27–36.

2. General Characteristics of the Solid Freeform Fabrication Process

Solid Freeform Fabrication (SFF) is a Rapid Prototyping (RP) technology that makes it possible to build the part from metal or ceramic powder instead of cutting it out of metal blocks larger than the part dimensions. The SFF process consists of two steps, as indicated by Manzur et al. [7] and Wang et al. [2].

- Design using CAD methods that will result in part STL, VRML, or other file that will allow the model to be sliced or processed otherwise to generate sections of the SFF product.

- Manufacturing the physical product by layering 2D or 3D sections with a small third dimension creating, step by step, stereometric shapes.

A wide variety of RP technologies have been proposed during the last decade; these technologies can be used for the purpose of rendering of 3D models in resin, plastic, or metal powder as described by Bzymek et al. [3], but only a few were successfully applied to building structurally strong parts. One of them is SFF technology, which allows for building objects from powders by thermal fabrication. The heat source in the SFF process is always a laser impulse in the infrared region of the electromagnetic spectrum.

3. Sintering Processes

The Selective Laser Sintering (SLS) process used in desk-top SFF leads to fabrication of nearly full density products from metal or ceramic powder layering, sintering or melting with a pulse induced by a CO_2, YAG, or diode laser (Figure 1). The part fabricated by such a process consist of slices (200-300 μm) of stacked and fused powder. Loose powder serves to support the part and is removed from the descending bin in which the part is sintered. In addition to SLS, SALD (Selective Area Laser Deposition) and SALDVI (Selective Area Laser Deposition with Vapor Infiltration) (Figure 2) processes are also used in SFF technology. Methods of parts and laser path design for desktop manufacturing can be also used in both SFF and SALD technologies as pointed out by Bzymek, Shaw and Marks [5].

4. Present State of Problems in Design for SFF manufacturing.

In spite of the fact that SLS, SALD and SALDVI are successfully used to sinter some sample parts in laboratories, they are still not quite ready for industrial applications. There is an entire group of questions still to be answered before the industrial application of SFF can take place. These questions address the use of CAD systems, representation of the part model, communication between CAD systems and the sintering setup, and effectiveness of the sintering process. There are several questions to be answered that concern the technology and design for manufacturing of the

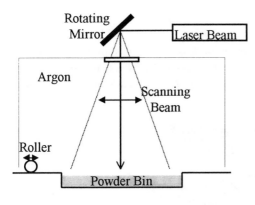

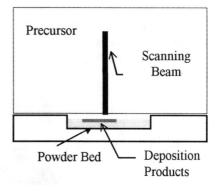

Figure 1. A schematic of the SFF process with stacionary laser.

Figure 2. A schematic of SALD process

sintered part. Some of them can be readily solved, such as the statement that dimension tolerance of the part cannot be smaller than the diameter of the laser beam, but other questions require further studies. The aspects that require studies are: orientation of the part, optimal shapes of the part for SFF processes, configuration of the part, and the relation between the part mechanical properties and the rendering patterns.

5. Present State of Knowledge on Design for SFF

The CAD design for SFF has two stages. The first stage does not differ from the design for conventional manufacturing and produces a file representing the geometry of the part. In the second stage, the file is processed for SFF. There are two basic questions: what should be the structure of the file and how should it be processed? The IGES (International Graphics Exchange Standard) is an international standard of output files used by most CAD vendors. It represents the solid model with Constructive Solid Geometry (CSG) primitives and Boundary Representation Support (BRS). The size of this format is very large and difficult to handle. But this format allows data exchanges among different CAD packages. Most of the literature on SFF design problems is an international standard of output files (Graphics Exchange Standard) used by most CAD vendors. It represents the solid model with Constructive Solid Geometry (CSG) primitives and boundary representation support. The size of this format is very large and difficult to handle. But this format allows data exchanges between different CAD packages. Most of the literature on SFF design problems concentrate on the output format and slicing. The STL file is the most widely used standard for SFF processes. Though several other structures for the file were proposed by Wang, Dong and Marcus [8], the STL and Virtual Reality Modeling Language (VRML) data format are becoming standards for SFF. VRML was created to put interconnected 3D worlds onto every desktop and has become the standard language for 3D World Wide Web. So the VRML format has its potential application for Tele-Rapid Prototyping (RP) and 3D-fax.

6. Desk -Top SFF System

The sintering apparatus used for the experiment in desktop SFF is shown Figure 3. The apparatus includes a diode laser power system as described by Chen, Roychoudhuri and Bana [1], a powder feeding system, an oxidation prevention system and laser scan control system.

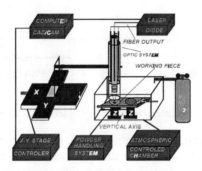

Figure 3. Schematic Diagram & Sketchof the Powder Feeding System

After sintering a new layer is laid down on top of the sintered one layer-by-layer to form a 3-D structure. Layer thickness ranges between 50 and 300 μm for various powders. The layers of powder are traced with a computer controlled laser scanning system. The laser beam diameter is of the order of 0.5 mm. Working output power ranges between 5 to 60 watts. The laser beam rasters the powder bed surfaces and raises the temperature of the powder, binding the loose powder. The laser scan rate is of the order of 0.1 to 3 cm/s for metals and ceramic powders. At the end of the first scan, the second layer of loose powder is deposited as described earlier and the process is repeated. An extra piston is used to pressure the powder to the sintered layer each time a new layer of powder is added.

7. Sintered Parts

7.1 PARTS FROM METAL /CERAMIC POWDER.

With the current rapid prototyping technologies, the majority of prototypes can only be made from wax, plastics, nylon, and polycarbonate materials. Extending current technologies or discovering new technologies to produce prototypes from metal and ceramic material is greatly desired by industry.

7.2 MECHANICAL QUALITIES OF PROTOTYPES.

Most prototype parts made from the current RP machines and/or research have very low structural strength and integrity. The quality of a prototype depends heavily on the

CAD data used to control laser beams, the materials, and the sintering processes. Examples shown in Figure 4 (a, b) are of preliminary SSF sintered parts with multiple sintered layers made from Fe/Ni-Bronze premix powders. A comparison of the parts in Figure 4 reveals that the part sintered from finer particle size powder has more dense packing. Figure 4 shows near net shape of fully dense functional parts. The SFF parts were sintered from 44 μm Fe powder with the aid of 810 nm diode laser. A set of simple plates was also rendered with different laser patterns (Figure 5) within one continuos session. The experiment has shown the positive influence of laser patterns on properties of rendered parts as noticed by Bzymek et al. [6]. The powder size and laser parameters were as described in Figure 4 a -b. The laser speed was 2 mm/sec. The values of density and strength of rendered plates are listed in Table 1. As can be seen, the plate No. 2 is one of the weakest as far as tensile strength was concerned. This was the reason for which it was a subject of further studies. In Figure 6, the density of the cross - section of that plate is shown. The surface roughness which is a function of laser pattern is shown in Figure 7. The tensile strength results are shown in Figure 8. In spite of the fact that it was almost the worst sample, the results shown are in agreement with predicted values and suggest methods of improvement.

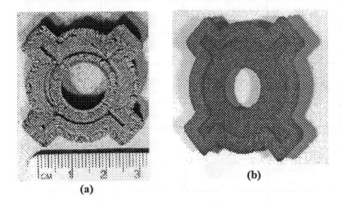

(a)   (b)

Figure 4(a-b) The particle size effects of sintered samples made from iron-bronze premix powder inan argon atmosphere. As powder size decreases from left to right the resolution of SLS parts increases. Figure 4(a) Parameters used: l = 980 nm, SS = 700±80 mm, SP = 1 mm/sec, Powder size = 44 μm, Power = 12 watts. Figure 4(b) Parameters used: l = 980 nm, SS = 700±80 mm, SP = 1 mm/sec, Powder size = 150 μm, Power = 15 watts.

8. CAD System Output and Interface Files

8.1 STL FORMAT

The samples shown in Figure 4 were produced by a slicing STL file. The STL format is the *de-facto* standard input for SFF machines. It is a list of the triangular facets that

TABLE 1. The density and strength of generated plates.

Sample #	Relative Density (%)	Tensile Strength (MPa)
1	36.2	7.6
2	36.7	4.4
3	40.8	5.9
4	37.5	3.5
5	36.8	6.9
6	37.3	7.1

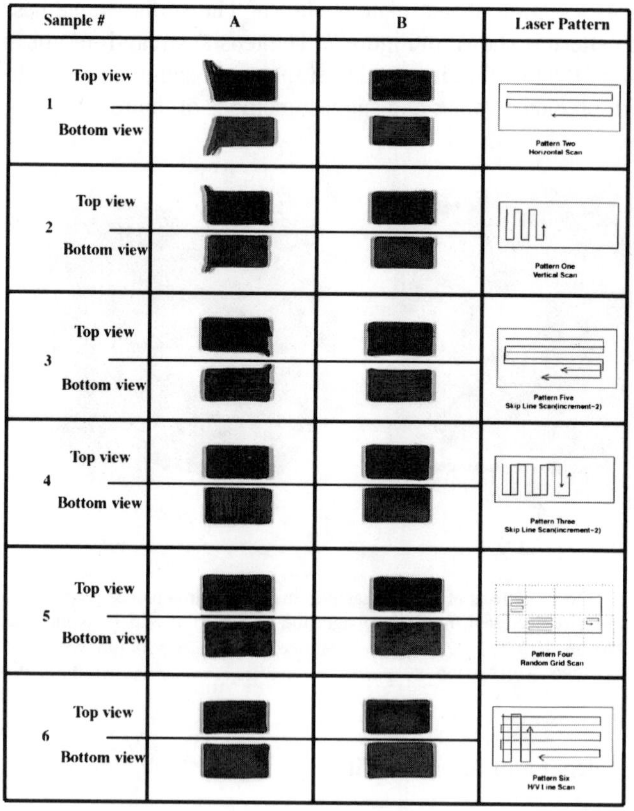

Figure 5. The set of plates rendered with different patterns.

describe a computer generated solid model. It has evolved from the explicit polygon mesh representation of a solid model and established by 3D System™. Two representation are commonly known; the ASCII representation and the binary representation. Both describe the coordinates of three points that form a triangle in space and its associated out-pointing normal. The ASCII STL file format must start with the lower case word solid and end with endsolid. Within these keywords are a listing of individual triangles that define the faces of the solid model. Each individual triangle description defines a single normal vector directed away from the solid's surface followed by the x, y, z components for all three of the vertices. These values are all in Cartesian coordinates and floating point numbers. STL is difficult to use in handling large curvature surfaces and surfaces with fine features, which may lead to errors of unwanted holes or gaps or give rise to non-manifold models of parts.

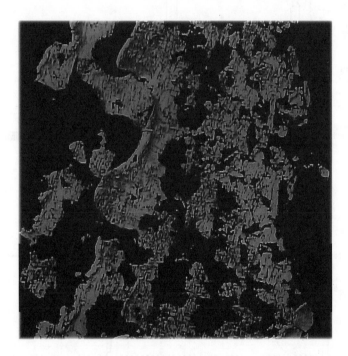

Figure 6. The density of the cross-section of the plate No. 2 along the shorter axis of symmetry (magnified 50 times).

8.2 VRML FORMAT

The origin of VRML dates back to the middle of 1994, to a European Web conference in which Tim Berners-Lee presented a 3D Web standard. Later a group of researchers joined the VRML mail list and started to produce the VRML specification. VRML is a subset of the Silicon Graphics Open Inventor file format for use in Internet

34

applications. It is a platform-independent language for virtual reality scene description. This particular format has become the standard Internet Modeling Language format for the World Wide Web. VRML is also a rapidly emerging industry standard for exchanging three-dimensional data over Internet. The VRML 97 format is an attractive method for data transmission for desktop telemanufacturing.

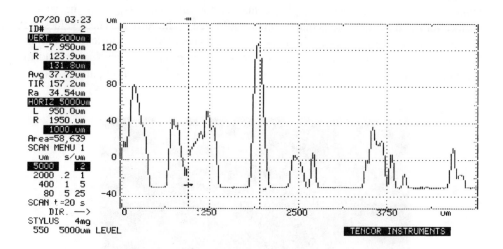

Figure 7. The surface roughness of the plate No. 2.

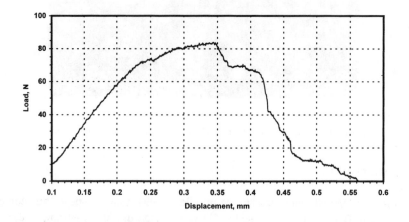

Figure 8. The tensile strength results of the plate No. 2.

9. Design for Desk - Top SFF

Process planning combines design information with manufacturing knowledge to set up the procedures and control parameters for fabrication of a designed part as described by Dong [4]. Using SFF processes to fabricate a part, the process planning usually includes: part orientation, geometric slicing, and path planning. The performance of SFF is influenced by the orientation in which the part is fabricated. Before geometric slicing, the part must be well oriented. There are several criteria for identifying the fabrication orientation, they are: the consideration of surface quality, building time and the complexity of the support structure. The goal of geometric slicing is to generate 2.5D sections from a 3D model. 2.5D as it is used here means the 2D layer contour with the layer thickness. The 3D model is usually represented in a specific 3D data format (for example, STL or VRML). To avoid going through a full manufacturing process in order to check part correctness, virtual modeling is recommended. In modern manufacture processes, virtual prototyping becomes more and more important to avoid as much physical prototyping as possible. Thus, it is necessary to process the virtual prototyping before physical prototyping. At the University of Connecticut, a virtual prototyping system has been developed using C, Inventor and OpenGL on an IRIX machine). The diode laser desktop SFF setup was developed from a conceptual idea at UConn and is unique in its concept. The concept of the SIC file and rendering by pillar, truss and skip methods are also unique concepts developed at the University of Connecticut and among others described by Bzymek et al. [3].

10. Conclusions

The goal of the above research is to form a foundation for Computer-Aided Sintering (CAS). This will create a path for further development of this new method of design for revolutionary manufacturing in which the material can be delivered easily in powder form and leave the factory in the form of solid stock. There are still some manufacturing problems to be solved i.e. warping, strength, and accuracy. However the results of the research in progress are very encouraging and prove that it is possible to obtain parts of a desired shape and structure.

Acknowledgments

The desk-top sintering device was developed at the Photonics Research Center, where the described samples were rendered. The research on design methods was conducted at the following University of Connecticut institutions: Photonics Research Center, the Institute of Material Science and the Department of Mechanical Engineering. The authors appreciate all the help and support provided by the leadership of these institutions and colleagues.

36

References

1. Chen, W., Roychoudhuri C. S., and Bana S. C., (1994) Design Approaches for Diode Laser Material Processing System, *Optical Engineering*, Vol. 33, 3662 - 3669.
2. Wang Y., (1997) Computer Interfacing and Virtual Prototyping for SFF, MS Thesis, University of Connecticut, J. Dong, H. Marcus, Z. M. Bzymek - advisors.
3. Bzymek, Z. M., Benson, S., Garrett, R. E. and Ramakrishnan, B. T., (1996) Thermo-Printing and Laser Sintering Slicing Simulation for Rapid Prototyping of Engine Parts, *ISATA 1996 Proceedings*, Florence, Italy, 291 - 298.
4. Dong, J., (1996) The Issues in Computer Modeling and Interfaces to Solid Freeform Fabrication, *Proceedings of ASME Winter Congress and Exposition*, Atlanta , GA.
5. Bzymek Z. M., Shaw L.L., Marks W., (1998) A Theoretical Model for Optimization of SALD Parameters, *Proceedings of the Solid Freeform Fabrication Symposium*, University of Texas at Austin, Austin, TX, August 11 - 13 1998, (in print).
6. Bzymek Z. M., Theis S., Manzur T., Roychaudhuri C., Sun L. and Leon L. Shaw L. L., (1998) Stereometric Design for Desk - Top SFF Fabrication, *Proceedings of the Solid Freeform Fabrication Symposium*, University of Texas at Austin, Austin, TX, August 11 - 13 1998, (in print).
7. Manzur T., Roychoudhure C. S., and Marcus, H. L., (1996) SFF Using Diode Lasers, *Proceedings of the Solid Freeform Fabrication Symposium*, University of Texas at Austin, Austin, TX, August 1996, 363 - 368.
8. Wang, Y., Dong, J. and Marcus, H. L., (1997) The Use of VRML to Integrate Design and Solid Freeform Fabrication, *Proceedings of the Solid Freeform Fabrication Symposium*, University of Texas at Austin, Austin, TX, August 1997, 669 - 676.

SMART STRUCTURES RESEARCH IN THE U. S.

K.P. CHONG
National Science Foundation
4201 Wilson Blvd., Arlington, VA 22230, U.S.A.

Abstract
Fundamental research and development in smart structures and materials have shown great potential for enhancing the functionality, serviceability and increased life span of our civil and mechanical infrastructure systems and as a result, could contribute significantly to the improvement of every nation's productivity and quality of life. The intelligent renewal of aging and deteriorating civil and mechanical infrastructure systems includes efficient and innovative use of high performance sensors, actuators, materials, mechanical and structural systems. In this paper some examples of NSF funded projects and research needs as well as a new initiative on Knowledge and Distributed Intelligence are presented.

Keyword: smart structures, structural control, solid mechanics, smart materials, composites, distributed intelligence.

1. Introduction

Recent explosive growth in computer power and connectivity is reshaping relationships among people and organizations, and transforming the processes of discovery, learning, and communication [see NSF 98-42 in Article 4], providing unprecedented opportunities for providing rapid and efficient access to enormous amounts of knowledge, data (including real-time data for smart structures) and information; for studying vastly more complex systems than was hitherto possible; and for advancing in fundamental ways our understanding of learning and intelligence in living and engineered systems. NSF's Knowledge and Distributed Intelligence (KDI) theme is a Foundation-wide effort to promote the realization of these opportunities. In the following Article 4, more discussion is presented on KDI.

In recent years, researchers from diverse disciplines have been drawn into vigorous efforts to develop smart or intelligent structures that can monitor their own condition, detect impending failure, control damage, and adapt to changing environments. The potential applications of such smart materials/systems are abundant -- ranging from

J. Holnicki-Szulc and J. Rodellar (eds.), Smart Structures, 37–44.
© *1999 Kluwer Academic Publishers. Printed in the Netherlands.*

design of smart aircraft skin embedded with fiber optic sensors to detect structural flaws; bridges with sensoring/actuating elements to counter violent vibrations; flying MEMS with remote control for surveying and rescue missions; and stealth submarine vehicles with swimming muscles made of special polymers. Such a multidisciplinary civil infrastructral systems (CIS) [1] research front, represented by material scientists, physicists, chemists, biologists, and engineers of diverse fields--mechanical, electrical, civil, control, computer, aeronautical, etc.--has collectively created a new entity defined by the interface of these research elements. Smart structures/materials are generally created through synthesis by combining sensoring, processing, and actuating elements integrated with conventional structural materials such as steel, concrete, or composites. Some of these structures/materials currently being researched or in use are [2,3]:

• piezoelectric composites, which convert electric current to (or from) mechanical forces;

• shape memory alloys, which can generate force through changing the temperature across a transition state;

• electro-rheological (ER) fluids, which can change from liquid to solid (or the reverse) in an electric field, altering basic material properties dramatically.

Current research activities aim at understanding, synthesizing, and processing material systems which behave like biological systems. Smart structures/materials basically possess their own sensors (nervous system), processor (brain system), and actuators (muscular systems)--thus mimicking biological systems. Sensors used in smart structures/materials include optical fibers, corrosion sensors, and other environmental sensors and sensing particles. Examples of actuators include shape memory alloys that would return to their original shape when heated, hydraulic systems, and piezoelectric ceramic polymer composites. The processor or control aspects of smart structures/materials are based on microchip, computer software and hardware systems. In the past, engineers and material scientists have been involved extensively with the characterization of given materials. With the availability of advanced computing, and new developments in material sciences, researchers can now characterize processes, design and manufacture materials with desirable performance and properties. One of the challenges is to model short term micro-scale material behavior, through meso-scale and, macro-scale behavior into long term structural systems performance (cf. Fig. 3). Accelerated tests [4] to simulate various environmental forces and impacts are needed. Supercomputers and/or workstations used in parallel are useful tools to solve this scaling problem by taking into account the large number of variables and unknowns to project micro-behavior into infrastructure systems performance, and to model or extrapolate short term test results into long term lifecycle behavior.

2. Research Needs in Smart Structures and Materials

One of the challenges is to achieve optimal performance of the total system rather than just in the individual components. Among the topics requiring study are energy -

absorbing and variable dampening properties as well as those having a stiffness that varies with changes in stress, temperature or acceleration. The National Materials Advisory Board has published [5] a good perspective on the materials problems associated with a high performance car and a civilian aircraft which develops the "values" associated with these applications. Among the characteristics sought in smart structures/materials are self healing when cracks develop and in-situ repair of damage to structures such as bridges and water systems in order that their useful life can be significantly extended. There is the associated problem of simply being able to detect (predict) when repair is needed and when it has been satisfactorily accomplished. The use of smart materials as sensors may make future improvements possible in this area. The concept of adaptive behavior has been an underlying theme of active control of structures that are subjected to earthquake and other environmental type loads. Through feedback control and using the measured structural response, the structure adapts its dynamic characteristics to meet the performance objectives at any instant. Fig. 1 is a sketch of a futuristic smart bridge system, illustrating some new concepts: including wireless sensors, optical fiber sensors, data acquisition and processing systems, advanced composite materials, structural controls, dampers, and geothermal energy bridge deck de-icing . An artist rendition of this bridge appeared in *USA Today*, 3/3/97.

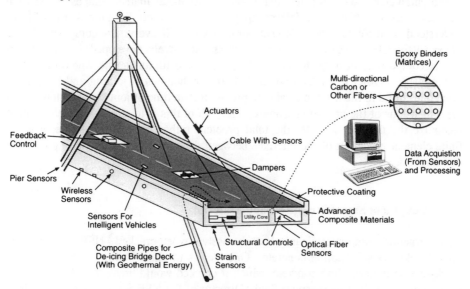

Figure 1. A Futuristic Smart Bridge System

The recent CIS activities [1] for example call for efforts in:
- deterioration science
- assessment technologies
- renewal engineering
- institutional effectiveness and productivity at the system level.

These are in addition to the implied needs in:
- Reliable accelerated tests for long term durability behavior
- improved computers, microprocessors and the information highway
- more accurate/complete modeling of lifetime predictions
- new sensors and control systems; NDT; new materials
- electro - rheological fluids.....shape memory alloys, etc.
- understanding corrosion better at the detail level
- life-cycle performance and costs.

This list is meant to illustrate rather than be complete. The research needs in the CIS may not be very different than in other applications such as the "clean car" or even in defense applications. What is different is the economic considerations as the problem changes from high cost / low volume to low cost / high volume applications. The need to be able to predict safe lifetime, for example, with incomplete data is generic. Presently we do not know all the parameters that need to be included in the material behavior models and in the prediction methodology in order that the predictions are both reliable and cost effective. The use of new (and therefore a moving target as the list of) materials adds to the need to provide a framework that leads to reliable closure in the parameter choice in a few iterations. Some metal matrix composites have an extensive research history in defense applications and their limitations are still not understood from a total life cycle cost point of view. If every new composite that is proposed for CIS or the clean car must be as extensively developed as these metal matrix composites, in order to be able to predict safe lifetimes; the whole enterprise may collapse. Other research needs include better methods of joining and fastening. Then one comes to the need to be able to predict the lifetime of joints and fasteners and the dimensions of the problem becomes great. All of this is to stimulate researchers to recognize the need to consider the total problem rather than, for example, just the finite element modeling of the interface between dissimilar materials, as important as that might be.

3. Examples of Smart Structures and Materials

Smart structures and materials may heal themselves when cracked. NSF grantees have been developing self-healing concrete. One idea is to place hollow fibers filled with crack-sealing material into concrete, which if cracked would break the fiber releasing the sealant, according to Surendra Shah, Director of the NSF Science and Technology (S&T) Center at Northwestern. Lehigh researchers at ATLSS, an NSF Engineering Research Center, are looking into smart paints, which will release red dye (contained in capsules) when cracked. Optical fibers which change in light transmission due to stress are useful sensors. They can be embedded in concrete or attached to existing structures. NSF-supported researchers at Rutgers University studied optical fiber sensor systems for on-line and real-time monitoring of critical components of structural systems (such as bridges) for detection and warning of imminent structural systems failure.   NSF grantees at Brown University and the University of Rhode Island investigated the fundamentals and dynamics of embedded optical fibers in concrete. Japanese researchers recently developed glass and carbon fiber reinforced concrete which provides the stress data by measuring the changes in electrical resistance in carbon fibers.   Under an NSF Small Grants for Exploratory Research Programs, researchers at the University of California-Berkeley recently completed a study of the application of electro-rheological (ER) fluids for the vibration control of structures. ER fluids stiffen up very rapidly (changing elastic and damping properties) in an electric field.   Other NSF supported researchers are studying shape memory alloys (University of Texas, Virginia Tech, and MIT); surface superelastic microalloying as sensors and microactuators (Michigan State); and magnetostrictive active vibration control (Iowa State, Virginia Tech); etc. Photoelastic experiments at Virginia Tech demonstrated that NiTiNOL shape memory alloy wires could be used to decrease the stress intensity factor by generating a compressive force at a crack tip. Other examples include:

• Semi-Active Vibration Absorbers (SAVA), W.N.Patten, et al, (Univ. of Oklahoma). This is part of a 5-year NSF Structural Control Initiative with Prof. Larry T. T. Soong as the grantees coordinator. The project was originally started by R.Sack

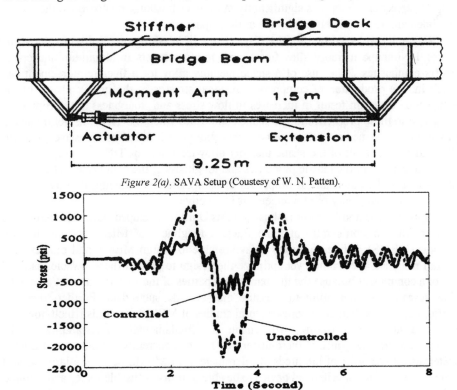

Figure 2(a). SAVA Setup (Coustesy of W. N. Patten).

Figure 2(b). Reduction of Stresses at Critical Locations, Controlled vs. Uncontrolled Conditions (Courtesy of W. N. Patten).

and W. N. Patten. A smart micro-controller coupled with hydraulic systems reduces large vibration amplitudes over 50% produced by heavy trucks passing through a highway bridge, adding 15% more load capacity and extending bridge life over 20 years. Figs. 2 (a) and (b) show the SAVA setup and the stress reduction respectively.

• Fiber-optic Sensors in Bridges, R. Idriss (New Mexico State University).

Fiberoptic cables are etched by laser with 5 mm long internal gauges, spaced about 2 m apart. These cables, strung under the bridge with epoxy, will be able to detect the stresses by sending light beams down the cable at regular intervals and by measuring the bending of the light beams. These gauges can also be used to monitor general traffic patterns. The snsors serves as a data collector as well as a wireless transmitter.

• Intelligent Structures and Materials, C.A. Rogers (Univ. of S. Carolina).

Specifically, this Presidential Young Investigator (PYI) project [6] includes: distributed sensing and health monitoring of civil engineering structures, active buckling control structures and active damage control of composites. Micron size magnetostrictive particles are mixed with the composite for health monitoring. Proof-of concept experiments have demonstrated that delaminations could be located using active magnetostrictive tagging. Low velocity impact damage resistance of composites can be improved by hybridizing them with shape memory alloy material (SMA). The graphite epoxy specimen has significantly more fiber breakage compared to the hybrid specimen and is completely perforated at the impact site.

• Research conducted by Prof. M. Taya (Univ. of Washington, Seattle) explored the use of TiNi shape memory alloy fibers and reinforcements in aluminum and epoxy matrix composites. The use of a shape memory alloy for a fiber reinforcement can result in the production of compressive residual stresses in the matrix phase of the composite. This can result in increases in flow stress and toughness of the composite compared to these properties in the same composites when the TiNi was not treated to produce a shape memory effect in the composite. The induced compressive stress in the matrix as a result of the shape memory alloy treatment of TiNi is responsible for the enhancement of the tensile properties and toughness of the composite. The concept of enhancing the mechanical properties of composites by inducing desirable stress states via shape memory reinforcements is innovative.

• Profs. Sharpe and Hemker (Johns Hopkins Univ.) developed new test methods to measure the Poisson's ratio and the fracture toughness of Microelectromechanical systems (MEMS). While engineers today are able to design MEMS and predict their overall response, they cannot yet optimize the design to predict the allowable load and life of a component because the mechanical properties of the material are not available. This research is attempting to measure and provide such data. Further, they are performing comprehensive microstructural studies of MEMS materials which, together with the mechanical testing, will enable a fundamental understanding of the micromechanical response of these materials. There is currently a trend toward thicker materials (of the order of hundreds of micrometers) in MEMS because a larger aspect ratio is needed for a mechanical device to be able to transmit usable forces and torques.

The mechanical testing techniques are being extended for such applications also. MEMS have great potential in smart structures.

- Synthesis of Smart Material Actuator Systems for Low and High Frequency Macro Motion: An Analytical and Experimental Investigation, G. Naganathan (University of Toledo). This project is to demonstrate that motions of a few centimeters can be performed by smart material actuator system. A program that combines theoretical investigations and experimental demonstrations will be conducted. Potential configurations made of piezoceramic and electrostrictive materials will be evaluated for providing larger motions at reasonable force levels for applications.

4.Knowledge and Distributed Intelligence [NSF 98-55 and NSF 98-60; Ref. 7]

According to this new initiative, NSF aims to achieve the next generation of human capability to generate, model, and represent more complex and cross-disciplinary scientific data from new sources and at enormously varying scales (see e.g. Fig. 3); to transform this

Materials		**Structures**	**Infrastructure**
Micro-level	*meso-level*	*macro-level*	*systems integration*
Atomic scale	**Microns**	**Meters**	**Up to km scale**
~micro-mechanics	~meso-mechanics	~beams	~bridge systems
~nanotechnology	~interfacial	~columns	~lifelines
smart materials		*smart structures*	*smart systems*

Figure 3. Scales in Materials and Structures

information into knowledge by combining and analyzing it in new ways; to deepen our understanding of learning and intelligence in natural and artificial systems; to collaborate in sharing knowledge and working together interactively; and others.

KDI have three foci: Knowledge Networking (KN); Learning and Intelligent Systems(LIS); and New Computational Challenges (NCC). KN focuses on the integration of knowledge from different sources and domains across space and time. One of the goals of KN research is to achieve new levels of knowledge integration, information flow, and interactivity among people, organizations, and communities, LIS seeks to stimulate multidisciplinary research that will promote the use and development of information technologies in learning and discovery across a wide variety of fields. LIS emphasizes research that advances basic understanding of learning and intelligence in natural and artificial systems, as well as research that supports the development of tools and environments to test and apply this understanding in real situations. NCC focuses on research and tools needed to discover, model, simulate, analyze, display, or understand complex phenomena, to control resources and deal with massive volumes of data in real time, and to predict the behavior of complex systems. These aims will require major advances in hardware and software to handle complexity, representation, and scale, to enable distributed collaboration, and to facilitate real-time interactions and control. These three foci could have important implications and applications on smart structure/material systems, e.g. in the control and reliability of highway bridges related to problems of scale (in space and time) and structure as well as to interplay between computations and data. For more information about the three foci and their particular emphases for FY 1998, please see the Program Solicitation (NSF 98-55; www.nsf.gov).

5. Summary and Conclusions

An overview of the state of the art and NSF engineering research in smart structures and materials are presented. The authors hope that this paper will act as a catalyst, sparking interest and further research in these areas. This paper reflects the personal views of the author (see also [8]), not necessarily those of the National Science Foundation.

Acknowledgments
Input and additions by his NSF grantees and colleagues, including Drs. C.S. Hartley, S. Saigal, S.C. Liu , O.W. Dillon, C. A. Rogers, W. N. Patten and L. Bergman are gratefully acknowledged.

Bibliography

Chong, K. P.,et al, eds (1990) *Intelligent Structures*, Elsevier, London.
Tzou, H. S. and Anderson, G. L., eds (1992) *Intelligent Structural Systems*, Kluwer Academic Publ.

References

1. *Civil Infrastructure Systems Research: Strategic Issues* (1993) NSF 93-5, 54 pp., National Science Foundation, Arlington, VA.
2. Chong, O.W. Dillon, J.B. Scalzi and W. A. Spitzig (1994) Engineering Research in Composite and Smart Structures, *Composites Engineering*, 4(8), 829-852.
3. Liu, K. P. Chong and M. P. Singh (1994) Civil Infrastructure Systems Research Hazard Mitigation and Intelligent Material Systems, *Smart Mater. Struct.*, 3, A169-A174.
4. Chong, K. P., et al, (1998) *Long Term Durability of Materials and Structures: Modeling and Accelerated Techniques*, NSF 98-42, NSF, Arlington, VA.
5. National Academy of Sciences, (1993), *Materials Research Agenda for the Automotive and Aircraft Industries*, NMAB-468.
6. Rogers, C. A., and Rogers R.C., eds., (1992), Recent Advances in adaptive and Sensory Materials and Their applications, Technomic Publishing, Lancaster, PA.
7. *Knowledge and Distributed Intelligence*, NSF 98-55 (1998), NSF, Arlington, VA
8. Boresi, A.P. and Chong, K. P., (1987), *Elasticity in Engineering Mechanics*, Elsevier, New York; Chinese edition published by Science Publishers (1995).

DAMAGE PREDICTION CONCEPT BASED ON MONITORING AND NUMERICAL MODELLING OF A MEDIEVAL CHURCH

W.CUDNY, M.SKŁODOWSKI, D.WIĄCEK, J.HOLNICKI-SZULC
Institute of Fundamental Technological Research
ul. Świętokrzyska 21, 00-049 Warsaw, Poland

1. Introduction

System of monitoring of historical buildings is presented. This system is under development in application to observation of the medieval St. John's church in Gdansk, Poland (Fig.1). The corresponding design of monitoring network with instrumentation on the one side, and the necessary numerical analysis tools for assessment and prediction of the damage development on the other, are presented.

The church of St. John in Gdańsk is dated from the 14th century. It has been founded on a soil being a mixture of sand with very high humidity and organic components. This is considered to be the origin of all the troubles.

Differential settlement of the foundation caused civil engineering problems almost a century later.

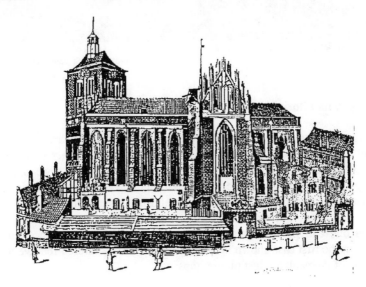

Figure 1. St. John's church in Gdansk

45

J. Holnicki-Szulc and J. Rodellar (eds.), Smart Structures, 45–52.

46

Today the church walls are cracked from the top to bottom in a few locations (Fig.4), 3 pillars of the presbytery and its walls are cracked (cf.Fig.5), at least 5 pillars are out of plumb with bending and torsional distortions at the same time.

Movements which occurred during last two years caused almost pulling of the anchor of the tie rod out of two pillars. Hence it is quite a challenge to build an effective monitoring and damage prediction system.

Figure 2. Cracked walls of the south-west part of St. John's church

Figure 3. Cracked wall of the north-east part of the presbytery of St. John's church

2. Developed monitoring system

Monitoring of the building deformations can usually be done in various ways, without significant influence on the usefulness of the data obtained. Another situation is in the case of development of the computer-based damage prediction system. Satisfactory monitoring system may not provide suitable data set for computer calculations.

As an example of the above problem, two monitoring systems for a part of a medieval church are described. The first one is very suitable for the assessment of vaults while the second one can provide data for numerical modelling and hence, the damage prediction. In the first case shown in the Fig. 4 there are monitored relative movements of the centre of the vault against pillars' tops.

In the St. John's church in Gdańsk, thanks to the suitable geometry, it can be done optically with laser beams directed from the pillars towards a Position Sensing Detector placed at the centre of the vault. The other solution is given in Fig. 5. Two displacement components (vertical and a horizontal one) for each pillar are monitored from the common place situated far from the measurement points. We can not measure reliably the other horizontal displacement in this manner. Thus it was decided to measure the tilt of each pillar in this direction. This set of data is useful not only for the assessment but also for the suggested numerical modelling of the church behaviour.

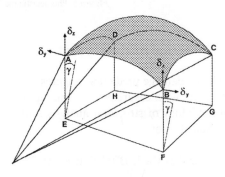

Figure 4. Schematic diagram of optical assessment of a vault.

Figure 5. Displacements of pillar tops and tilt monitoring schema

A more general set of measuring equipment has been found to be of a great interest and is being under study. In addition to the last set of displacement and tilt data, the total change of the distance between two pillar tops (e.g. points A and B) can be monitored by measuring the change of the length of a wire stretched between the measurement points (like the method used in ISMES-Italy [4]). As an alternative method for such a measurement, an optical triangulation is considered, with two laser diodes placed on one pillar and a sensor on the other one.

To complete the data set, one needs to measure the settlement of the pillars. For this purpose it is possible to use the optical system shown in the Fig. 5 or a hydraulic net of communicated vessels equipped with a liquid level sensors [5]. During our preliminary research, two types of such sensors were developed. One with LVDT transducers, and the other for measuring conductivity changes between two electrodes partly immersed in the electrolyte being the self-levelling medium. Sensors of this hydraulic system are mounted at the base of pillars at the points E, F, G, H. Thus the described monitoring system can provide data of vertical and horizontal displacement and allows us to calculate the tilt and translation components (the last ones being hardly probable to occur).

Figure 6 shows a simple, inexpensive yet reliable method of measurement of crack opening. A photoelastic plate of suitable dimensions is spanned across the crack and glued to the construction. Changes in the crack opening change the material birefringence measured with a specially designed compensator with stress-frozen isochromatic fringes. It is also possible to calculate the opening and sliding mode components if a complementary measurement of isoclinic parameter is made.

48

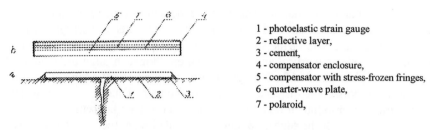

1 - photoelastic strain gauge
2 - reflective layer,
3 - cement,
4 - compensator enclosure,
5 - compensator with stress-frozen fringes,
6 - quarter-wave plate,
7 - polaroid,

Figure 6 Photoelastic measurement of crack opening

3. Numerical Analysis System

Reliable damage prediction requires a numerical model of entire structure updated due to current measurements coming from the monitoring system. The applied methodology of structural analysis settlement identification and damage prediction is presented below.

3.1. VDM BASED BRITTLE-PLASTIC PROGRESSIVE DAMAGE ANALYSIS

Brittle material response to tension and elasto-plastic to compression is assumed, while the damage development is simulated through fictitious field of the so-called *virtual distortions*. In the consequence, the final nonlinear structural behaviour is composed of the linear response and non-linear part caused by virtual distortions. The self-equilibrated stresses and compatible (except crack discontinuities) deformations induced by virtual distortions are calculated with use of pre-computed influence vectors describing relations between the locally provoked unit distortion (e.g. dilatation) and the global stress/strain fields. Numerical efficiency of the proposed method is based on the fact that there is no need for re-meshing and modification of the global stiffness matrix in the simulation process of propagation of crack and plastic zones (cf.Ref.2). In continuation, application of the proposed approach to identification of settlement modification and prediction of the provoked damaged zones development is proposed.

The VDM based approach gives a chance to handle the progressive damage analysis of a large numerical model at a relatively low computational cost. Taking into account application to analysis of brick-made historical buildings it was assumed, in the first approximation, the simple Nadai's type of material characteristic, since cracks provoked by local tension are the major cause of damage development.

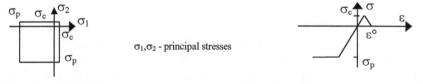

σ_1, σ_2 - principal stresses

a) Stress limits in2D Model

b) One-dimensional stress-strain characteristics

Figure 7. Assumed material characteristics

Let us consider material properties (for simplicity for a 2D continuum case) characterised by the brittle-plastic surface shown in Fig.7a, where x_1 and x_2 are local principal directions and the uni-axial behaviour shown in Fig.5b σ^c denotes the critical local stress (for tension) causing fracture, and σ^p denotes the plastic yield stress value (for compression).

Applying the VDM concept of brittle-plastic progressive damage analysis (cf. Ref.1) we postulate that local stresses σ as well as deformations ε for the *modified* structure (with non-linear material properties $\mathbf{C}^*$ shown in Fig.7) and for the *distorted* structure (with original linear properties $\mathbf{C}$ and with introduced virtual distortions ε°) are identical:

$$\sigma = \mathbf{C}^*\varepsilon = \mathbf{C}\,(\,\varepsilon - \varepsilon^\circ\,),\tag{1}$$

where ε can be decomposed:

$$\varepsilon = \varepsilon^L + \mathbf{D}\,\varepsilon^\circ,\tag{2}$$

and where: ε^L - strains caused by external load, ε - global strains caused by external load as well as virtual distortions, $\mathbf{D}$ - the influence matrix denotes strains caused by unit virtual distortions, $\mathbf{C}$ - constitutive matrix.

Note that the virtual distortions ε° (analogously to the deformation state ε) are determined by 3 independent components ε°_{11}, ε°_{22}, ε°_{12} for each finite element. Composing the actual influence matrix $\mathbf{D}$ gradually, from vectors describing structural deformations caused by unit local virtual distortions $\varepsilon^\circ_{11}=1$, $\varepsilon^\circ_{22}=1$, $\varepsilon^\circ_{12}=1$ in each overloaded finite element detected in the loading process, the following requirements satisfying local brittle or plastic behaviour has to be postulated due to the particular case of the stress state:

- for the fracture case (tension)   - for the plastic case (compression)

$$
\begin{aligned}
\sigma_n &= 0 \\
\sigma_{nt} &= 0 \\
\sigma_t &= E\,\varepsilon_t
\end{aligned}
\tag{3}
\qquad\qquad
\begin{aligned}
\sigma_n &= \sigma^p \\
\sigma_{nt} &= 0 \\
d\varepsilon^\circ\,\sigma &= 0
\end{aligned}
\tag{4}
$$

where E denotes the Young's modulus and t, n denote direction, respectively, along the crack and normal. To the crack condition $(4)^3$ describes orthogonality of plastic distortions increment to the plastic surface. Expressing local stresses as superposition of linear solution σ^L and influence of the virtual distortions (cf. Eqs.1, 2):

$$\sigma = \mathbf{C}\,[\varepsilon^L + (\mathbf{D} - \mathbf{I})\,\varepsilon^\circ]\tag{5}$$

and substituting (5) to (3) and (4), we can get a system of the linear equations (6) determining the virtual distortions simulating the cracks (in over-tensioned areas, Eqs.3) and plastic deformations (in over-compressed areas, Eqs.4)

$$\mathbf{B}\,\varepsilon^\circ = F(\sigma^L, \mathbf{CD}^u \mathbf{u}) , \qquad\qquad (6)$$

where $\mathbf{D}^u$ - strains caused by unit settlements, and $\mathbf{u}$ - the settlements.

In this way, solving the system of linear equations (with the size equal to the number of observed damaged finite elements times three), the piecewise-linear brittle-plastic response can be determined. If no new damaged element is detected, thé analysis can be finished. Otherwise, the new, overloaded elements has to be included into the damaged area and the simulation process (determining virtual distortions) must be repeated.

3.2. STRUCTURAL MONITORING AND ASSESMENT

Important cause of damage development in historical buildings is non-homogeneous settlement and modification of support conditions. Therefore, identification of modified loading provoked by these changes is crucial for damage prediction. Assuming installation of the sensor systems monitoring the crack propagation as well as deformation of distances between the selected l'' pairs of locations, the problem of selection of a limited number l' of sensors providing the most accurate identification of unknown components of the expected modes of deformation can be formulated. For example, if we expect two components of unknown settlement u^1 and u^2 of the wall shown in Fig.8, then l''≥2 possible locations of the deformation sensor can be considered (cf.Fig.8). Calculation of the matrix determines sensitivity of each sensor to a unit modification of loading component S^{ij} (cf. Ref.3), and the problem of optimal selection of the limited number l' (e.g. l'=2 in our example) of sensors can be formulated. To this end, a selection of a [l'x2] submatrix $S' \subset S$ with the best condition number can be proposed.

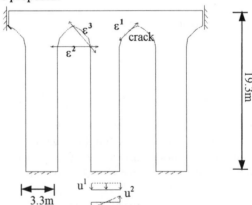

Figure 8.. Example, possible sensor location

Having the identification problem solved (virtual distortions modelling settlement and cracks determined), the current deformation and stress states for the entire structure can be updated as follows:

$$\varepsilon = \varepsilon^L + \mathbf{D}\varepsilon^\circ + \mathbf{D}^u\mathbf{u} \,,$$
$$\sigma = \mathbf{C}\,\varepsilon^L + (\mathbf{D}\text{-}\mathbf{I})\,\varepsilon^\circ + \mathbf{D}^u\mathbf{u}] \,, \tag{7}$$

where $\mathbf{u} = \{u^1, u^2\}$, $\mathbf{D}^u$ are strains caused by unit settlements.

3.3 PREDICTION OF DAMAGE DEVELOPMENT

Finally, having the stress field (7) determined, the gradient

$$\frac{d\sigma}{dt} = \frac{\partial\sigma}{\partial u}\frac{du}{dt} + \frac{\partial\sigma}{\partial\varepsilon^\circ}\frac{d\varepsilon^\circ}{du}\frac{du}{dt} \tag{8}$$

denoting stress response for the settlement modification can be calculated. Taking into account the formula $(7)_2$, the following derivatives can be obtained

$$\frac{\partial\sigma}{\partial u} = \mathbf{D}^u\,, \qquad \frac{\partial\sigma}{\partial\varepsilon^\circ} = \mathbf{C}(\mathbf{D}-\mathbf{I})\,,$$

$$\frac{\partial\sigma}{\partial u} = \mathbf{D}^u\,, \qquad \frac{\partial\sigma}{\partial\varepsilon^\circ} = \mathbf{C}(\mathbf{D}-\mathbf{I})\,. \tag{9}$$

On the other hand, ε° is determined from equations (6), hence

$$\mathbf{B}\frac{\partial\varepsilon^\circ}{\partial u} = \frac{\partial F}{\partial u}\,. \tag{10}$$

Solving this set of linear equations is the main numerical cost necessary to determine the gradients (8), and allowing to predict development of the stress concentration in vicinity of the crack. In the consequence, crack's growth can be predicted.

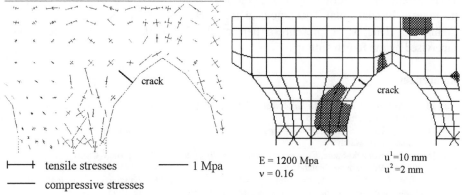

├───┤ tensile stresses ──── 1 Mpa	E = 1200 Mpa u^1=10 mm
──── compressive stresses	v = 0.16 u^2=2 mm

Figure 9. Principal stresses in crack's vicinity   *Figure 10*. Maximal gradients $d\sigma_1$ / du zones

The current distribution of principal stresses (due to settlement and crack generation) for the numerical example shown in Fig.8 is demonstrated in Fig.9. On the other hand, the zones of maximal derivatives of positive principal stresses are shown in Fig.10. This information allows to predict the most probable location of the next crack generation, assuming continuation of the settlement process.

4. Conclusions

The DAmage MONitoring system (DAMON) specialized in prediction of the damage development in historical buildings is under development. It includes the following modules:
- identification of the global structural model on the basis of the measured response to thermal loading (climatic day cycles),
- optimal location of a limited number of sensors focused on detection of the expected variations of loading conditions
- detection of the current state of settlement and other variation of loading conditions;
- updating strains and stresses due to the current, modified loading conditions;
- prediction of the damage development and eventual step back to modification of the sensor system.

It was found that more precise identification of the current loading conditions is very important for reliable evaluation of the structural response. Usually, the number of sensors measuring the variation of loading conditions (e.g. settlement) is limited and therefore, the missing information has to be determined indirectly. The DAMON system enables the progressive damage analysis curried for relatively large buildings, thanks to the VDM based FEM simulation of crack propagation, without, need for remeshing and stiffness matrix reformations in the analysis process.

Acknowledgement

This work was supported by the grant No. KBN 7T07E01096C/3049. from the Institute of Fundamental Technological Research, funded by the National Research Committee. This paper presents a part of the Ph.D. thesis of the third author, supervised by the fourth author.

References

1. J.Holnicki-Szulc, J.T.Gierliński *Structural Analysis, Design and Control by the VDM Method*, J.Wiley & Sons, Chichester, 95
2. D.Wiacek, J.Holnicki-Szulc, *Progressive Damage Analysis in Historical Buildings*, Proc. Complas Barcelona,17-19 March 97
3. D.Wiacek, J.Holnicki-Szulc, *Damage predictions in Historical Buildings*, Proc. WCSMO-2, 26-30 May - 97 Zakopane, Poland
4. P. P. Rossi *Possibilities of the experimental techniques for the structural analysis of historical buildings*, Proc. Structural Analysis of Historical Constructions, Barcelona 8-10 Nov. 1995
5. ISMES - Italy: personal communications

FEW REMARKS ON THE CONTROLLABILITY OF AN AEROACOUSTIC MODEL USING PIEZO-DEVICES

PHILIPPE DESTUYNDER
CNAM-IAT 15, rue Marat, 78210 Saint-Cyr l'École, France

1. Introduction

Let us consider a two-dimensional airfoil as shown in Figure 1. For small angles of attack, a turbulent boundary layer appears in the rear part of the airfoil. The sub-layer can produce transverse vortices. The mechanical advantage of this phenomenon is that it enables a drag-reduction. More precisely the airfoil, or at least its rear part, can "roll" over these vortices. But the instable feature of the sub-layer requires an active control of the vortices. Several aerodynamicists have suggested that this is the way the shark skin works. In other words, it seems necessary to energize the vortices through a resonance strategy (I have been told that shark skin would have small resonators which are magically adjusted at the right frequency!). Our strategy is to use piezo-electric actuators in order to energize the micro-vortices. But at first, modelling seems to be necessary in order to understand what happens. Let us also underline that experiments (see Wojciechowski - Szuster - Pietrucha [1998]) have proved the efficiency of the strategy. But the location and the frequency of the actuator should be adjusted carefully. Let us now sketch the method that we suggest in order to improve the efficiency of the piezo-actuators. Near the airfoil, there is a so-called "bubble" as shown in Figure 1. The mean value of the velocity is U and it is represented by a vector parallel to the boundary of the airfoil. Then the perturbations due to the acoustic field can be approximately represented by the following equation :

$$\begin{cases} \dfrac{\partial^2 \varphi}{\partial t^2} + 2U \dfrac{\partial^2 \varphi}{\partial x_1 \partial t} + U^2 \dfrac{\partial^2 \varphi}{\partial x_1^2} - C^2 \Delta \varphi = 0 \text{ in} \Omega, \\[2mm] \varphi = 0 \text{ on} \Gamma_o, \dfrac{\partial \varphi}{\partial \nu} = 0 \text{ on} \Gamma_{1nc}, \dfrac{\partial \varphi}{\partial \nu} = g \text{ on} \Gamma_{1c}, \Gamma_1 = \Gamma_{1c} \cup \Gamma_{1nc}. \end{cases} \qquad (1)$$

The notation $"\dfrac{\partial .}{\partial \nu}"$ is used for the normal derivative of a function along the boundary

of Ω. Furthermore, x_1 is the coordinate parallel to the boundary of the airfoil as shown in Figure 1. Finally φ is the velocity potential traducing the acoustic perturbation. The sound velocity which is denoted by C, is assumed to be constant. The function $g(x,t)$ is defined on a part Γ_{1nc} of the airfoil. This is the control. It should be defined such that at a finite time T, one has :

J. Holnicki-Szulc and J. Rodellar (eds.), Smart Structures, 53–62.

$$\varphi(x,T)=\varphi_0(x), \frac{\partial\varphi}{\partial t}(x,T)=\varphi_1(x) \text{ in}\Omega. \tag{2}$$

Hence one can formulate the controllability problem as follows : *"can we find g(x,t) such that if :*

$$\varphi(x,0)=\frac{\partial\varphi}{\partial t}(x,0)=0 \tag{3}$$

then φ, *a solution of (1), satisfies (2) for an arbitrary couple* (φ_0,φ_1) *chosen in a right space of functions".* If the answer to the previous question is yes, then the model (1)-(3) is said to have the exact controllability property. The next point is then : *"how to prescribe a velocity g on the part* Γ_{1nc} *of the airfoil".*

2. Controllability of the acoustic model

First of all, let us notice that a function g which characterizes the solution φ of (1) and such that (2) would be satisfied is not unique (if one, at least, exists !). Moreover, the existence and uniqueness of a solution φ_g of (1)-(3) is classical as far as $g \in H^1\left(]0,T[;L^2(\Gamma_{1c})\right)$ (see J.L. Lions - E. Magenes, Vol. 2 [1968]). Then:

$$\varphi_g \in C^0\left([0,T];V\right)\cap C^1\left([0,T];L^2(\Omega)\right) \tag{4}$$

where (see Figure 1 for the definition of Γ_0) :

$$V=\left\{v\in H^1(\Omega),v=0 \text{ on } \Gamma_0\right\} \tag{5}$$

Let us now introduce a bilinear and positive form defined by :

$$\forall g,h \in H^1\left(]0,T[;L^2(\Gamma_{1c})\right) \to a(g,h) =\int_\Omega \varphi_g(x,T)\varphi_h(x,T) +\int_\Omega \frac{\partial\varphi_g}{\partial t}(x,T)\frac{\partial\varphi_h}{\partial t}(x,T)$$

and a linear form by (we assume that $(\varphi_0,\varphi_1)\in V\times L^2(\Omega)$) :

$$\forall h \in H^1\left(]0,T[;L^2(\Gamma_{1c})\right) \to \pi(h)=\int_\Omega \varphi_0(x)\varphi_h(x,T) +\int_\Omega \varphi_1(x)\frac{\partial\varphi_h}{\partial t}(x,T).$$

Then the optimal control we are looking for is defined as the unique solution of the following optimization problem $\left(\Sigma_{1c}^T=\Gamma_{1c}\times]0,T[\right)$:

$$\begin{cases} \text{find} g^\varepsilon \in H^1\left(]0,T[;L^2(\Gamma_{1c})\right), \\ \forall h \in H^1\left(]0,T[;L^2(\Gamma_{1c})\right), \quad a(g^\varepsilon,h)+\varepsilon\int_{\Sigma_{1c}^T}\left(g^\varepsilon h+\frac{\partial g^\varepsilon}{\partial t}\frac{\partial h}{\partial t}\right)dxdt=\pi(h), \end{cases} \tag{6}$$

ε being a small positive parameter. The existence and uniqueness of g^ε is very standard. This is the way the optimal control is defined and computed. In practical application, a stabilization can be obtained using Ricatti equations. But the most important point is the asymptotic analysis of g^ε when ε tends to zero. The mathematical procedure is the following one. First of all we define the closed set :

$$K_0=\left\{g\in H^1\left(]0,T[;L^2(\Gamma_{1c})\right);a(g,g)=0 \right\}. \tag{7}$$

Then we introduce the completed space of the quotient space : $H^1\left(\left]0,T\right[;L^2\left(\Gamma_{1c}\right)\right)/K_0$, which is denoted by $C^{\#}$, and defined with respect to the norm induced by the bilinear form a $(\bullet,\bullet)$. From the classical mathematical analysis, one can prove that g^ε converges to an element g of a class g^0 of $C^{\#}$ such that :

$$\forall h \in K_0, <g,h>=0 \qquad (8)$$

where $< , >$ denotes the duality between $H^1\left(\left]0,T\right[;L^2\left(\Gamma_{1c}\right)\right)$ and its dual space.

But the most important point is certainly to decide whether φ_0 and φ_1 are reached or not. The classical strategy in order to answer to this question is to use the adjoint state - say $p\,(x,t)$ - defined as the solution of the following wave equation.

$$p(x,T)=\frac{\partial\varphi_g}{\partial t}(x,T)-\varphi_1(x), \qquad \frac{\partial p}{\partial t}(x,t)=-\left(\varphi_g(x,T)-\varphi_0(x)\right)\text{in}\,\Omega, \qquad (9)$$

$$\begin{cases} \dfrac{\partial^2 p}{\partial t^2}+2U\dfrac{\partial^2 p}{\partial x_1 \partial t}+U^2\dfrac{\partial^2 p}{\partial x_1^2}-C^2\Delta p=0\,\text{in}\,\Omega\times\left]0,T\right[, \\[2mm] p=0\,\text{on}\,\Sigma_0^T=\Gamma_0\times\left]0,T\right[,\dfrac{\partial p}{\partial\nu}=0\,\text{on}\,\Sigma_1^T=\Gamma_1\times\left]0,T\right[. \end{cases} \qquad (10)$$

The existence and uniqueness of a solution to (9) and (10) are similar (and even more simple due to the homogeneous boundary conditions) to those of φ_g. For instance, if:

$$g\in H^1\left(\left]0,T\right[;L^2\left(\Gamma_{1c}\right)\right) \text{ then}$$

$$\begin{cases} \varphi_g \in C^0\left(\left[0,T\right];V\right)\cap C^1\left(\left[0,T\right];L^2\left(\Omega\right)\right) \text{and therefore:} \\[2mm] p(x,T)\in L^2\left(\Omega\right),\dfrac{\partial p}{\partial t}(x,T)\in V' \text{ (dual space of V).} \end{cases} \qquad (11)$$

Consequently the solution of (9)-(10) is such that (see J.L. Lions-E. Magenes [1968]) :

$$p(x,t)\in C^0\left(\left[0,T\right];L^2\left(\Omega\right)\right)\cap C^1\left(\left[0,T\right];V'\right) \qquad (12)$$

and the boundary conditions (10) are satisfied in a weak sense by $p(x,t)$. Furthermore, because g is solution of

$$\begin{cases} \forall h\in\dot{h}\in C^{\#},a(g,h)=\pi(h) \\[2mm] g\in\dot{g}\in C^{\#} \end{cases} \qquad (13)$$

we deduce by a simple computation that p satisfies the equation

$$\forall h\in H^1\left(\left]0,T\right[;L^2\left(\Gamma_{1c}\right)\right)\rangle p,h\langle=0. \qquad (14)$$

This relation implies (in a weak sense !) that :

$$p = 0 \quad \text{on} \quad \Gamma_{1c} \times \left]0,T\right[. \qquad (15)$$

Our goal now is to characterize p satisfying both (10) and (14). Then this will give information on the quantities $\frac{\partial\varphi_g}{\partial t}(x,T)-\varphi_1(x)$ and $\varphi_g(x,T)-\varphi_0(x)$. Obviously, if we prove that (10)-(14) imply that $p = 0$ then we could conclude that

$$\frac{\partial \varphi_g}{\partial t}(x,T)= \varphi_1(x) \quad \text{and} \quad \varphi_g(x,T)=\varphi_o(x) \text{in} \Omega, \tag{16}$$

and therefore the aeroacoustic system can be controlled. The method used is identical to the one of J.L. Lions [1988] and known as the H.U.M. method. The first step is to prove the regularity of several boundary expressions of p. The second one is the derivation of an inverse inequality. The third step consists in deducing that $p = 0$.

2.1. THE INVARIANTS OF THE AEROELASTIC MODEL

Let us first assume that $p\,(x,t)$ and $\frac{\partial p}{\partial t}\,(x, t)$ are smooth (for instance, elements of the space $C^\infty(\Omega)\cap V$) ; then the following computations are justified.

i) Conservation of the energy. Multiplying (10) by $\frac{\partial p}{\partial t}$ and integrating over $\Omega\times]0,T[$, we deduce that ($\nabla \bullet$ is the gradient)

$$E(t)=\left[\frac{1}{2}\int_\Omega \left(\frac{\partial p}{\partial t}\right)^2 +\frac{C^2}{2}\int_\Omega |\nabla p|^2 -\frac{U^2}{2}\int_\Omega \left|\frac{\partial p}{\partial x_1}\right|^2\right](t)=E(T)=\text{constant.} \tag{17}$$

ii) The Maupertuis Principle . Let us multiply (10) by p . We obtain

$$\left[\int_\Omega \left(\frac{\partial p}{\partial t}+U\frac{\partial p}{\partial x_1}\right)p\right]_0^T =\int_o^T \int_\Omega \left(\frac{\partial p}{\partial t}+U\frac{\partial p}{\partial x_1}\right)^2 -C^2|\nabla p|^2. \tag{18}$$

iii) The domain derivative invariant . Let us set $v = \nabla p \bullet \theta$ and let us multiply (10) by this term. The vector θ is arbitrary but with smooth components. It represents a virtual movement of the points of the open set Ω. Using several integrations by parts and because of the boundary conditions satisfied by p, one obtains ($v = (v_1, v_2)$ is the unit outwards normal to $\partial\Omega$):

$$\int_o^T \int_{\Gamma_o}\left(\frac{C^2-U^2 v_1^2}{2}\right)\left(\frac{\partial p}{\partial v}\right)^2 \theta \bullet v+ \int_o^T \int_{\Gamma_{inc}}\left[\frac{1}{2}\left(\frac{\partial p}{\partial t}+U\frac{\partial p}{\partial x_1}\right)^2 -\frac{C^2}{2}\left(\frac{\partial p}{\partial x_1}\right)^2\right]\theta \bullet v$$

$$=\frac{1}{2}\int_0^T \int_\Omega \left[\left(\frac{\partial p}{\partial t}+U\frac{\partial p}{\partial x_1}\right)^2 -C^2|\nabla p|^2\right]div\theta+C^2\int_0^T \int_\Omega \partial_i p\partial_j p\partial_i\theta_j \tag{19}$$

$$-U\int_0^T \int_\Omega \left(\frac{\partial p}{\partial t}+U\frac{\partial p}{\partial x_1}\right)\nabla p\bullet \frac{\partial\theta}{\partial x_1}+\left[\int_\Omega \left(\frac{\partial p}{\partial t}+U\frac{\partial p}{\partial x_1}\right)\nabla p \bullet \theta\right]_0^T,$$

where $\partial_i p=\frac{\partial p}{\partial x_i}$, $i = 1, 2$ (implicit summation is assumed over repeated indices).

2.2. THE REGULARITY RESULTS

First of all, let us notice that we assume $U < C$, and therefore by choosing θ at purpose, one deduces that

$$\text{"}\ \text{if } p(x,T) \in C^0\big([0,T]; V\big) \cap C^1\big([0,T]; L^2(\Omega)\big)\text{"} \quad \text{then} \tag{20}$$

$$\alpha) \frac{\partial p}{\partial \nu} \in L^2\big([0,T[; L^2(\Gamma_0)\big) \text{ and } \beta)\left(\frac{\partial p}{\partial t} + U\frac{\partial p}{\partial x_1}\right)^2 - C^2\left(\frac{\partial p}{\partial x_1}\right)^2 \in L^1\big(0T; L^1(\Gamma_{1nc})\big)$$

Therefore (17)-(18) and (19) can be extended to functions p with the regularity (20).

2.3. INVERSE INEQUALITY (J.L. LIONS [1988])

Let us set in (19) $\theta = (x - x_o)$, where x_0 is an arbitrary point in R^2. Then noticing that :

$div\theta = 2, \partial_i p\, \partial_j p\, \partial_i \partial_j = |\nabla p|^2$, and combining (18) and (19), we deduce that

$$\int_0^T \int_{\Gamma_0}\left(\frac{C^2 - U^2 v_1^2}{2}\right)\left(\frac{\partial p}{\partial \nu}\right)^2 \theta \bullet v + \int_0^T \int_{\Gamma_{1nc}}\left[\frac{1}{2}\left(\frac{\partial p}{\partial t} + U\frac{\partial p}{\partial x_1}\right)^2 - \frac{C^2}{2}\left(\frac{\partial p}{\partial x_1}\right)^2\right]\theta \bullet v$$

$$= TE(T) + \left[\int_\Omega\left(\frac{\partial p}{\partial t} + U\frac{\partial p}{\partial x_1}\right)\left(\frac{p}{2} + \nabla p \bullet \theta\right)\right]_0^T, \tag{21}$$

$$\geq TE(T) - 2\alpha \sup_{t \in]0,T[}\left\{\frac{1}{2}\int_\Omega\left(\frac{\partial p}{\partial t} + U\frac{\partial p}{\partial x_1}\right)^2 + \frac{1}{2\alpha^2}\int_\Omega\left(\frac{p}{2} + \nabla p \bullet \theta\right)^2\right\}(t), (\forall\, \alpha > 0) \text{ where} \quad \text{we}$$

have set D $= \max_{x \in \Omega}(x - x_0)$. But from (Cf. J.L. Lions [1988]) :

$$\int_\Omega\left(\frac{p}{2} + \nabla p\theta\right)^2 \leq \int_{\Gamma_{1nc}}\frac{p^2}{2}\theta \bullet v + D^2\int_\Omega|\nabla p|^2 \leq \left(D^2 + \frac{DC_0^2}{2}\right)\int_\Omega|\nabla p|^2 \tag{22}$$

(C_0 is the constant of continuity of trace operator from $H^1(\Omega)$ into $L^2(\Gamma_{1nc})$). Let us

set $\alpha = \dfrac{D}{C}\sqrt{1 + \dfrac{C_0^2}{D}}$. Then the inequality (21) leads to

$$\int_0^T \int_{\Gamma_0}\left(\frac{C^2 - U^2 v_1^2}{2}\right)\left(\frac{\partial p}{\partial \nu}\right)^2 \theta \bullet v + \int_0^T \int_{\Gamma_{1nc}}\left[\frac{1}{2}\left(\frac{\partial p}{\partial t} + U\frac{\partial p}{\partial x_1}\right)^2\right]\theta \bullet v \tag{23}$$

$$\geq TE(T) - \frac{2D}{C}\sqrt{1 + \frac{C_0^2}{D}}\sup_{t \in]0,T[}\left\{\frac{1}{2}\int_\Omega\left(\frac{\partial p}{\partial t} + U\frac{\partial p}{\partial x_1}\right)^2 + \frac{C^2}{2}\int_\Omega|\nabla p|^2\right\}(t) \ .$$

Finally one can check by a simple computation that

$$\frac{E(t)}{1+\dfrac{U}{C}} \leq \frac{1}{2}\int_{\Omega}\left(\frac{\partial p}{\partial t}+U\frac{\partial p}{\partial x_1}\right)^2 + \frac{C^2}{2}\int_{\Omega}|\nabla p|^2 \leq \frac{E(t)}{1-\dfrac{U}{C}} \tag{24}$$

and therefore : $2E(T)(T-T_0) \leq$

$$\int_0^T (\int_{\Gamma_0}\left(C^2-U^2 v_1^2\right)\left(\frac{\partial p}{\partial v}\right)^2 \theta\bullet v + \int_{\Gamma_{1nc}}\left[\left(\frac{\partial p}{\partial t}+U\frac{\partial p}{\partial x_1}\right)^2 -C^2\left(\frac{\partial p}{\partial x_1}\right)^2\right]\theta\bullet v)$$

with

$$T_0 = \frac{2D}{C}\sqrt{1+\frac{C_0^2}{D}\left(\frac{1}{1-\dfrac{U}{C}}\right)}. \tag{25}$$

2.4. THE EXACT CONTROLLABILITY RESULTS

If the point x_0 is chosen as in Figure 2, then $\theta\bullet v\leq 0$ on Γ_0 and $\theta\bullet v=H$ on Γ_{1nc}. The inequality (24) becomes :

$$\int_0^T \int_{\Gamma_{1nc}}\left[\frac{1}{2}\left(\frac{\partial p}{\partial t}+U\frac{\partial p}{\partial x_1}\right)^2 -\frac{C^2}{2}\left(\frac{\partial p}{\partial x_1}\right)^2\right] \geq \frac{E(T)}{H}(T-T_0). \tag{26}$$

The first consequence is that if $\Gamma_{1nc} = \phi$, and for $T>T_0$, then $E\ (T) = 0$, and therefore

$$\varphi_g(x,T)=\varphi_0(x),\frac{\partial\varphi_g}{\partial t}(x,T)=\varphi_1(x). \tag{27}$$

In such a situation the aeroacoustic surroundings can be exactly controlled. But unfortunately it seems more realistic to consider the situation where $\Gamma_{1nc}\neq\phi$. Then the sign of the left-hand side of (26) is not obvious. For instance let $p(x,t)$ be a solution of (see Figure 2 and L is the length of Γ_{1nc}):

$$\begin{cases}\dfrac{\partial^2 p}{\partial t^2}+2U\dfrac{\partial^2 p}{\partial x_1\partial t}+\left(U^2-C^2\right)\dfrac{\partial^2 p}{\partial x_1^2}=0\text{ on }\Gamma_{1nc}\times]0,T[\\ p(0,t)=p(L,t)=0\ \forall t\in\ ]0,T[.\end{cases} \tag{28}$$

Then the left-hand side of (26) is zero. A simple calculus enables one to characterize the solutions of (28). The physical meaning of these solutions is certainly close to the one of the Rayleigh or Stoneley waves. Their velocities are $U+C$ and $U-C$. In our case it is possible to improve the controllability results. First of all let us notice that for arbitrary functions $p(x,t)$ one has (we use Cauchy-Schwarz and Poincaré inequalities) :

$$\int_0^T \int_{\Gamma_{1nc}}\left(\frac{\partial p}{\partial t}+U\frac{\partial p}{\partial x_1}\right)^2 -C^2\left(\frac{\partial p}{\partial x_1}\right)^2 \leq \frac{C}{U}\int_0^T \int_{\Gamma_{1nc}}\left[\left(\frac{\partial p}{\partial t}\right)^2 -\left(\frac{\Pi U}{L}\right)^2\left(\frac{C^2}{U^2}-1\right)p^2\right]. \tag{29}$$

Let us now consider the subspace V^N of V spanned by the N first eigenmodes of the aeroelastic model. The corresponding frequencies are denoted by η_j $j = 1,N$. Then the aeroelastic model can be exactly controlled in a time T_0 (defined at (23)), if the term φ_0, φ_1 belongs to the space $(V^N)^2$ and as soon as

$$(2\Pi\eta_N)^2 \leq \left(\frac{\Pi C}{L}\right)^2 \left(1 - \frac{U^2}{C^2}\right). \tag{30}$$

The interesting point is that the control time T_0 is independent of the eigenmodes concerned. We underline that the information given here by the multiplier method are more precise than those deduced from the Bellman controllability theorem, (see Ph. Destuynder, A. Saidi [1994]).

3. Controllability of the structural model

The flexible part of the airfoil which was denoted by Γ_{1c} in the previous section, is modelled by a beam. The deflection is denoted by $w(x_1, t)$ and the motion is driven by a system of piezo-devices as shown in Figure 3. The effect is equivalent to a pointwise force and a pointwise bending moment. The equations of the beam are:

$$\begin{cases} \forall t: w(0,t) = w(\ell,t) = \dfrac{\partial w}{\partial x_1}(0,t) = \dfrac{\partial w}{\partial x_1}(\ell,t), w(x_1,0) = \dfrac{\partial w}{\partial t}(x_1,0) = 0 \, \mathrm{on} \, \Gamma_{1c}, \\[2mm] \rho_s \dfrac{\partial^2 w}{\partial t^2} + D\dfrac{\partial^4 w}{\partial x_1^4} = -\rho_F\left(\dfrac{\partial \varphi}{\partial t} + U\dfrac{\partial \varphi}{\partial x_1}\right) + V_1(t)\delta_a(x_1) + V_2(t)\delta_a'(x_1) \, \mathrm{on} \, \Gamma_{1c}, \end{cases} \tag{31}$$

where ρ_s (or ρ_F) is the mass density of the beam (or of the fluid), D is the bending modulus and V_1, V_2 are the voltages applied to the two piezo-devices. Let us introduce the solution w^0 of

$$\begin{cases} w^0(0,t) = w^0(\ell,t) = \dfrac{\partial w^0}{\partial x_1}(0,t) = \dfrac{\partial w^0}{\partial x_1}(\ell,t) = w^0(x_1,0) = \dfrac{\partial w^0}{\partial t}(x_1,0) = 0, \\[2mm] \rho_s \dfrac{\partial^2 w^0}{\partial t^2} + D\dfrac{\partial^4 w^0}{\partial x_1^4} = -\rho_F\left(\dfrac{\partial \varphi_g}{\partial t} + U\dfrac{\partial \varphi_g}{\partial x_1}\right), \mathrm{on} \, \Gamma_{1c} \times \,]0,T[, \end{cases} \tag{32}$$

where φ_g is the aeroacoustic solution corresponding to the optimal control g^ε discussed in Section 2. Then we set $\widetilde{w}(x,t) = w(x,t) - w^0(x,t)$ and we define the following control problem $(T' \ll T)$:

$$find V_1(t), V_2(t) \, \mathrm{such\,that} : \frac{\partial \widetilde{w}}{\partial t}(x_1,T') = -g(x_1,T'), \widetilde{w}(x_1,T') = 0 \, \forall x_1 \in \Gamma_{1c}, \tag{33}$$

Furthermore let us point out that $\widetilde{w}(x_1, T')$ is not really important in the modelling. This is why we suggest a homogeneous condition. But the physical aspect of the previous strategy can be justified only if the control delay T' can be chosen small enough compared to T. We know that this is possible but with increasing amplitudes of the controls V_1 and V_2. The problem (33) has been discussed in Ph. Destuynder-A. Saidi [1994] when only one piezo-device is used. We proved that no controllability result was realistic. The application of the mulyiplier method leads to an inverse inequality for the adjoint state $z(x_1,t)$, (a is the abscissa of the extremities of the piezo-devices):

$$\begin{cases} \rho_s \dfrac{\partial^2 z}{\partial t^2} + D\dfrac{\partial^4 z}{\partial x_1^4} = 0 \ on \ \Gamma_{1c} \times \,]0,T'[,\ z(a,t) = \dfrac{\partial z}{\partial x_1}(a,t) = 0,\ \forall t \in \,]0,T'[, \\ \\ z(0,t) = z(\ell,t) = \dfrac{\partial z}{\partial x_1}(0,t) = \dfrac{\partial z}{\partial x_1}(\ell,t) = 0 \,\forall t \in \,]0,T'[. \end{cases} \tag{34}$$

Then multiplying (34) by $\dfrac{\partial z}{\partial x_1}x_1$ (assuming the regularity of z as we did for the aeroacoustic model), we obtain the two following inequalities (Ph. Destuynder [1998]):

$$\left(T' - \frac{a}{\lambda_a}\right)E^{sa}(T) \le \frac{aD}{2}\int_o^{T'}\left[\frac{\partial^2 z}{\partial x_1^2}(a,t)\right]^2 dt \le \left(T' + \frac{a}{\lambda_a}\right)E^{sa}(T') \tag{35}$$

where $E^{sa}(T)$ is the energy of the structure located between 0 and a and defined by

$$E^{sa}(T) = \left\{\frac{\rho_s}{2}\int_0^a\left(\frac{\partial z}{\partial t}\right)^2 dx + \frac{D}{2}\int_0^a\left(\frac{\partial^2 z}{\partial x_1^2}\right)dx\right\}(T'), \tag{36}$$

and λ_a is the smallest eigenvalue of the Poisson model on $]0,T[$, defined by

$$\lambda_a = \left(\frac{\Pi}{a}\right)\sqrt{\frac{D}{\rho_s}}.$$

From (35) we deduce that

$$T'\left|\frac{E^{sa}(T')}{a} - \frac{E^{s(\ell-a)}(T')}{\ell-a}\right| \le \frac{E^{sa}(T')}{\lambda_a} + \frac{E^{s(\ell-a)}(T')}{\lambda_{\ell-a}}$$

and finally, the only movements which can not be exactly controlled would correspond to those satisfying the condition

$$\frac{E^{sa}(T')}{a} = \frac{E^{s(\ell-a)}(T')}{\ell-a}. \tag{37}$$

This result can be improved using the analytical expressions of the eigenmodes of the beam. For instance, for a simply supported beam the exact controllability can be easily obtained because of the orthogonality with respect to the time of the coefficients of the eigenmodes. For further details we refer to Destuynder [1998].

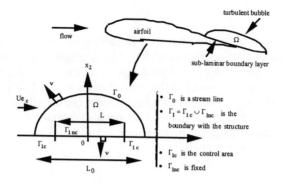

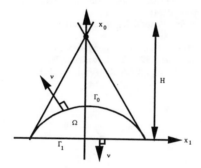

Figure 1 The geometry to be considered

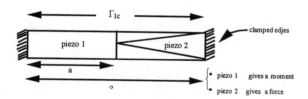

Figure 2 The position of x_0

Figure 3. The piezo-devices on the airfoil

4. Conclusion

A simple aeroacoustic model have been used in this paper for describing the dynamic behaviour of the waves near a subsonic airfoil. The decoupled controllability problem has been discussed. Several rules have been pointed out. They aim at a better efficiency of the strategy. An important aspect of the method suggested is that piezo-devices are used. Basing upon the very good experimental results obtained by J. Wojciechowski, T Szuster and J. Pietrucha [1998], who used a harmonic function for the voltage of the piezo-device, the present analysis gives a few rules which should enable one to improve the efficiency of the method.

References

1. Ph. Destuynder, [1998]. Mathematical aspects of a fluid-structure controllability problem, to appear.
2. Ph. Destuynder, A. Saidi [1997]. Smart materials and flexibles structures. Control and cybernetics, vol. 26, n° 2, p. 161-205, (IAT internal report [1994].
3. J.L. Lions [1988]. Contrôlabilité exacte. Perturbation et stabilisation de systèmes distribués. RMA n° 8-9 - MASSON - Paris.
4. J.L. Lions - E. Magenes [1968]. Problèmes aux limites non homogènes. Tomes 1 et 2, Dunod, Paris.
5. J. Wojciechowski, T. Szuster, J. Pietrucha [1998]. Étude expérimentale de la réduction de traînée d'un profil laminaire à l'aide d'actionneurs piezo électriques. CRAS - Acad. Sci., Paris t 326, série IIb, p.85-90.

SUPERVISION, MAINTENANCE AND RENOVATION OF REINFORCED AND PRESTRESSED CONCRETE BRIDGES

GY. FARKAS
Technical University of Budapest, Dept. of Reinforced Concrete Structures
H-1521 Budapest, Hungary

1. Introduction

Considering that the volume of road transport has multiplied in the course of the past few years, the maintenance of more than the 5700 road bridges in Hungary is an actual problem with considerable technical and economical consequences. Therefore, the development of an adequate Bridge Management System (BMS) is a common interest.

The renovation and, if necessary, the strengthening works of these bridges are carried into effect in function of the results of their regular supervisions and the possibilities of the financial background. The determination of the optimal moment of the renovation is very difficult because of the subjectivity of the visual supervisions that we generally made, as well as the uncertainties around the predictions of the material and mechanical characteristics of the construction. However, the development of a regular control system based on the "smartness" of the structures can conduce to establish an efficient and economical method to help the decision concerning the necessity of the strengthening of a bridge. The smartness of a structure in this regard signifies the variation of its response to various excitations, from which conclusions about the actual state of the construction can be drawn.

After a short presentation of the general state of the road bridges in Hungary, the description of a proposed method for monitoring, based on the measurement of the dynamic characteristics of the structure with some applications, will be presented in this paper.

2. Actual state of the road bridges in Hungary

Principal characteristics of more than 5200 Hungarian road bridges are registered in the National Road Data Register (cf.[1]). Their distribution according to the

J. Holnicki-Szulc and J. Rodellar (eds.), Smart Structures, 63–70.

64

theoretical load carrying capacity corresponding to the Hungarian Road Bridge Codes is shown in Table 1.

TABLE 1. Distribution of the Hungarian road bridges according to their load carrying capacity

Load carrying capacity in kN (Hungarian Code)	800	600	400	360	max 200
Number of bridges	1810	383	2417	429	168
Percentage (%)	34	7	47	8	4

Based on the results of the yearly due visual supervisions, the bridges are classified in five categories. The different categories that characterize the general state of a bridge are specified as follows:

Category 1. good general state (without visual deterioration),
2. starting of the deterioration (only small superficial faults detected),
3. middle importance deteriorations (not only superficial faults found),
4. considerable deterioration (advanced corrosion traces detected),
5. dangerous state (structural stability in danger).

The distribution of the bridges according to their state category is shown in Fig.1.

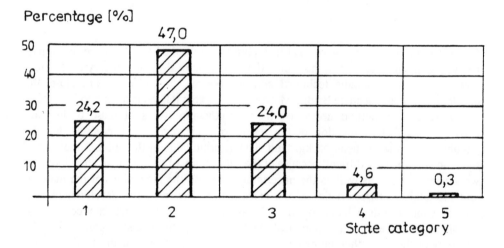

Figure 1. Distribution of the Hungarian road bridges according to their state category

In their publication, S. B. Chase and L. Gáspár (cf. [2]) developed a solution to estimate the relation between the state category and the load carrying capacity of a bridge. The method was based on the application of the regression analysis and the theory of Markoff - chains. The results concerning the Hungarian road bridges due to the application of this method are shown in Fig. 2. The bridges in state category 5 were excluded from the study because of their relatively small number. The relation developed with the use of the method of minimum squares to estimate the load carrying capacity of a bridge is given by the following expression,

Load carrying capacity [kN] = 610 - 64,6 (State category)   (1)

with the coefficient of correlation equal to 0,97.

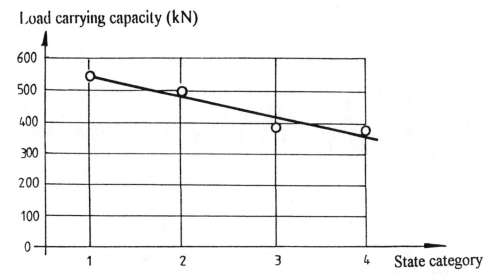

Figure 2. Relation between the load carrying capacity and the state category of Hungarian road bridges

Experiences show that only simple visual supervisions are not always sufficient to estimate the real state of a bridge with sufficient precision and to take decisions concerning the necessity of its renovation or strengthening. Such decisions must be based on more adequate considerations. Principles of a monitoring system, based on the variation of the dynamic characteristics of the structures, are shown in the next chapters.

3. Dynamic characteristics of the reinforced and prestressed concrete structures

The continuous surveying of a bridge can be based on the measurement of the variation of its dynamic characteristics. The main characteristics, from which conclusions can be taken concerning the change of the general state (cracking or variation of the prestressing force for example) of a reinforced or prestressed concrete structure are the natural frequency and the damping coefficient. Experiences shown (cf. [3]) that the damping coefficient of an uncracked reinforced concrete bridge deck is around D = 0,05 and it changes to proximity of D = 0,15 if considerable cracks appear on the structure. Also, the natural frequencies of a structure must theoretically decrease with diminution of the intensity of the prestressing force applied to it (cf. [4]).

The determination of the dynamic characteristics of a structure mentioned above may be based on the registration of the time-dependent A (t) amplitude variation of the vibration of the construction submitted to an excitation. The diagram of the movement of a tested bridge deck is shown in Fig. 3.

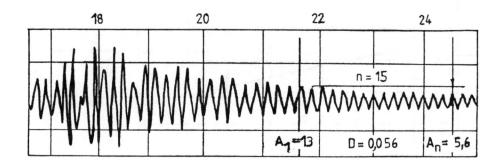

Figure 3. Vibration diagram of a tested bridge deck

The dynamic characteristics of the structure can be determined corresponding to the free vibration range. The damping coefficient is given by the following expression:

$$D = \frac{1}{n} \ln \frac{A_1}{A_n},$$

(2)

where A_1 and A_n are the first and the last amplitude of the free vibration in the examined range, and n is the number of the waves in this range (generally $n = 15$ is proposed).

The examination of the first three vibration modes are satisfactory to have sufficient information about the dynamic properties of a construction. The measurement of the characteristics of the higher vibration modes is generally very difficult or, in many cases, impossible.

The facility of the determination of the dynamic properties of a structure highly depends on the method of the excitation applied on it. The ideal excitation in this respect is called "white noise", in which the amplitude of the excitation is almost the same for all frequencies, namely $A(\omega)$ = const. The influence of the wind in several frequency ranges approximates the white noise excitation. To the realization of the regular supervision of a reinforced concrete road bridge deck, the excitation due to the normal road traffic can be used.

The registration of the response to the excitation of a structure can be based on the measurement of the time-dependent amplitude $A(t)$ or acceleration $a(t)$ functions of the movement. Because of its higher intensity due to the second derivative of the function of the movement, the measurement of the acceleration is technically easier. The disadvantage of this method is that the range of measurement must be known and adjusted previously.

The registration of the time-dependent amplitude variation of the movement due to the vibration can be based on the continual measurement of the displacements at different points of the construction with use of electrical displacement traducers. The displacements can be continually recorded by an instrumentation tape recorder. The position of the measuring points on the bridge deck must be chosen appropriately to be able to determine the different characteristics of the flexural and torsional vibration modes, too. The registration of the time-history of deflections in six symmetrically positioned points on the bridge deck is generally sufficient.

The evaluation of the results is based on the determination of the time-dependent amplitude functions $A(t)$ in the examined points of the structure. Generally from these functions the first natural frequency of the bridge deck may be directly determined and the damping coefficient must be calculated from the equation (2). The estimation of the exact values of the natural frequencies of the structure is based on the application of a Fourier transformation from which the amplitude spectrum $A(\omega)$ of the vibration can be determined. A special computer program was developed to execute this transformation.

4. Applications for testing of bridge decks

Some applications of the dynamic response method to evaluate the efficiency of the strengthening, and the proposition to develop a monitoring system of a bridge will be shown in this chapter.

4.1. TESTING OF THE EFFICIENCY OF THE STRENGTHENING OF TWO REINFORCED CONCRETE BRIDGES

Two reinforced concrete bridges were repaired and strengthened by additional post-tensioning. Both were continuous two-span structures of 29,0 and 38,7 m between abutments with beam-slab superstructures. The strengthening was necessary because of the results of a long-time deterioration. The load carrying capacities of the continuous beams were increased by additional post-tensioning. Static and dynamic load testing were performed on the bridges both before and after the strengthening works to verify its efficiency. The longitudinal sections of the two bridges are shown in Fig. 4.

Dynamic load tests were performed to determine the dynamic deflection factors and the natural frequencies of the bridges before and after retrofitting. One 20 ton truck was used to vibrate the bridge. The truck crossed each bridge several times at speed of 5, 30 and 60 km/h. The measurement was realised by the use of six displacement transducers. Each transducers measured the time-history of deflections, which were recorded by a tape recorder. The dynamic deflection factors were calculated as the ratio of the maximum deflection at 60 km/h speed to the maximum deflexion under the 5 km/h run. The results are shown in Table 2. It is seen that the dynamic deflection factors after strengthening are considerably lower than before retrofitting.

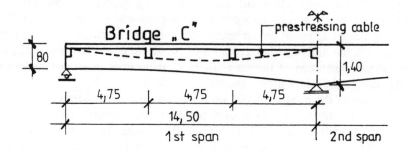

Figure 4a. Longitudinal section of the bridge C

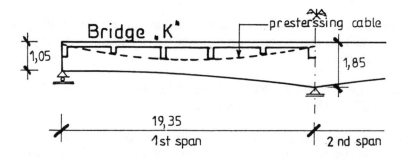

Figure 4b. Longitudinal section of the bridge K

TABLE 2. Dynamic deflection factors of bridges C and K

Bridge	Dynamic factors				Hungarian Code
	Before strengthening		after strengthening		
	1st span	2nd span	1st span	2nd span	
C	1,41	1,67	1,12	1,1	1,3
K	1,28	1,14	1,14	1,05	1,26

The natural frequencies of the bridges were determined by the so-called "microtremor" method. The excitations in these tests were the normal highway traffic. The measurements were made by accelerometers. The average response spectrum of the measured accelerations gives the natural frequency. Results of these tests are shown in Table 3. It is seen that there is a considerable increase in the natural frequencies of the bending and torsional modes of vibrations as a result of the strengthening.

TABLE 3. Natural frequencies of bridges C and K

Bridge	Mode	Natural frequencies [Hz]		Change [%]
		before strengthening	after strengthening	
C	bending	8,15	8,74	7,2
	torsional	11,62	11,87	2,2
K	bending	4,20	4,39	4,5
	torsional	6,59	7,08	7,4

4.2 MONITORING OF A PRESTRESSED CONCRETE BRIDGE

Considerable corrosion of the prestressing tendons was detected in the Bodrog bridge near Alsóberecki in Hungary. The structural system of the three-span reinforced concrete arche bridge with prestressed slab bridge deck is shown in Fig. 5.

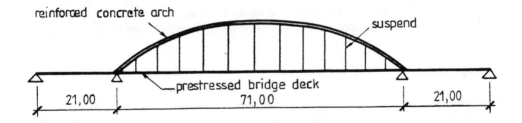

Figure 5. Structural system of the Bodrog bridge near Alsóberecki

Static and dynamic load tests indicate that immediate strengthening of the bridge is not necessary, mainly because it was originally designed for uncracked condition. The damping coefficient given by test was $D = 0.056$. However, further corrosion of prestressing tendons is not excluded therefore, monitoring of the bridge behaviour is required. The aspects of selection of the method were that they could be used on a long run under varying environmental conditions, without the continuous presence of a measuring staff. For this reason the following methods were studied.

a) Gluing of hybrid non-metallic (composed of lower and higher failure strength fibers) reinforcement to the surface, from which the failure of the lower failure strength fiber may give an indication of the increasing strains of the structure.

b) Glass fiber optic sensors could be glued to the surface of the bridge giving a possibility for continuous control of stresses.

c) Continuous measurement of the deflections, even at the mid-span section above the river, would be possible by a laser. Measurement results could be transferred by telephone line to the required location.

d) A continuous registration of deformations in the characteristic points of the bridge would be possible with strain gauges. Deformation measurements could be stored on site or transferred by telephone line to the required location.

e) Accelerometers could be adjusted to characteristic points of the structure in order to register the frequency distributions. As it was shown above, modifications in frequency spectra would give an indication to modifications in the characteristics of the bridge. Frequency signals also could be stored in site or transferred to the required location.

After examination of the possible methods, the parallel use of the last two was accepted to realisation. The installation of the necessary instruments is in progress.

5. Conclusion

Necessities and possibilities of the use of the methods based on the "smartness" of the structures to know more about their real behaviour is shown in this paper. Examples show that the measurement and the evaluation of the dynamic characteristics of a construction may be a useful method to verify the efficiency of its strengthening or develop a system of monitoring which is in the focus of interest worldwide.

References

1. *Directory of the Hungarian Road Bridges*, (1993) Edited by the Ministry of Transportation, Communication and Hydraulics, Budapest (in Hungarian)

2. Chase, S. B. and Gáspár, L. (1998): Modelling of the diminution of the load carrying capacity of road bridges, *Scientific Revue of Civil Engineering* Vol. XLVIII. no. 1. pp.9-15. (in Hungarian)

3. Farkas, Gy. and Szalai, K. (1998): Supervision of the Bodrog Bridge by Alsóberecki, *Scientific Revue of Civil Engineering* Vol. XLVIII. no. 4. pp (in Hungarian)

4. Dalmy, D. (1995): Static and Dynamic Load Testing of Concrete Bridges Before and After Retrofitting, *Proceeding of the Congress Restructuring XIII., America and Beyond*, Vol. 2. pp. 1369-1373.

NUMERICAL AND EXPERIMENTAL INVESTIGATIONS OF ADAPTIVE PLATE AND SHELL STRUCTURES

U. GABBERT, H. KÖPPE, F. LAUGWITZ
Institut für Mechanik
Otto-von-Guericke-Universität Magdeburg
D-39106 Magdeburg, Universitätsplatz 2

1. Introduction

In many branches of engineering lightweight design has become very important to reduce the mass and the energy consumption and to increase simultaneously the safety,
integrity and environmental compatibility of a system. To meet these opposite objectives adaptive structural concepts have attracted increasing attention. These concepts are
characterized by a synergistic integration of active (smart) materials, structures, sensors,
actuators, and control electronic to an adaptive system. The potential uses of such concepts cover the entire range from mechanical engineering, aerospace engineering and
civil engineering to manufacturing, transportation, robotics, medicine etc. The use of
plate and shell structures as basic components of such adaptive systems is very common. Such smart material systems often consist of different layers of passive and active
materials, e.g. steel or aluminium sheets attached with piezoceramics, fiber reinforced
composites with embedded piezoelectric wafers etc. [1], [2], [4], [6]. The global behavior of piezoelectric smart structures can be modeled with sufficient accuracy by the
linearized coupled electromechanical constitutive equations. However, it is generally
recognized that analytical solutions of the coupled electro-mechanical field equations
are limited to relatively simple geometry and boundary conditions. In practical applications, finite element (FE) techniques provide the versatility in modeling, simulation,
analysis and optimal design of real engineering adaptive structures.

The paper gives an overview about our finite element simulation tool. Then two
examples are presented from a number of test examples which we have investigated recently by numerical as well as experimental methods.

2. Governing Equations of the Electromechanical Problem

The following derivation of the electromechanical fundamental relations is based on
conventional formulations and notations of the theory of elasticity. In the sense of the
principle of virtual work extended by the electrical part we can multiply these equations
with a virtual displacement $\delta\mathbf{u}$ and with a virtual electric potential $\delta\Phi$, respectively. In-

J. Holnicki-Szulc and J. Rodellar (eds.), Smart Structures, 71–78.

tegration over the entire domain and the surface with applied loads, respectively, provides the coupled electromechanical functional

$$\chi = \int_V \delta\mathbf{u}^T\left(\mathbf{B}_u^T\sigma + \overline{\mathbf{p}} - \rho\ddot{\mathbf{u}}\right)dV + \int_V \delta\Phi\left(\mathbf{B}_\Phi^T\mathbf{D}\right)dV - \int_{O_q} \delta\mathbf{u}_q^T\left(\overline{\mathbf{t}} - \mathbf{t}\right)dO - \int_{O_Q} \delta\Phi\left(\overline{Q} - Q\right)dO = 0 , \quad (1)$$

with the stress vector σ, the body force vector $\overline{\mathbf{p}}$, the electrical displacement vector $\mathbf{D}$, the vector of applied surface traction $\overline{\mathbf{t}}$, the vector of applied surface charge $\overline{Q}$ and the density ρ. The linear coupled electromechanical constitutive equations have the form

$$\sigma = \mathbf{C}\varepsilon - \mathbf{e}\mathbf{E} , \qquad \mathbf{D} = \mathbf{e}^T\varepsilon + \kappa\mathbf{E} , \qquad\qquad (2a,b)$$

with the (6x6) elasticity matrix $\mathbf{C}$, the (6x3) piezoelectric matrix $\mathbf{e}$, the (3x3) dielectric matrix κ, the strain vector ε and the electric field vector $\mathbf{E}$. In a preliminary polarised piezoelectric ceramic, where the direction of the polarisation at each point of the body is assumed to be direction 3 of a local Cartesian co-ordinate system, the material tensors are reduced to $c_{11}, c_{12}, c_{13}, c_{33}, c_{44}, c_{66}, e_{31}, e_{33}, e_{15}, \kappa_{11}, \kappa_{33}$, where the five independent elastic constants are measured under constant (or vanishing) electric field, and the three piezoelectric constants and the two dielectric constants are measured under constant (or vanishing) deformation.

Introducing equations (2) into equation (1) and using partial integration, the Gaussian integral theorem, the strain-displacement relationship $\varepsilon = \mathbf{B}_u\mathbf{u}$ and the analogous relationship for the electric field $\mathbf{E} = -\mathbf{B}_\Phi\Phi$, we get a suitable form of the functional (1) to derive finite element matrices as

$$\chi = \int_V \delta\mathbf{u}^T\rho\ddot{\mathbf{u}}\,dV + \int_V \left(\mathbf{B}_u\delta\mathbf{u}\right)^T\mathbf{C}\mathbf{B}_u\mathbf{u}\,dV + \int_V \left(\mathbf{B}_u\delta\mathbf{u}\right)^T\mathbf{e}\mathbf{B}_\Phi\Phi\,dV$$

$$+ \int_V \left(\mathbf{B}_\Phi\delta\Phi\right)^T\mathbf{e}^T\mathbf{B}_u\mathbf{u}\,dV - \int_V \left(\mathbf{B}_\Phi\delta\Phi\right)^T\kappa\mathbf{B}_\Phi\Phi\,dV \qquad (3)$$

$$- \int_V \delta\mathbf{u}^T\overline{\mathbf{p}}\,dV - \int_{O_t} \delta\mathbf{u}^T\overline{\mathbf{t}}\,dO - \int_{O_Q} \delta\Phi\overline{Q}\,dO = 0$$

3. Finite Element Formulation and Piezoelectric Element Library

For finite element discretization we approximate the mechanical displacements and the electric potential in a finite element by using the interpolation functions G_L

$$u_i = \sum_{L=1}^N G_L u_{iL} , \qquad \Phi = \sum_{L=1}^N G_L\Phi_L , \qquad\qquad (4a,b)$$

where N is the number of element nodes and L is the nodal index. For an electromechanical finite element we extend the mechanical element nodal vector $\mathbf{q}_e$ by adding the electric potential Φ to each node

$$\mathbf{q}_e^T = \left[\mathbf{u}_1^T \;\vdots\; \Phi_1 \;\vdots\; \cdots \;\vdots\; \mathbf{u}_L^T \;\vdots\; \Phi_L \;\vdots\; \cdots \;\vdots\; \mathbf{u}_N^T \;\vdots\; \Phi_N\right]. \qquad\qquad (5)$$

To express **e** and **E** we need the derivatives of the shape functions G

$$\begin{bmatrix} \varepsilon \\ E \end{bmatrix} = \begin{bmatrix} B_{uG_1} & 0 & \vdots & B_{uG_2} & 0 & \cdots & B_{uG_N} & 0 \\ 0 & B_{\Phi G_1} & \vdots & 0 & B_{\Phi G_2} & \cdots & 0 & B_{\Phi G_N} \end{bmatrix} q_e = B_{u\Phi G} \, q_e \qquad (6)$$

where

$$B_{uG_L} = B_u \, G_L, \quad B_{\Phi G_L} = B_\Phi \, G_L. \qquad (7a,b)$$

Introducing these approximations into equation (3) and using the fundamental lemma of the calculus of variation provide the relations for one finite element (index e) in the form of a differential equation system

$$M_e \ddot{q}_e + R_e \dot{q} + K_e q_e = f_e, \qquad (8)$$

where M_e is the element mass matrix, K_e is the element stiffness/electric matrix and f_e is the element load vector, respectively.

$$M_e = \int_{V_e} \rho G^T G dV, \quad K_e = \int_{V_e} B_{u\Phi G}^T \begin{bmatrix} C & e \\ e^T & -\kappa \end{bmatrix} B_{u\Phi G} dV, \quad f_e = \begin{bmatrix} \int_{V_e} G^T \overline{p} dV + \int_{O_{eq}} G^T \overline{t} dO \\ \int_{O_{eQ}} G^T \overline{Q} dO \end{bmatrix}$$

$$(9a,b,c)$$

In addition, a rate-dependent damping matrix R_e can be included to take into account damping effects as well.

Based on the theoretical background given above a library of piezoelectric finite elements has been developed (Figure 1) and testwise implemented in the general purpose finite element code COSAR. The shape functions of the elements can be linear or quadratic, and the isoparametric element concept has been used to approximate the element geometry. The solid element family consists of a basic brick element (hexahedron) and some special degenerate elements which have been derived by collapsing nodes [8]. The quadrilateral and triangular multilayered shell elements shown in Figure 1 have been developed on the basis of a triangular piezoelectric shell element by H.S.Tzou & R.Ye [3]. We added a quadrilateral element and extended the geometry approximation by an isoparametric description, and consequently, complex laminated structures can be modelled by these elements [11]. Thin shell assumptions can be included for the shell as a whole or several layers, and consequently, the number of degrees of freedom can be reduced by constraint conditions. Global and local effects such as the transfer behavior of active to passive parts of a structure or delamination propagation can also be investigated effectively by these elements [12].

Piezoelectric elements and conventional mechanical elements can be combined in one model. This technique provides an efficient way to analyse complex adaptive structures which contain in general only a few special piezoelectric actuators or sensors located at special parts. The finite element code has the capability to use a substructure technique, and consequently, it is possible to separate mechanical and piezoelectric structures, and only the hyperstructure has merged DOF's.

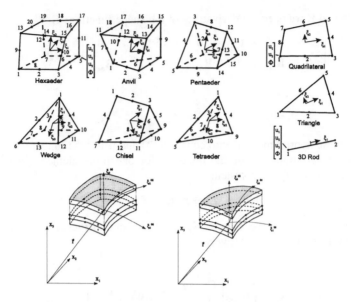

Figure 1. Piezoelectric finite element library

4. Comparison of Numerical and Experimental Investigations

In this section two examples of smart structures are presented which have been analysed by numerical as well as experimental methods. In the experimental investigations the voltage supply for the piezoceramic patches was realized with piezodrivers having an adjustable offset voltage 0 V to +100 V. With respect to the low stiffness of the structures a non contact method of displacement measurements using a laser triangulation sensor was used. The measurement range of this sensor is ± 1 mm and the resolution is 0.5 μm. The driving voltage was switched between 0 V and 100 V several times before accepting the readings of displacement in order to normalize the influence of the hysteresis behaviour of the piezoceramic patches.

4.1 ADAPTIVE BEAM STRUCTURE

Figure 2 gives the geometry, the position of the actuators and the finite element model of the adaptive beam structure. The material of the base structure is aluminium. The bending deformation results from the contraction and the stretch of the piezoelectric actuators which are glued at the upper and the lower side of the beam. For this example we used a control voltage of 100 V. The piezoceramic layers of 0.2 mm thickness consist of PZT material PIC 151 made by PI Ceramic GmbH. The static deflections induced by externally applied voltages are calculated using 640 3D finite elements with 20 nodes per element. For the numerical simulation we used two sets of data for the PIC 151material with different electric material parameters. One data set (Mat. 2 in Figure 3), which is given in Figure 2, was taken from an official data sheet of the supplier, and another data set (Mat. 1 in Figure 3) is from an internal source of PI Ceramic GmbH. Figure 3 compares the FE solution for the static beam deflection along the beam length with laboratory experiments.

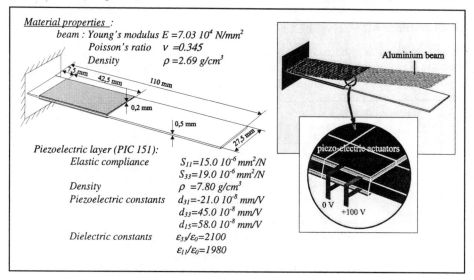

Figure 2. Model of a clamped beam with piezoelectric actuators

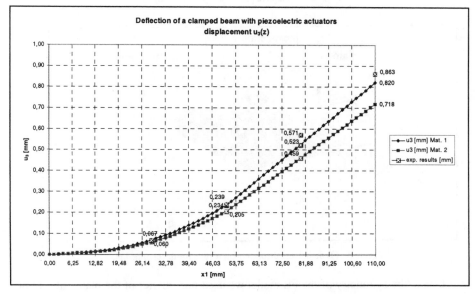

Figure 3. Comparison between numerical and experimental results

The experimental results are measured by an optical measuring method. Figure 3 demonstrates that the measured displacements are greater than the results obtained from the FE calculation. Finally, the two different sets of material parameters result in a maximum error of the tip deflection with respect to the experimental results of 16.8% (Mat. 2) and 4.9% (Mat. 1).

4.2 CLAMPED ADAPTIVE PLATE

The second example is an adaptive plate attached with piezoceramic patches (Figure 4). The plate is made from steel coated with eight piezoelectric layers (PIC 151, see Figure 1) on the top and on the bottom surfaces of the plate.

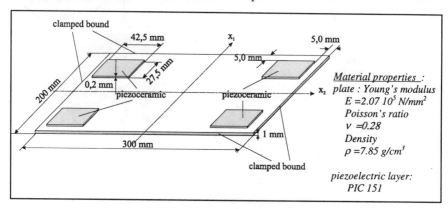

Figure 4. Adaptive plate, clamped at all edges

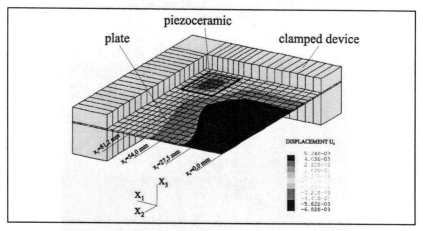

Figure 5. FE model and displacement results for the clamped plate

Due to the symmetry of the FE model we used one quarter of the plate only. In order to get a better agreement between the numerical and experimental results a coarse model of the boundary conditions was used (Figure 5). The finite element model consists of 506 3D elements with 20 nodes per element. First, the eigenvalues and eigenmodes were determined in order to check the correspondence of the numerical model and the laboratory experiment. The first frequency of the experimental analysis is f_{exp}= 163.5 1/s, and the numerical result is f_{num}=167.2 1/s. Figure 6 and Figure 7 show the displacements along line x_1=0 and x_1=81.2 cm, respectively.

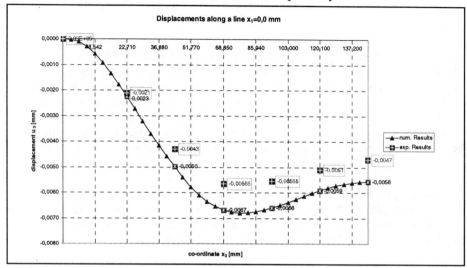

Figure 6. Comparison of numerical and experimental results at x_1=0 mm

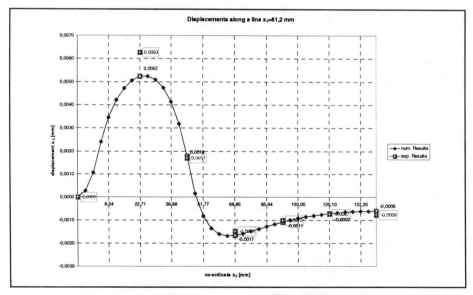

Figure 7. Comparison of numerical and experimental results at x_1=81.2 mm

The experimental data in Figures 6 and 7 are mean values obtained from several tests. It can be seen that the best agreement between the numerical and experimental displacements is reached in the centerline (x_1=0 mm, 0.0 mm $\leq x_2 \leq$ 150.0 mm) of the plate. The greatest differences in the deflections we observed near the clamped edge $(x_1 = 81.2 \text{ mm})$, especially where the piezoceramic layers are connected with the plate.

5. Conclusion

The paper presents a general concept for the development of piezoelectric finite elements. Based on this concept a library of 1D, 2D and 3D elements as well as multilayered composite shell elements were developed and implemented in an existing finite element code. The options to simulate adaptive structures include both static and dynamic analysis. Several beam, plate and shell structures were investigated experimentally to verify the software and get more insight information about the behaviour of adaptive structures. Two examples of an adaptive beam structure and a plate structure are reported in the paper.

Acknowledgement

This work is part of the project *Innovationskolleg Adaptive mechanische Systeme - ADAMES* supported by the German Research Foundation (DFG) under the project number INK 25/A1-1. This support is gratefully acknowledged.

References

1. Rao, S. S.; Sunar, M.: Analysis of Distributed Thermopiezoelectric Sensors and Actuators in Advanced Intelligent Structures. *AIAA Journal*, Vol. 31, No.7, 1993, pp. 1280 - 1286.
2. Hwang, W.-S.; Hyun, C. P.: Finite Element Modeling of Piezoelectric Sensors and Actuators. *AIAA Journal*, Vol. 31, No. 5, 1993, pp. 930 - 937.
3. Tzou, H. S.; Ye, R.: Piezothermoelasticity and Precision Control of Piezoelectric Systems: Theory and Finite Element Analysis. *Journal of Vibration and Acoustics*, Vol. 116, 1994, pp. 489 - 495.
4. Lammering, R.: The Application of a Finite Shell Element for Composites Containing Piezo-Electric Polymers in Vibration Control. *Comp. & Struct.*, Vol. 41, No. 5, pp. 1101 - 1109.
5. Lin, M. W.; Abatan, A. O., Rogers, C. A.: Application of commercial finite element codes for the analysis of induced strain-actuated structures. *Proc. of 2nd Int. Conf. on Intell. Materials*, June 5 - 8, 1994, Williamsburg (USA), pp. 846 - 855.
6. Ha, S. K.; Keilers, C.; Chang, F.-K.: Finite Element Analysis of Composite Structures Containing Distributed Piezoceramic Sensors and Actuators. *AIAA Journal*, Vol. 30, No. 3, 1992, pp. 772 - 780.
7. Gabbert, U.; Zehn, M.: Universelles FEM-System COSAR - ein zuverlässiges und effektives Berechnungswerkzeug für den Ingenieur. *Berichte zur III. COSAR-Konferenz*, 24./25. Sept. 1992, TU Magdeburg.
8. Altenbach, J.; Berger, H.; Gabbert, U.: Numerical Problems in 3D Finite Element Analysis Based on Degenerated Elements. in J.R.Whiteman(Ed.): *The Mathematics of Finite Elements and Applications V.* Academic Press, Inc. 1985, pp. 459 - 467.
9. International Standard IEC 483, 1st edition, 1976, "Guide to dynamic measurements of piezoelectric ceramics with high electromechanical coupling".
10. Bahrami, H.; Tzou, H. S.: Precision Placement Analysis of a New Multi-DOF Piezoelectric End-Effector via Finite Elements. *Proceedings of DETC '97, 1997 ASME Design Engineering Technical Conferences*, September 14-17, 1997, Sacramento, California.
11. Köppe, H.; Gabbert, U.; Tzou, H. S.: On Three-Dimensional Layered Shell Elements for the Simulation of Adaptive Structures. *Proc. of*
    the EUROMECH 373 Colloquium - Modelling and Control of Adaptive Structures, Madgeburg, 11 - 13 March 1998, pp. 103-114
12. Cao, X.; Gabbert, U., Poetzsch, R.: Delamination modelling and analysis of adaptive composites. *Proc. of the 39th AIAA Conference, Adaptive Structure Forum*, April, 20-23, 1998, Long Beach, CA, paper AIAA-98-2046.

LAYERED PIEZOELECTRIC COMPOSITES:
MACROSCOPIC BEHAVIOUR

A. GALKA AND R. WOJNAR

Institute of Fundamental Technological Research
00-049 Warsaw, Świ/etokrzyska 21, Poland

1. Introduction

The knowledge of artificial composite materials (French acronym - *Science de* M*atériaux* ART*ificiels*) reaches a prehistory. Architecture, textile craftmanship, jewellery have used such materials extensively. For example, in mid-way between Pu/ltusk and Warsaw, in Wieliszew in 1992, in an archeological excavation, a V-IV century dish made of clay ceramics with 3 brass circles inserted was found.

Intelligent (English "smart") materials denote objects which are manufactured in such a way that they perform certain planned actions. A number of smart materials are related to discovery of a piezoelectric phenomenon by the brothers Jacques and Pierre Curie in 1880, [1]. A composite known as a "Curie strip (bilame de quartz)" invented in 1889 and consisting of two X-cut quartz plates cemented together with polarities opposed, and vibrating in a flexural mode, is one of such smart materials, [2, 4, 5]. The other composite material is known as Langevin's quartz-steel "sandwich", [3, 5]. It has the form of a few millimeters thick quartz plate cemented between two massive slabs of steel that serve as electrodes in a transducer.

The aim of this contribution is to analyze the formulae for macroscopic moduli of layered piezoelectric composites in terms of anisotropy of the layers. We limit ourselves to the study of a composite made of the periodically arranged quartz layers with different directions of crystal axes in each layer.

First, by using the general homogenization formulae derived in [7], cf. also [8-11], the effective elastic, piezoelectric and dielectric moduli for a microscopic layered composite are obtained. These formulae are derived in a particular reference frame. Thus a natural question arises: how coud we obtain the effective moduli in arbitrary direction of lamination. Moreover, a macroscopic behaviour of layered piezoelectric composites, in which

J. Holnicki-Szulc and J. Rodellar (eds.), Smart Structures, 79–88.
© *1999 Kluwer Academic Publishers. Printed in the Netherlands.*

anisotropy axis of each layer constituting the basic cell do not concide, is also of some interest. By choosing particular piezoelectric materials it is shown that depending on the anisotropy axis of layers, the effective moduli differ from one composite to another, though the layers are made of the same materials.

2. Basic relations

Let $\Omega \subset \mathbf{R}^3$ be a bounded, sufficiently regular domain and $(0, \tau)(\tau > 0)$ – a time interval. The elastic, piezoelectric and dielectric moduli are denoted by c_{ijkl}, g_{ijk} and ϵ_{ij}, respectively; ρ is the density. We identify Ω with the underformed state of the piezoelectric composite with a microperiodic structure. Thus for $\varepsilon > 0$ the material functions just introduced are εY-periodic, where $Y = (0, Y_1) \times (0, Y_2) \times (0, Y_3)$ is the so-called basic cell. We write

$$c_{ijkl}^\varepsilon(x) = c_{ijkl}(\frac{x}{\varepsilon}), \quad g_{ijk}^\varepsilon(x) = g_{ijk}(\frac{x}{\varepsilon}), \quad \epsilon_{ij}^\varepsilon(x) = \epsilon_{ij}(\frac{x}{\varepsilon}), \quad \rho^\varepsilon(x) = \rho(\frac{x}{\varepsilon}),$$

where $x \in \Omega$ and the functions $c_{ijkl}^\varepsilon, g_{ijk}^\varepsilon$, etc are εY-periodic, where $\varepsilon > 0$ is a small parameter.

For a fixed $\varepsilon > 0$ the basic relations describing a linear piezoelectric solid with the microperiodic structure are:

(i) *Field equations*

$$\sigma_{ij,j}^\varepsilon + b_i^\varepsilon = \rho \ddot{u}_i^\varepsilon, \qquad D_{i,i}^\varepsilon = 0 \qquad in \quad \Omega \times (0, \tau).$$

(ii) *Constitutive equations*

$$\sigma_{ij}^\varepsilon = c_{ijkl}^\varepsilon e_{kl}(\mathbf{u}^\varepsilon) - g_{kij}^\varepsilon E_k(\varphi^\varepsilon), \qquad D_i^\varepsilon = g_{ikl}^\varepsilon e_{kl}(\mathbf{u}^\varepsilon) + \epsilon_{ik}^\varepsilon E_k(\varphi^\varepsilon).$$

(iii) *Geometrical relations*

$$e_{kl}(\mathbf{u}^\varepsilon) = u_{(k,l)}^\varepsilon = \frac{1}{2}(u_{k,l}^\varepsilon + u_{l,k}^\varepsilon), \qquad E_k(\varphi^\varepsilon) = -\varphi_k^\varepsilon.$$

Here $(\sigma_{ij}^\varepsilon)$, (u_i^ε), (b_i^ε), ρ^ε, (E_i^ε), φ and (D_i^ε) are the stress tensor, the displacement vector, the body force vector, the mass density, the electric field vector, the electric potential and the electric displacement vector, respectively. Moreover $\mathbf{b}^\varepsilon$ is $\varepsilon - Y$ periodic. The tensors of material functions satisfy the usual symmetry conditions

$$c_{ijmn}^\varepsilon = c_{mnij}^\varepsilon = c_{ijnm}^\varepsilon = c_{jimn}^\varepsilon, \qquad g_{kij}^\varepsilon = g_{kji}^\varepsilon, \qquad \epsilon_{ij}^\varepsilon = \epsilon_{ji}^\varepsilon.$$

We make the usual assumption: there exists a constant $\alpha > 0$ such that for almost every $x \in \Omega$, the following conditions are satisfied:

$$c_{ijmn}^\varepsilon(x)e_{ij}e_{mn} \geq \alpha \mid \mathbf{e} \mid^2, \qquad \epsilon_{ij}^\varepsilon(x)a_i a_j \geq \alpha \mid \mathbf{a} \mid^2,$$

for each $\mathbf{e} \in \mathbb{E}_s^3$ and each $\mathbf{a} \in \mathbb{R}^3$; here $\mathbb{E}_s^3$ is the space of symmetric 3×3 matrices. The basic equations in terms of the displacement and electric potential read

$$(c_{ijmn}^\varepsilon u_{m,n}^\varepsilon + g_{kij}^\varepsilon \varphi^\varepsilon{}_{,k})_{,j} + b_i^\varepsilon = \rho^\varepsilon \ddot{u}_i^\varepsilon, \qquad (g_{imn}^\varepsilon u_{m,n}^\varepsilon - \epsilon_{ik}^\varepsilon \varphi^\varepsilon{}_{,k})_{,i} = 0.$$

3. Homogenization

In order to find the effective or macroscopic coefficients we employed the method of two-scale asymptotic expansions. We make the following *Ansatz*, cf. [8],

$$\mathbf{u}^\varepsilon(x,t) = \mathbf{u}^0(x,y,t) + \varepsilon \mathbf{u}^1(x,y,t) + \varepsilon^2 \mathbf{u}^2(x,y,t) + \ldots$$

$$\varphi^\varepsilon(x,t) = \varphi^0(x,y,t) + \varepsilon \varphi^1(x,y,t) + \varepsilon^2 \varphi^2(x,y,t) + \ldots$$

where $y = \frac{x}{\varepsilon}$. The functions $\mathbf{u}^0(x,.,t), \mathbf{u}^1(x,.,t), \ldots, \varphi^0(x,.,t), \varphi^1(x,.,t), \ldots$ are Y-periodic. The main steps of the asymptotic analysis are outlined in the paper [2].

The homogenized form of the field equations is, cf. [8,9],

$$<\rho> \frac{\partial^2 u_i^0}{\partial t^2} = c_{ijmn}^h \frac{\partial^2 u_m^0}{\partial x_j \partial x_n} + g_{kij}^h \frac{\partial^2 \varphi_m^0}{\partial x_j \partial x_k} + <b_i>,$$

$$g_{ikj}^h \frac{\partial^2 u_k^0}{\partial x_j \partial x_i} - \epsilon_{ij}^h \frac{\partial \varphi^0}{\partial x_j \partial x_i} = 0,$$

where $<f> = \frac{1}{|Y|} \int_Y f(y) dy$. We see that the displacement field $\mathbf{u}^0(x,t)$ and electric potential field $\varphi^0(x,t)$ do not depend on the local variable $y \in Y$.

The effective coefficients are given by the following expressions:

$$c_{ijmn}^h = < c_{ijmn} + c_{ijpq} \frac{\partial \chi_p^{(mn)}}{\partial y_q} + g_{pij} \frac{\partial \psi^{(mn)}}{\partial y_p} >,$$

$$g_{kij}^h = < g_{kij} + c_{ijmn} \frac{\partial \Phi_m^{(k)}}{\partial y_n} + g_{mij} \frac{\partial R^{(k)}}{\partial y_m} >,$$

$$\epsilon_{im}^h = < \epsilon_{im} - g_{ipq} \frac{\partial \Phi_p^{(m)}}{\partial y_q} + \epsilon_{ik} \frac{\partial R^{(m)}}{\partial y_k} >,$$

where the *local* functions $\chi_i^{(mn)}$, $\psi^{(mn)}$, $\Phi_i^{(m)}$ and $R^{(m)}$ are Y-periodic. They are solutions to the *local* problems:

$$\frac{\partial}{\partial y_j}\left[c_{ijmn}(y) + c_{ijpq}(y)\frac{\partial\chi_p^{(mn)}}{\partial y_q} + g_{kij}(y)\frac{\partial\psi^{(mn)}}{\partial y_k}\right] = 0$$

$$\frac{\partial}{\partial y_i}\left[g_{imn}(y) + g_{ipq}(y)\frac{\partial\chi_p^{(mn)}}{\partial y_q} - \epsilon_{ik}(y)\frac{\partial\psi^{(mn)}}{\partial y_k}\right] = 0,$$

$$\frac{\partial}{\partial y_j}\left[g_{mij}(y) + g_{kij}(y)\frac{\partial R^{(m)}}{\partial y_k} + c_{ijpq}(y)\frac{\partial\Phi_p^{(m)}}{\partial y_q}\right] = 0$$

$$\frac{\partial}{\partial y_i}\left[\epsilon_{im}(y) + \epsilon_{ik}\frac{\partial R^{(m)}}{\partial y_k} - g_{ipq}\frac{\partial\Phi_p^{(m)}}{\partial y_q}\right] = 0.$$

4. Microperiodic layered composite

Now the basic cell reduces to an interval $(0,1)$. We assume that the material coefficients of such a composite are piecewise constant; for the lamination in the direction y_3 they are specified by, cf. [9],

$$C_{\alpha\beta\mu\nu}(y) = \begin{cases} C_{\alpha\beta\mu\nu}^{(1)} & \text{for } y_3 \in (0,\xi), \\ C_{\alpha\beta\mu\nu}^{(2)} & \text{for } y_3 \in (\xi,1). \end{cases} \tag{1}$$

Here $\alpha, \beta, \mu, \nu = 1, 2, 3, 4$, and $C_{ijmn} = c_{ijmn}$, $C_{4jmn} = g_{jmn}$, $C_{4j4n} = -\epsilon_{jn}$, $(i,j,m,n = 1,2,3)$.

It means that the composite is made of two materials. After calculations the local functions can be found in a closed form; they are piecewise linear and we find

$$C_{\alpha\beta\mu\nu}^h = < C_{\alpha\beta\mu\nu} > -\xi(1-\xi)(\tilde{B}^{-1})^{\kappa\lambda}[\![C_{\lambda3\alpha\beta}]\!][\![C_{\mu\nu\kappa3}]\!], \tag{2}$$

where

$$< C_{\alpha\beta\mu\nu} >= \xi C_{\alpha\beta\mu\nu}^{(1)} + (1-\xi)C_{\alpha\beta\mu\nu}^{(2)}, \qquad [\![C_{\alpha\beta\mu\nu}]\!] = C_{\alpha\beta\mu\nu}^{(2)} - C_{\alpha\beta\mu\nu}^{(1)}$$

$$[\tilde{B}_{\alpha\beta}] = [\xi C_{\alpha3\beta3}^{(2)} - (1-\xi)C_{\alpha3\beta3}^{(1)}].$$

By employing these formulae we determine the homogenized coefficients for layered composite for various mutual orientations of crystallographic axes of components.

5. Rotation of crystallographic axes

The rotation matrices are given by

$$Q_1(\varphi) = \begin{bmatrix} 1 & 0 & 0 \\ 0 & \cos(\varphi) & -\sin(\varphi) \\ 0 & \sin(\varphi) & \cos(\varphi) \end{bmatrix}, \quad Q_2(\psi) = \begin{bmatrix} \cos(\psi) & 0 & \sin(\psi) \\ 0 & 1 & 0 \\ -\sin(\psi) & 0 & \cos(\psi) \end{bmatrix}$$

$$Q_3(\theta) = \begin{bmatrix} \cos(\theta) & -\sin(\theta) & 0 \\ \sin(\theta) & \cos(\theta) & 0 \\ 0 & 0 & 1 \end{bmatrix}. \tag{3}$$

We consider a composite made of two kinds of quartz layers. The layers differ by orientation of crystal axes.

In piezoelectricity we are often concerned with α-quartz (low-quartz) to which we shall refer as "quartz". It crystallizes below 573 $°$C and belongs to trigonal system D_3 (32). If cristallization takes place between 573 and 870 $°$C, the form known as β-quartz is produced of hexagonal D_6 (622) instead of the trigonal structure, cf. [5]. The dielectric, piezoelectric and elastic coefficients for quartz are given by

$$\epsilon^{(1)} = \begin{bmatrix} 0.392 & 0 & 0 \\ 0 & 0.392 & 0 \\ 0 & 0 & 0.41 \end{bmatrix}, \tag{4}$$

$$\mathbf{g}^{(1)} = \begin{bmatrix} 0.171 & -0.171 & 0 & -0.04 & 0 & 0 \\ 0 & 0 & 0 & 0 & 0.04 & -0.171 \\ 0 & 0 & 0 & 0 & 0 & 0 \end{bmatrix}, \tag{5}$$

$$\mathbf{c}^{(1)} = \begin{bmatrix} 8.674 & 0.699 & 1.191 & -1.791 & 0 & 0 \\ 0.699 & 8.674 & 1.191 & 1.791 & 0 & 0 \\ 1.191 & 1.191 & 10.72 & 0 & 0 & 0 \\ -1.791 & 1.791 & 0 & 5.794 & 0 & 0 \\ 0 & 0 & 0 & 0 & 5.794 & -1.791 \\ 0 & 0 & 0 & 0 & -1.791 & 3.9875 \end{bmatrix}, \tag{6}$$

respectively, cf. [6]. These coefficients are taken in the reference frame $\{x_i\}$. Performing a rotation according to (3) we write

$$\epsilon_{ij}^{(2)} = Q_{im}Q_{jn}\epsilon_{mn}^{(1)}, \quad g_{ijk}^{(2)} = Q_{im}Q_{jn}Q_{kp}g_{mnp}^{(1)},$$

$$c_{ijkl}^{(2)} = Q_{im}Q_{jn}Q_{kp}Q_{lq}c_{mnpq}^{(1)},$$

84

For example, the dielectric coefficients $\epsilon_{ij}^{(1)}$, after rotation through the angle φ about the x_1-axis, are given by

$$\epsilon^{(2)} = \begin{bmatrix} 0.392 & 0 & 0 \\ 0 & 0.401 - 0.009\cos(2\varphi) & -0.009\sin(2\varphi) \\ 0 & -0.009\sin(2\varphi) & 0.401 + 0.009\cos(2\varphi) \end{bmatrix}.$$

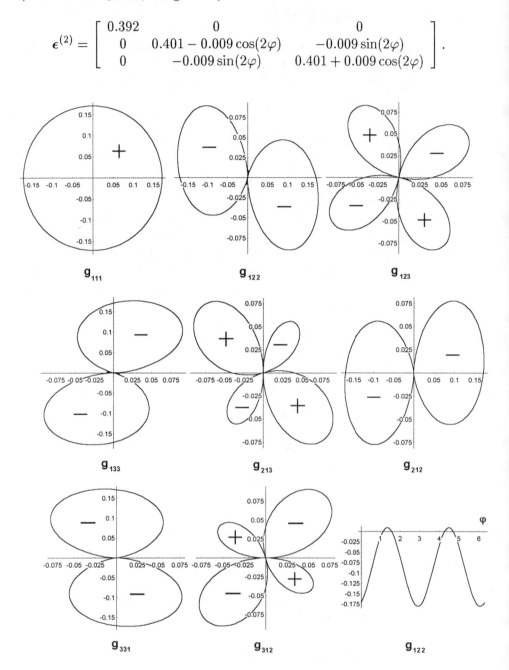

Figure 1. Polar diagrams of piezoelectric coefficients: x_1-rotation.

Nonzero piezoelectric coefficients *versus* angle of rotation about the x_1-axis are depicted in Fig. 1. The coefficient g_{122} is shown twice; the angle intervals of positive values are distinctly seen in the Cartesian coordinate set. Some particular elastic coefficients after rotation about the same axis are shown in Fig. 2.

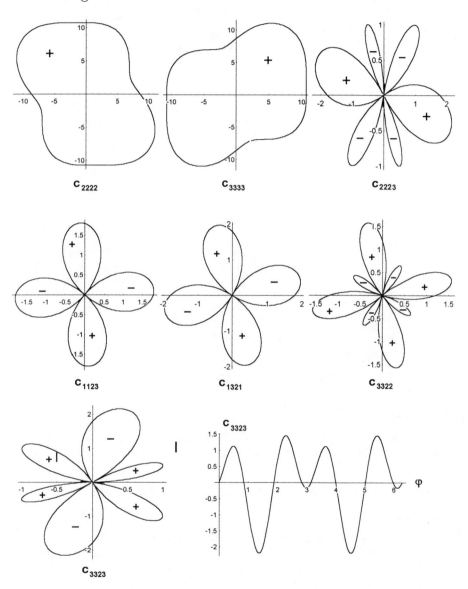

Figure 2. Polar diagrams of some elastic coefficients: x_1-rotation.

Twofold symmetry of quartz about the x_1 axis is visible.

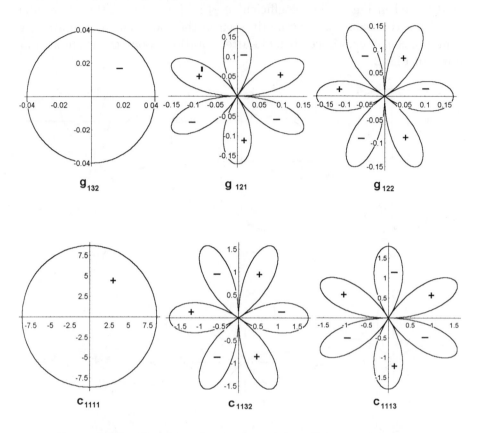

*Figure 3.*Polar diagrams of some material coefficients: x_3-rotation.

In Fig. 3, some selected piezoelectric and elastic moduli after rotation about the x_3-axis are presented. The moduli g_{132}, g_{133}, c_{1111}, c_{2222} do not change under the rotation. Other components reveal a symmetry of D_3-group.

6. Some particular cases of homogenized coefficients

Example 1

We consider a case in which one layer is described by matrices (5) and (6), and the adjacent layer is obtained by rotation with respect to the first layer through angle $\pi//3$ about the x_3-axis.

Volume fraction of the first layer is ξ, cf. Eq. (1). By using Eq. (2) we obtain the following effective coefficients:

$$\mathbf{g}^h =$$

$$\begin{bmatrix} -0.171 + 0.342\xi & 0.171 - 0.342\xi & 0 & -0.04 & 0 & 0 \\ 0 & 0 & 0 & 0 & 0.04 & 0.171 - 0.342\xi \\ 0 & 0 & 0 & 0 & 0 & 0 \end{bmatrix}$$

$$\mathbf{c}^h =$$

$$\begin{bmatrix} 8.674 - 2.214(1-\xi)\xi & 0.699 + 2.214(1-\xi)\xi & 1.191 & 1.791 - 3.582\xi \\ 0.699 + 2.214(1-\xi)\xi & 8.674 - 2.214(1-\xi)\xi & 1.191 & -1.791 + 3.582\xi \\ 1.191 & 1.191 & 10.72 & 0 \\ 1.791 - 3.582\xi & -1.791 + 3.582\xi & 0 & 5.794 \\ 0 & 0 & 0 & 0 \\ 0 & 0 & 0 & 0 \end{bmatrix}$$

$$\begin{bmatrix} 0 & 0 \\ 0 & 0 \\ 0 & 0 \\ 0 & 0 \\ 5.794 & 1.791 - 3.582\xi \\ 1.791 - 3.582\xi & 3.9875 - 2.214(1-\xi)\xi \end{bmatrix}$$

We see that for $\xi = 1//2$, *i.e.* for the equal volume fractions of both layers we obtain a material with symmetry D_6. Such a symmetry exists in a β quartz.

The terms proportional to $\xi(\xi - 1)$ in the matrix $\mathbf{c}^h$ result from homogenization.

Example 2

In this example, one layer is the same as in the previous example, and another layer is obtained by rotation through angle π about the x_2-axis (the rotation through angle π about the x_3-axis gives the same result). Volume fractions of both layers are the same ($\xi = 1//2$). The coefficients $\mathbf{g}^{(1)}$ and $\mathbf{c}^{(1)}$ of the first layer are given by (5) and (6). The coefficients of the second layer $\mathbf{g}^{(2)}$ and $\mathbf{c}^{(2)}$ are given below. By using Eq. (2) we obtain the effective coefficients denoted by $\mathbf{g}^h$ and $\mathbf{c}^h$.

$$\mathbf{g}^{(2)} = \begin{bmatrix} -0.171 & 0.171 & 0 & -0.04 & 0 & 0 \\ 0 & 0 & 0 & 0 & 0.04 & 0.171 \\ 0 & 0 & 0 & 0 & 0 & 0 \end{bmatrix},$$

$$\mathbf{g}^h = \begin{bmatrix} 0 & 0 & 0 & -0.0047 & 0 & 0 \\ 0 & 0 & 0 & 0 & -0.0368 & 0 \\ 0 & 0 & 0 & 0 & 0 & 0 \end{bmatrix},$$

$$
\mathbf{c}^{(2)} =
\begin{bmatrix}
8.674 & 0.699 & 1.191 & 1.791 & 0 & 0 \\
0.699 & 8.674 & 1.191 & -1.791 & 0 & 0 \\
1.191 & 1.191 & 10.72 & 0 & 0 & 0 \\
1.791 & -1.791 & 0 & 5.794 & 0 & 0 \\
0 & 0 & 0 & 0 & 5.794 & 1.791 \\
0 & 0 & 0 & 0 & 1.791 & 3.9875
\end{bmatrix},
$$

$$
\mathbf{c}^{h} =
\begin{bmatrix}
8.74859 & 0.624406 & 1.191 & 0 & 0 & 0 \\
0.624406 & 8.74859 & 1.191 & 0 & 0 & 0 \\
1.191 & 1.191 & 10.72 & 0 & 0 & 0 \\
0 & 0 & 0 & 5.4242 & 0 & 0 \\
0 & 0 & 0 & 0 & 4.98957 & 0 \\
0 & 0 & 0 & 0 & 0 & 3.43388
\end{bmatrix}.
$$

The symmetry of $\mathbf{c}^{h}$ is D_6. As far as the symmetry of $\mathbf{g}^{h}$ is concerned, we note that only two components of the matrix $\mathbf{g}^{h}$ do not vanish, and these components have the same sign and differ by one order of magnitude. Hence, a composite with $\mathbf{g}^{h}$ symmetry is not a crystal of the D_6 class.

Acknowledgement. The authors were supported by the State Committee for Scientific Research (Poland) through the grant No. 7 T07 A016 12.

References

1. Curie, J., Curie, P., (1880) Développment, par pression, de l'électricité polaire dans les cristaux hémièdres à faces inclinées, *C. R. Acad. Sci. Paris* **91**, pp. 294-295
2. Curie, P., (1908) Oeuvres, Gauthier-Villars, Paris
3. Langevin, P., (1921) "Improvements relating to the emission and reception of submarine waves", British Patent No. 145 691, 28 July 1921
4. Gramont, A. de, Béretzki, D., (1936) Sur la génération d'ondes acoustiques au moyen de quartz piézoélectriques, *C. R. Acad. Sci. Paris* **202**, pp. 1229-1232
5. Cady, W.G., (1964) Piezoelectricity, Dover Publications, New York
6. Hellwege, K.-H., (1979) (ed.), Landolt-Börnstein Numerical Data and Functional Relationships in Science and Technology, Springer, Berlin; Vol. **11**
7. Telega, J.J., (1991) Piezoelectricity and homogenization. Application to Biomechanics, in Continuum Models and Discrete Systems, vol 2, ed. by G.A. Maugin, pp. 220-229, Longman, Scientific and Technical, Harlow, Essex
8. Ga/lka, A., Telega, J.J., Wojnar, R., (1992) Homogenization and thermopiezoelectricity, *Mech. Res. Comm.*, **19** pp. 315-324
9. Ga/lka, A., Telega, J.J., Wojnar, R., (1996) Some Computational aspects of homogenization of thermopiezoelectric composites, *Comp. Assisted Mech. and Eng. Sci.*, **3**, pp. 133-154
10. Telega, J.J., Ga/lka, A., Gambin, B., (1998) Effective properties of physically nonlinear piezoelectric composites, *Arch.Mech.* **50**, pp. 321-340
11. Ga/lka, A., Gambin, B., Telega, J.J., (1998) Variational bounds on the effective moduli of piezoelectric composites, *Arch.Mech.*, in press.

SMART STRUCTURES USING CARBON FIBRE REINFORCED CONCRETE (CFRC)

J. M. GONZALEZ & S. JALALI
University of Minho,
Campus de Azúrem,
4800 Guimarães, Portugal

1. Introduction

Carbon-Fibre-Reinforced Concrete (CFRC) demonstrates attractive features such as high flexural strength and toughness, resistance to earthquake damage, low drying shrinkage, ability to shield electromagnetic interference, pseudo-strain hardening (PSH) behaviour, i.e. the ability to sustain higher loads even after the concrete cracks.

However, the feature that makes the CFRC a promising material for smart structures is the ability to conduct electricity and most importantly, capacity to change its conductivity with mechanical stress. This feature is of special interest because it can lead to a relatively easy self – monitoring of the state of stress in structures.

Results of a research work investigating the changes of conductivity with the applied mechanical stress are presented. The potential applications and requirements in civil engineering structures are discussed. Furthermore, some of the difficulties encountered and possible means of bridging them are indicated. This research work is in progress and it is expected to generate data for a more detailed knowledge of the mechanisms producing the change in electrical conductivity with stress in CFRC.

2. Carbon Fibre Reinforced Concrete

The use of carbon fibres in construction has become a reality and many applications have been reported so far [1]. The use of carbon fibres has become economically viable due to the production of pitch carbon fibres and the consequent cost reduction. The use of carbon fibres is known to enhance the physical and mechanical properties of concrete. The lowering of drying shrinkage, increasing the flexural strength and the ability to sustain higher loads even after the concrete has cracked, i.e. pseudo-strain hardening (PSH), are but a few interesting features.

The electrical conductivity of pitch carbon fibres is also interesting for the construction material. The conductivity of concrete increases drastically when it is reinforced with such fibres. One of the major applications of such carbon-reinforced concretes (CFRC) is in cathodic protection of degraded structures [2].

89

J. Holnicki-Szulc and J. Rodellar (eds.), Smart Structures, 89–95.

The interesting feature of the electrical conductivity of CFRC is that it changes with the applied mechanical stress. It has been reported that CFRC when subjected to mechanical stress changes its conductivity [3]. This characteristic enables the direct measurement and monitoring of the actual stresses applied to structures. Hence, it can be visualised that CFRC can be used as a sensor to monitor its own state of stress. This self-monitoring capacity has been investigated in this research work and the results of the laboratory tests are presented. Cycles of loading and unloading, in the range of stresses, were applied, and the conductivity and the deformation were measured.

Although so far the laboratory results show that CFRC is a promising material for smart structures, certain difficulties for its wide-spread application exist. The difficulties foreseen are related to:

- control of the water content in CFRC,
- distribution of fibres in CFRC, and finally
- the repeatability of the electrical response of CFRC.

Possible solutions or ways of bridging these difficulties are also investigated.

In normal concrete the conductivity changes with the degree of saturation and also with the type of the ions present in the liquid phase. However, the conductivity of CFRC is a combined effect of the liquid phase and that of the carbon fibres. Hence, continuity of the liquid phase, liquid-fibre system and the fibre-fibre contact is essential for the conductivity. When this continuity is altered, the intensity of the electrical current changes.

When the CFRC is subjected to mechanical stresses in the elastic phase, it is accepted that the existing micro-cracks are modified, therefore the conductivity changes. When the stress is removed, i.e. the deformations are removed, the micro-cracks retain their original state, and original conductivity is re-established.

However, when an element is subjected to stresses that it has not experienced before, the conductivity may change in an irreversible manner. It can be predicted that the reversibility may occur only up to the level of stresses that the material has experienced before. The cyclic loading and unloading, on the other hand, can produce micro-cracks that will change the intensity of the current passing through CFRC.

3. Materials and Equipment Used

The carbon fibres were small pitch-type manufactured by 'Kurhea Carbon Fiber', with 0.018 mm in diameter and 3 mm in length. Cubic concrete specimens of 100x100x100 mm were moulded. A metallic mesh was placed on both faces of each specimen connected to wires for measuring the intensity of current. A 2.5 DC voltage was used for normal concrete and a 6 DC voltage was used for CFRC specimens. Current intensities were registered when the loading was maintained constant for about 4 seconds at each stage. The rate of loading was 1.5 kN per second.

Materials and the mix used for normal and carbon-fibre-reinforced concretes are presented in Table 1.

TABLE 1.   The mix composition of the normal and carbon-fibre-reinforced concrete

Material	Normal Concrete (NC)	CFRC
Ordinary Portland cement (Class 42.5)	300 kg/m³	350 kg/m³
Sand	1000	1150
Coarse aggregate (½"maximum size)	1250	1500
Water/Cement ratio	0.5	0.5
Super plasticiser	1.5 lit/m³	1.5 lit/m³
Carbon fibre	None	1% Cement weight
Silica fume	10% Cement weight	10% Cement weight

4. Results Obtained

Figure 1 shows the current intensity versus applied force up to 200 kN, i.e. 20 N/mm², which is approximately 2/3 of the final strength. It is seen that intensity increases with increasing the load when applied for the first time. However, in the following cycle of loading, the intensity of current does not change until the previous peak is achieved. Only after reaching the peak of the previous stress, the intensity of current increases. This behaviour is in accordance with the predicted behaviour for normal concrete.

Hence, this material does not show the capacity for self –monitoring of the state of stress it is subjected to. Fig.2 shows the loading and current intensity versus deformations. It is interesting to note that up to around 0.32 mm deformation, the intensity of current decreases and thereafter it increases rapidly to achieve its maximum value around the peak of loading. The use of current intensity shows clearly that rapid growth of micro-cracks starts before the crushing load is reached.

Results of a similar test on CFRC for 12 cycles show that the intensity of current increases with increasing stress. Figs. 3 and 4 show the change of load and deformations versus current intensity for cycle 11 and cycle 12. Results obtained can be explained well by linear relationships having a high correlation coefficient (R2 = 0.97).

Figs. 5 and 6 show the results of loads and deformations versus current intensity for both cycles. It can be seen that the best linear fits have the same slope for both cycles (see Table 2).

5. Conclusions

Results obtained clearly indicate that in the range of cycles and stresses applied, CFRC has a self-monitoring capacity, i.e. the current intensity changes with the applied stress and deformation. Furthermore, this change is seen to be linear and the best fit of the data resulted in a high correlation coefficient ($R^2 = 0.99$). For repeated loading in cycles 11 and 12 it is seen that the best linear fit of data has the same slope while the constants obtained are slightly different.

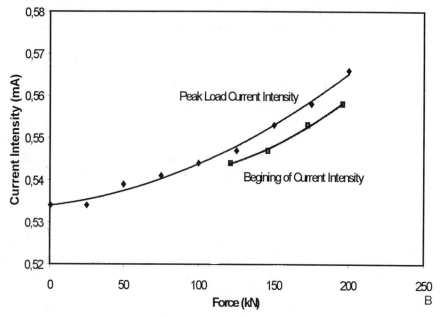

Figure 1. Current intensity versus applied force for normal concrete (NC)

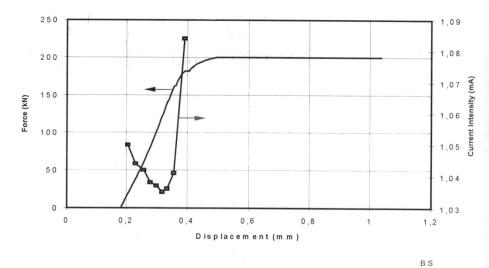

Figure 2. Current intensity versus displacement for normal concrete

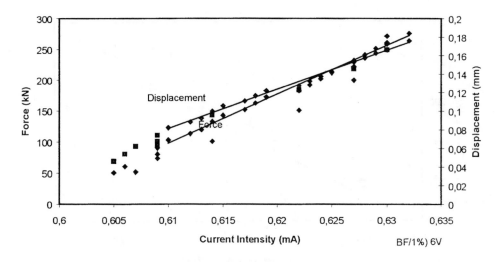

Figure 3. Force and displacement versus current intensity for CFRC during Cycle 11

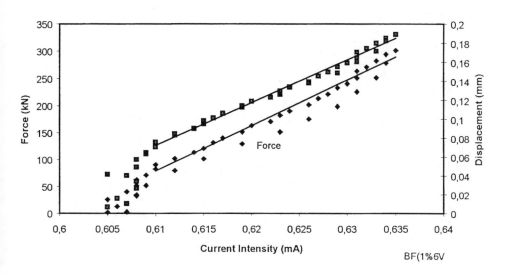

Figure 4. Force and displacement versus current intensity for CFRC during cycle 12

94

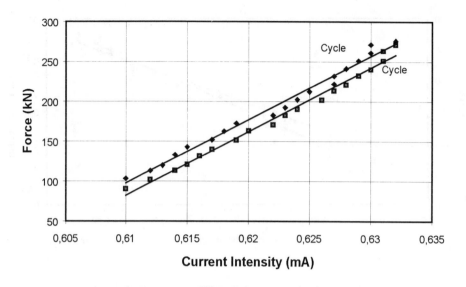

Figure 5. Data and the best fit for force versus current intensity for cycles 11 and 12

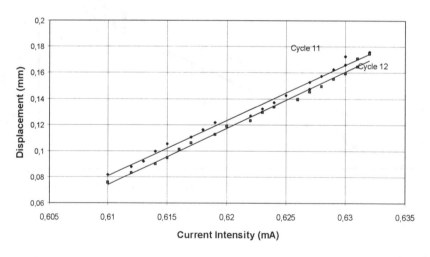

Figure 6. Data and the best fit for displacement versus current intensity for cycles 11 and 12

This difference, which is assumed to be due to the effect of repeated loading, is very small and does not seem to lead to large errors in prediction. Further, it is found that simple concrete does not have this self-monitoring capacity indicating that this behaviour is due to the presence of carbon fibres.

This research work is under way to determine the effect of higher cycles of loading and possibility of its application in full-scale structures.

TABLE 2. Parameters of the best linear fit for load-current intensity and deformation-current intensity

Parameters of best linear fit	Slope	Constant	Correlation coefficient (R2)
Load – current intensity			
Cycle 11	7935	4742	0.99
Cycle 12	8002	4799	0.99
Deformation–current intensity			
Cycle 11	4.237	2.504	0.99
Cycle 12	4.338	2.572	0.99

Acknowledgement

The authors would like to express their appreciation to Professor C. A. Bernardo for his encouragement and support during this research work.

References

1. Meier, U.(1997) *Bright Future for Carbon Fibre in Construction,* RILEM Seminar, Zurich.
2. Xuli Fu and DDL Chung (1995) Carbon Fiber Reinforced Mortar as an Electrical Contact Material for Cathodic Protection, *Cement and Concrete Res.*, **25**, No. 4, 689-694.
3. Pu Woei, C. (1994), *Carbon Fiber-Reinforced Concrete as a Strain/Stress Sensor and High Performance Civil Structure Material,* PhD Thesis, State University of New York at Buffalo.

OPTIMAL DESIGN OF ADAPTIVE STRUCTURES

J. HOLNICKI-SZULC
Institute of Fundamental Technological Research,
Polish Academy of Sciences,
Swietokrzyska 21, 00-048 Warsaw, Poland

1. Introduction. VDM Approach

The optimal design of actively controlled structures consists of the following (usually combined) problems: a) design of sensor's location for the best structural observability, b)design of actuator's location for the best structural controllability, c) design of the best control strategy, d) remodelling of material distribution.

Let us concentrate here on the points b) and c), assuming that the structural geometry and the sensor system for real time identification of the structural behaviour are defined. So-called *Virtual Distortion Method (VDM)* will be applied to solve the mentioned problems for several structural control applications. Especially, there is a need for new methods to solve the problem of optimal location of actuators leading usually to time-consuming discrete optimisation formulations (cf.Ref.1).

The VDM method makes use of the so-called *virtual distortion* (or *active distortions* in our case) field ε°_j describing the action of actuators (e.g. thermal distortions or local incompatibilities caused by hydraulic, piezo-electric, or other devices) causing compatible state of deformations ε^R_i and self-equilibrated state of stresses σ^R_i in the structure, where:

$$\varepsilon^R_i = \Sigma_j D_{ij}\, \varepsilon^{\circ}_j, \qquad\qquad \sigma^R_i = \Sigma_j E_i(D_{ij}-\delta_{ij})\, \varepsilon^{\circ}_j. \qquad (1)$$

The truss structure model is taken here into account for simplicity of presentation. However, the concept can be generalised to other elastic structures equipped with actuators. The *influence matrix* D_{ij} describes strains ε_i caused in the truss member i by the unit *virtual distortion* $\varepsilon^{\circ}_j=1$ generated in the member j (cf.Ref.1). This matrix stores information about all the structure properties (including boundary conditions in calculation of structural response). It can be demonstrated (making use of the Betti's reciprocal theorem) that multiplying each row of the matrix $D_{ij}{}^{\sigma}= Ei(Dij-\delta_{ij})$ by $A_i l_i$ (where A_i, l_i denote the cross-sectional area and length of the member i, respectively) it becomes symmetric. The rank of the n x n, non-symmetric matrix $D_{ij}{}^{\sigma}$, where n denotes the number of all members, is equal to the redundancy of the structure. It means, that only k=rank($D_{ij}{}^{\sigma}$) linearly independent states of *initial stresses* σ^R_i can be generated. In the consequence, rank(D_{ij})=n-k. It can be demonstrated (cf.Ref.1, p.8, corollary 1.2)

J. Holnicki-Szulc and J. Rodellar (eds.), Smart Structures, 97–106.
© 1999 *Kluwer Academic Publishers. Printed in the Netherlands.*

that an arbitrary state of distortions can be uniquely decomposed into two orthogonal components: a compatible part of deformations ε^R_i and the part $-\sigma^R_i/E$ corresponding to the provoked stress state:

$$\varepsilon^o_i = \varepsilon^o_{ci} + \varepsilon^o_{ri} = \varepsilon^R_i - \sigma^R_i/E, \qquad \text{where} \quad \Sigma_i A_i l_i E_i \varepsilon^o_{ci} \varepsilon^o_{ri} = \Sigma_i A_i l_i \sigma^R_i \varepsilon^R_i = 0. \qquad (2)$$

Assuming now strains ε^L_i and stresses σ^L_i caused by external loads, the final states take the form:

$$\varepsilon_i = \varepsilon^R_i + \varepsilon^L_i = \Sigma_j D_{ij} \varepsilon^o_j + \varepsilon^L_i, \qquad \sigma_i = \sigma^R_i + \sigma^L_i = \Sigma_j D_{ij}^\sigma \varepsilon^o_j + \sigma^L_i, \qquad (3)$$

and the following constitutive relation takes place: $\sigma_i = E_i (\varepsilon_i - \varepsilon^o_j)$. Note that σ_i and σ^R_i satisfied the equilibrium equations while ε_i and ε^R_i fulfil the compatibility conditions.

The *adaptive structure* can be defined now as an actively controlled structure requiring no external energy sources (except a small amount energy necessary for operation of the controller) in the control process:

$$\sigma_i \Delta \varepsilon^o_i \geq 0 \quad \text{for all } i=1,2,...n. \qquad (4)$$

It means that actuators dissipate only the potential strain energy accumulated in the structure during the loading process. Besides the advantage of energetically cost-less actuation, *adaptive structures* do not cause any problems of control stability.

Strains ε^L_i and stresses $\sigma^L_i = E_i \varepsilon^L_i$ caused by external loads can be decomposed (analogously to decomposition (2)) into two, orthogonal in the sense of scalar product, (3) components:

$$\sigma^L_i = \sigma^L_{ci} + \sigma^L_{ri} = E_i(\varepsilon^L_{ci} + \varepsilon^L_{ri}), \qquad (5)$$

where component σ^L_{ci} denotes the stresses caused by the force type of loads (e.g.Fig.1) while σ^L_{ri} denotes the stresses generated by displacement type of loading (e.g.Fig.2). From orthogonality of states $(...)_r$ and $(...)_c$ it follows that the global (integral) measure of components of strain fields ε^L_{ci} and stress fields σ^L_{ri} can be reduced (even to zero if no side constraints are imposed) by applying actions of actuators (through fields ε^o_{ci}, and ε^o_{ri} respectively). On the other hand, the components of strain fields ε^L_{ri} and stress fields σ^L_{ci} can be only redistributed (minimising maximal values) by applying actions of actuators (through fields ε^o_{ri} and ε^o_{ci}, respectively). In particular, the following shape and stress control problems can be considered as well defined.

a) Minimisation of the global strain measure for the force type of loading:
   min $\Sigma_i A_i l_i E_i \varepsilon_{ci} \varepsilon_{ci}$ subject to constraints on location of the actuators.
b) Minimisation of the maximal stress value $|\sigma_i|$ subject to constraints $|\varepsilon_i| \leq \varepsilon^u$.
c) Minimisation of the global stress measure for displacement type of loading:
   min $\Sigma_i A_i l_i \sigma^L_{ri} \sigma^L_{ri}/E_i$ subject to constraints on location of the actuators.
d) Minimisation of maximal strain value $|\varepsilon_i|$ subject to constraints $|\varepsilon_i| \leq \varepsilon^u$.

The above problems will be discussed in the next section on the basis of the VDM concept with algorithms leading to automatic generation of optimal location for actuators as well as their optimal performance. *Adaptive* realisation of the control strategy will be also examined in each case.

2. Shape and Stress Control

2.1. GLOBAL SHAPE CONTROL FOR THE FORCE TYPE OF LOADING

The problem a) of minimisation of the global measure of structural deformation can be formulated alternatively in the following way:

$$\min f_u = \min \Sigma_i u_i^2 \quad \text{where} \quad u_i = u_i^L + \Sigma_j D_{ij}^u \varepsilon_j^o; \tag{6}$$

the influence matrix D_{ij}^u denotes displacements u_i^R caused in the truss member i by the unit *virtual distortion* $\varepsilon_j^o = 1$ generated in the member j, and the number of available locations for actuators is limited to m. ($j=1,2,..m$). Additionally, the constraints on active distortions can be imposed:

$$|\varepsilon_i^o| \le \varepsilon^u \tag{7}$$

Substituting $(6)_2$ to $(6)_1$ and calculating the derivative of the objective function with respect to ε_j^o (located in the element j), one can obtain:

$$d\, f_u / d\varepsilon_k^o = 2\, u_i\, D_{ik}^u = 2(u_i^L + \Sigma_j D_{ij}^u \varepsilon_j^o)\, D_{ik}^u . \tag{8}$$

The problem of optimal design (to minimise the objective function f_u) of the sequence of most effective locations for actuators and their performance can be determined through the following algorithm.

1) Initiation. M - the number of available actuators, $A= \{0\}$ the set of actuator's location.
2) Find *not active* location j' (j'$\notin$A) with the highest sensitivity (8) and include it to the set $A= A \cup \{j'\}$.
3) Solve the best performance problem. Determine m ε_j^o from m linear eqs where m is the number of actually determined locations:

$$2(u_i^L + \Sigma_j D_{ij}^u \varepsilon_j^o)\, D_{ik}^u = 0 , \qquad j,k \in A . \tag{9}$$

4) If m.<M. Go to 2).
5) Calculate the objective function value $f_u = \Sigma_i (u_i^L + \Sigma_j D_{ij}^u \varepsilon_j^o)^2$.

The solution of the optimal shape control problem for the simply suported beam equipped with 3 actuators and exposed to the force type of loading is shown in Fig.1a. The solution reached is not adaptive in the sense of definition (4). However, it can be realised through the pre-shaped action (Fig.1b) with the same final result without the need for external power sources applied to actuators in the loading process. Having more shape modes to correct, the pre-shaping technique can be applied to some average shape mode, but the further shape corrections have to be adjusted actively (with conditions (4) not satisfied in a general case).

Application of the above strategy to the shape control of antenna backup truss structure (Fig.2) has been performed. The resulting best distribution of actuators to control the first mode of deformation is shown in Fig.3, while the distribution corresponding to the second mode is shown in Fig.4. Note that it follows from Fig.3

that a reasonable number of actuators to control the first mode are: 6,12,18, ...
Analogous numbers corresponding to the second mode are: 6,12,18,24,17,...

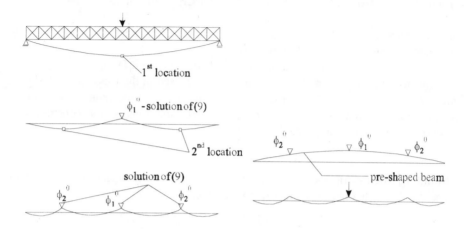

a)   b)

Figure1. Shape control of truss-beam structure

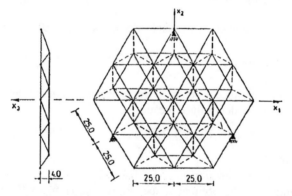

Figure.2 Antenna backup truss structure

2.2. MINIMISATION OF STRESS CONCENTRATIONS

The problem b) of stress redistribution to minimise the stress concentration can be
formulated as:

$$\min f_\sigma = \sigma^u ,\qquad (10)$$

subject to the following constraints describing elasto-plastic adaptation of the structure
to the external load and the yield limit σ^u :

$$\min \Sigma_i E_i (\varepsilon^\circ_i)^2 ,\qquad (11)$$

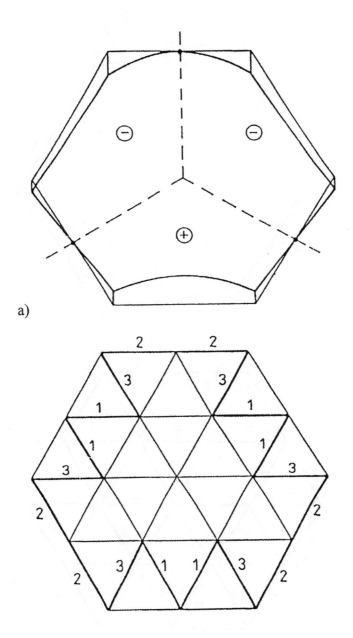

a)

b)

Figure 3. Location of actuators to control the first mode: (a) nodal lines and (b) optimal sequence of locations

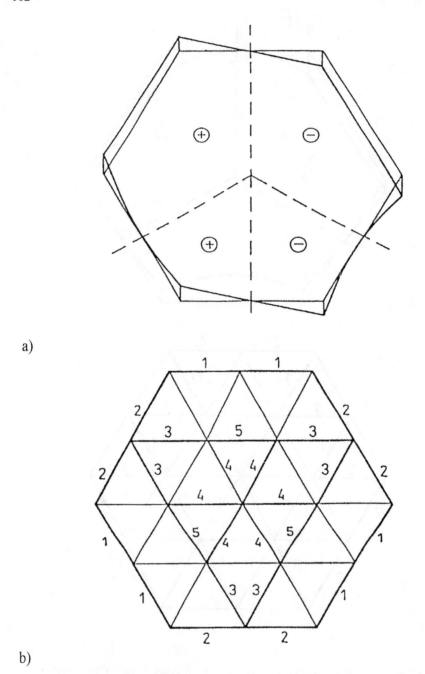

Figure.4. Location of actuators to control the second mode: (a) nodal lines and (b) optimal sequence of locations

subject to the considerations:

$$\sigma_i \varepsilon^o_i \geq 0, \qquad |\sigma_i| \leq \sigma^u, \qquad \text{where (cf. (3)}^2) \qquad \sigma_i = \Sigma_j D_{ij}{}^\sigma \varepsilon^o_j + \sigma^L_i \qquad (12)$$

The algorithm to determine the optimal solution (location of actuators as well as they performance) is written below:

1) Initiation: $\sigma_i = \sigma^L_i$, $\Delta\sigma$, $\sigma^u = \max|\sigma^L_i|$.
2) Solve elasto-plastic analysis problem (11),(12) determining $\dot{\varepsilon}^o_j$,
3) If the simulation matrix $B_{ij} = \cup_j D_{ij}{}^\sigma$ becomes singular STOP.
4) $\sigma^u = \sigma^u - \Delta\sigma$ and go to (2).

      The optimal result describes dissipative adaptation of elasto-plastic structure (cf. Fig.6c with the lowest possible yield stress limit σ^u) to external load. Naturally, it can be applied to the variable loading process. Optimal location for plastic-like dissipaters as well as their performance mimics the natural elasto-plastic adaptation of the structure.

2.3. GLOBAL STRESS CONTROL FOR THE DISPLACEMENT TYPE OF LOADING

The problem c) of minimisation of the global measure of structural stresses can be formulated in the following way:

$$\min f_\sigma = \min \Sigma_i (\sigma_i)^2 / E_i \quad \text{where} \quad \sigma_i = \Sigma_j D_{ij}{}^\sigma \varepsilon^o_j + \sigma^L_i \qquad (13)$$

and additionally, the constraint (7) can be also taken into account. Substituting $(13)^2$ to $(13)^1$ and calculating the derivative:

$$d f_\sigma / d\varepsilon^o_j = (\sigma_i / E_i) D_{ij}{}^\sigma, \qquad (14)$$

the sequence of most effective locations for actuators and their performance can be determined through an algorithm analogous to that discussed in the Section 1a. For illustration, the best location of the first dissipater reducing the global stress measure in a continuous beam subjected to non-homogeneous settlement of supports is shown in

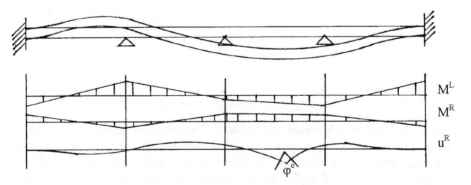

Figure.5 Stress control in continuous beam

104

Fig.5. Note that action of the dissipater obeys constraint (4) and the above procedure can be used to determine location of the so-called *structural fuses* protecting the structure against local overloading.

2.4. MINIMISATION OF STRAIN CONCENTRATIONS

The problem d) of strain redistribution to minimise the strain concentration can be formulated (analogously to (10)-(12)) as:

$$\min f_\varepsilon = \varepsilon^u . \tag{15}$$

subject to the following constraints describing behaviour of the structure made of so-called *locking material* (Ref.4)

$$\min \Sigma_i E_i (\varepsilon^o{}_i)^2 , \tag{16}$$

subject to the conditions:

$$\sigma_i \varepsilon^o{}_i \leq 0, \qquad |\varepsilon_i| \leq \varepsilon^u, \qquad \text{where (cf. (3)}_1) \quad \varepsilon_i = \Sigma_j D_{ij} \varepsilon^o{}_j + \varepsilon^L{}_i . \tag{17}$$

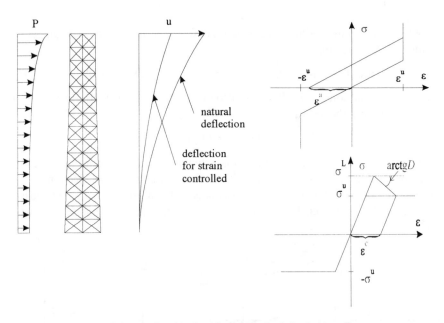

Figure 6. Local strain redistribution simulating locking effect

Applying the algorithm analogous to that described in the Section 1b, the overall structural behaviour as for a locking material (Fig.6b) can be reached. Location of locking elements as well as their performance can be automatically determined in this way. An example of application to a flexible skeletal structure of tall building with local strain control (e.g. to protect brittle crystal curtain walls) is shown in Fig.6a. The overall effect of locking the local extremal deformations can be reached.

Note that the locking process is not dissipative (cf.$(17)_1$). To the contrary, it requires an external source of energy in the actuator or some special device causing blockage of local strains wherever constraint $(17)_2$ is violated.

3. Design for Extremal Loads. Energy Accumulation/Release Strategy

The problem of structural adaptation to extremal loads can be formulated as the requirement of maximal dissipation of the strain energy in order to prevent structural elements against uncontrolled damage development.

$$\max \Sigma_i \sigma_i \varepsilon^\circ_i , \tag{18}$$

subject to: $\qquad \sigma_i \varepsilon^\circ_i \geq 0, \qquad |\sigma_i| \leq \sigma^u, \qquad |\varepsilon_i^\circ| \leq \varepsilon^u , \tag{19}$

where (cf.$(3)_2$): $\qquad\qquad \sigma_i = \Sigma_j D_{ij}{}^\sigma \varepsilon^\circ_j + \sigma^L_i . \tag{20}$

Substituting (20) to (18) we can decompose the objective function into a non-linear part $\Sigma_i \varepsilon^\circ_i \Sigma_j D_{ij}{}^\sigma \varepsilon^\circ_j$, where $\mathrm{rank} \mathbf{D}^\sigma = k$ (k is the structural redundancy) and the linear part is $\Sigma_i \sigma^L_i \varepsilon^\circ_i$

Let us now postulate: 1) first, introduction of k actuators to control the non-linear part of the objective function and then, 2) another k elements allowing generation of k independent modes of stress-less movements. The first stage of the solution is constrained to generation of distortions in k overloaded elements only, and the problem can be equivalently formulated through the stress limits σ^*_i as control parameters:

$$\max \Sigma_i \sigma^*_i \varepsilon^\circ_i , \tag{21}$$

subject to: $\quad \sigma^*_i \varepsilon^\circ_i \geq 0, \quad |\sigma^*_i| \leq \sigma^u, \quad |\varepsilon_i^\circ| \leq \varepsilon^* , \quad$ for $i=1,2,..k . \tag{22}$

The solution corresponds to the result for elasto-plastic structure designed to minimise the stress concentrations (cf. Problem 1b), however, with stress limits adjusted to each particular member. It follows from the decomposition (2),(3) that for the force type of loading, the above problem is equivalent to the problem of maximisation of the potential energy accumulation in the structure $\max \Sigma_i \sigma_i \varepsilon_i$, where $\varepsilon_i = \varepsilon^L_i + \varepsilon^\circ_{ci}$.

It means that in the first stage of action, the potential strain energy $\Sigma_i \sigma_i \varepsilon_i$ is maximised simultaneously with generation of distortions ε°_i in k overloaded members. Then, in the second stage of action, generation of k independent modes of stress-less movements $\varepsilon^R_i = \varepsilon^\circ_{ci}$ (up to the active constraints $(22)^3$) can be postulated. Formally, this means solving the problem (21),(22) for all i, with fixed stresses.

The above *accumulation/release* strategy has been applied to the truss cantilever beam (Fig.7, Ref.3) with location of dissipaters 1, 2, 3, 4 determined through solution of the problem (21),(22) and the location of further dissipaters 5, 6, 7, 8 chosen to form 2 dissipaters for each section of the beam. The amount of dissipated energy has grown from U=5.760x10^{2}J (for passive elasto-plastic behaviour Fig.7a) up to U=8.640x10^{2}J for the controlled elasto-plastic behaviour. The first stage of action dissipates only U=0.018x10^{2}J of energy, but it causes redistribution of stresses in such a way that in the second stage of action, higher stresses have greater contribution to the dissipation mentioned above.

The proposed *compromise solution* with 2k dissipaters allows for an active adaptation (pre-setting desired local stress limits σ^*_i, e.g. in valves of hydraulic actuators simulating plastic-like behaviour) to the detected loads.

Generalising the proposed approach to dynamic problems with inertial forces taken into account, decomposition (2), (3) is not longer valid as the dynamic influence matrix $\mathbf{D}_d$ describes the dynamic structural response to virtual distortion impulses and depends on the mass matrix $\mathbf{M}$ as well as on the stiffness matrix $\mathbf{K}$. In the consequence, $\mathbf{D}_d$ is not singular in general. However, the general strategy of *accumulation/release* technique can be also applied to dynamic problems (cf. Ref.5).

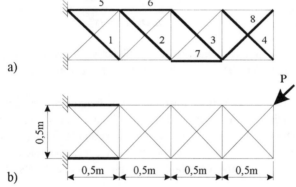

Figure 7. The best location of dissipaters (σ^u = 120 Mpa, P=0,0907 N, A=1,6 cm^2 , ε^u = 0,03 , E=100 Gpa)

Acknowledgement

This paper presents a part of the results of the COPERNICUS 263 Project „Feasibility Study of Active Railway Track Support" funded through a financial contribution from the Commission of the European Communities DGXII.

References

1. P.Albrecht, W.Zhang, B.Michaells, *3D Shape Measurement of Smart Mechaniocal Systems*, Proc. of the EUROMECH 373 Colloquium on Modelling and Control of Adaptive Mechanical Structures, March 1998, Magdeburg
2. J.Holnicki-Szulc, J.T.Gierliński *Structural Analysis, Design and Control by the VDM Method*, J.Wiley & Sons, Chichester, 95
3 J.Holnicki-Szulc, A.Mackiewicz and P.Kolakowski, *Design of Adaptive Structures for Improved Load Capacity*, AIAA Journal, vol.36, No.3, March 1998
4. A.Phillips, *The Theory of Locking Materials*, Trans.Soc.Rheology 3, pp.13-26, 1959
5. L.Knap and J.Holnicki-Szulc, *Dynamic Analysis of Adaptive, Visco-Plastic Structures*, Proc. Complas 5, Barcelona, 17-19 March 97

MAGNETOELASTIC METHOD OF STRESS MEASUREMENT IN STEEL

A. JAROSEVIC

Comenius University, Faculty of Mathematics and Physics
Mlynska dolina F2, 842 15 Bratislava, Slovakia

1. Introduction

The DYNAMAG measuring system has been developed as a means of contactless, reliable and precise determination of the stress or force in steel components. The system uses these "smart" steel components as direct measuring elements. Information about inner stress, temperature, fatigue etc. is hidden inside the steel and it is possible to obtain this information by non-contact method. The magnetic properties of the steel components changes under stress and temperature and may be measured.

This system has currently been operating for more than 10 years, monitoring the stress in reinforcement of bridges, prestressed concrete envelopes of reactors in nuclear power stations, rock and soil anchors and many other concrete and steel structures. Development and innovation of magnetoelastic sensors, measuring techniques and measuring systems is still in progress simultaneously with actual applications in civil and mechanical engineering.

2. Physical principle

The DYNAMAG system [2],[3] is based on the magnetoelastic phenomenon, i.e. the modification of the magnetic hysteresis loop of the ferrous material by static or dynamic mechanical stress. This phenomenon (discovered in 1862) where the magnetic properties of the steel change under the influence of mechanical load [3],[4],[7] is analogous to the electric resistance of conductors used in strain gauges. The magnetic characteristics of amplitude permeability and incremental permeability at a properly chosen working point are about 100 times more stress-sensitive than the electrical resistance effects. The relative change of the magnetic incremental permeability of steel is up to 10^{-3}/MPa, while the relative change of a strain gauge electric resistance is about 10^{-5}/MPa. The magnetoelastic method, therefore, enables the measurement of stress changes under 1MPa in noisy industrial environments and over a wide temperature range, also in cases when use of other methods is impossible.

Ferromagnetic material contains microscopic magnetic domains, in thermal equilibrium randomly oriented, so that no macroscopic magnetisation occurs. Under the external magnetic field domains rotate towards the direction of external magnetic

J. Holnicki-Szulc and J. Rodellar (eds.), Smart Structures, 107–114.

field and macroscopic magnetisation of the sample occurs. Motion of the magnetic domains is strongly affected by the internal state of the ferromagnetic material, inhomogenities, stresses etc. Magnetic properties of the ferromagnetic materials are characterised in a much more complicated way then, for example, conductivity and for appropriate application of the magnetoelastic method at least the brief description of these characteristics and their measurement is necessary.

The main magnetic characteristic of a ferromagnetic material is the relation between external (magnetising) magnetic field strength **H** and inner (induced) magnetic field flux density **B**, the so-called hysteresis loop. Experimentally we can measure the magnetic flux ϕ=**A**.**B** through the cross-section A. But the magnetic flux is not like electrical current - difference between conductivity of conductor and insulator is about 10^{10} and electrical current flows practically only through the conductor. Difference between the permeability of air and steel is from 2 to about 1000, so the magnetic flux flows not only through the ferromagnetic material. Only for narrow toroidal sample it is possible to calculate the magnetic field intensity and measure precisely the magnetic flux. In practice, in any case, the measured magnetic flux is a combination of flux, depending on the measured magnetic material and flux depending on sensor the arrangement and magnetic surrounding. For precise measurement, the magnetic flux must be closed inside the sensor, practically it means magnetic shielding of the sensor.

The main magnetic characteristics are shown in Fig.1. Amplitude permeability is defined as the ratio B/H, and incremental permeability defined as the ratio $\Delta B/\Delta H$. In both cases permeability depends also on "working point" in which it is measured. Only for very high field strength, where no hysteresis occur (technical saturation), these characteristics are not affected by the magnetic history of material.

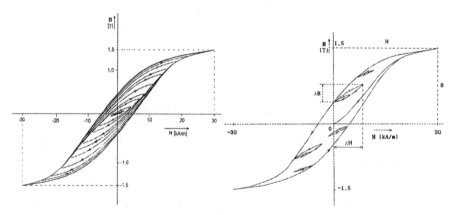

Figure1. Hysteresis loops, amplitude and incremental permeability

3. Measuring method

Ferromagnetic material with cross-section A_f is placed into the external magnetic field with strength $H(t)$ and passes through the sensor coil with N turns and area A_0.

According Faraday induction, low voltage induced in the coil is given by time change of the total magnetic flux flowing through the coil,

$$U_{ind}(t) = -\frac{\partial \Phi(t)}{\partial t} = -\left[NA_f \frac{dB(t)}{dt} + N(A_0 - A_f)\mu_0 \frac{dH(t)}{dt} \right]. \tag{1}$$

After integration of the induced voltage by electronic integrator with time constatnt RC, in the time interval from t_1 to t_2 we obtain the output voltage

$$U_{out} = \frac{1}{RC} NA_f \left[\Delta B + \left(\frac{A_0}{A_f} - 1 \right)\mu_0 \Delta H \right], \tag{2}$$

where

$$\Delta B = \int_{t_1}^{t_2} \frac{dB}{dt}dt \qquad \Delta H = \int_{t_1}^{t_2} \frac{dH}{dt} dt \tag{3}$$

Output voltage without the ferromagnetic material is equal to

$$U_0 = \frac{1}{RC} NA_0 \mu_0 \Delta H , \tag{4}$$

and the ratio U_{out}/U_0 is equal to

$$\frac{U_{out}}{U_0} = \frac{1}{\mu_0} \frac{A_f}{A_0} \frac{\Delta B}{\Delta H} + \left(1 - \frac{A_f}{A_0} \right) = \mu_{rel} \frac{A_f}{A_0} + \left(1 - \frac{A_f}{A_0} \right), \tag{5}$$

where

$$\mu_{rel} = 1 + \frac{A_0}{A_f} \left(\frac{U_{out}}{U_0} - 1 \right) \tag{6}$$

is the relative permeability of the measured material.
This simple model assumes the homogenity of magnetic field but illustrates the measuring principle.

4. Magnetoelastic sensor

4.1. CYLINDRICAL MAGNETOELASTIC SENSOR

It takes the form of a hollow cylinder in the middle of which the measured element (bar, wire, strand, cable) passes through. It must be slipped onto the measured element beforehand, during the construction. This type of sensor consists of primary, secondary and compensating windings, mounted in a protective steel shield and sealed with an insulating material. This cylindrical magnetoelastic sensor has no mechanical contact with the measured element so it can not be overloaded, it is resistant to water and mechanical injury, its characteristics does not change with time and its lifetime is unlimited (i.e. more than 50 years). These magnetoelastic sensors enable the stress measurement in strands and cables protected by thin-wal steel tube or plastic tube, without the need to remove them.

4.2. SENSOR WOUND ON THE TENDON IN-SITU

This sensor enables force measurement in external tendons or stay cables of bridges, mast stays etc. without the necessity to install the gauge during the construction period. The uncertainty of the measurement is, however, limited to about one third of that of the sensors mounted during construction (the zero stress state of the measured element is assumed to be unknown).

4.3. DIFFERENTIAL SENSOR

It was developed to give higher sensitivity and long-term stability. It contains a measured as well as a compensating steel element, both in the same magnetic field and at the same temperature. The accuracy and sensitivity are thereby increased to more than 1MPa and the temperature influence is essentially reduced.

4.4. U-SHAPED SENSOR

Such a sensor with windings on a U-shaped steel yoke enables the measurement of force and stress in rods, wires and strands up to $\varnothing$20mm. This U-shaped sensor can be attached to the element to be measured. In the case of embedded reinforcement it is necessary to remove the concrete cover from the area of about 200x50mm. Uncertainty after calibration on this steel specimen is about ±50MPa (for prestressing steel it is about ±5% of the residual stress). The main source of the error the air gap between the yoke and the measured element. In many cases, for example reconstruction, this may be the only type of sensor is able to measure the existing stress.

5. Measuring techniques, Accuracy and Reliability

5.1. STRESS ESTIMATION

Both the amplitude and the incremental permeability are stress and temperature-dependent and we can use them for stress estimation.
On Fig.2 is typical dependence of amplitude permeability on the stress and temperature for strand 15.5mm /1800MPa at different working points. It is obvious that precise measuring of steel temperature is necessary for precise measurement of the stress.

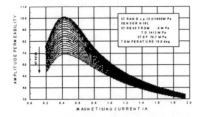 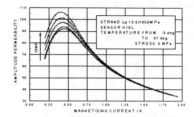

Figure2. Stress and temperature dependence of amplitude permeability

Fig.3 shows typical dependence of incremental permeability on the stress and temperature for the tendon 9x15.5mm /1800MPa. The strong dependence on temperature is present again. During the long-time measurements the temperature error caused by sensor heating may be significant.

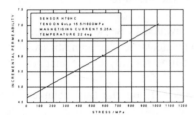 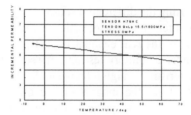

Figure3. Stress and temperature dependence of incremental permeability

This drawback was eliminated by a recently developed pulse method of measuring the incremental permeability. The incremental permeability is measured during the duration of short and high current pulse so that the average energy dissipation is very low and no heating of sensor occurs.

5.2. MEASURING TECHNIQUES AND MEASURING SYSTEM

Several measuring techniques were used during developing of the magnetoelastic method:
1. Continuous measurement of the amplitude permeability.
2. Continuous measurement of the incremental permeability.
3. Pulsed measurement of the incremental permeability.
4. Continuous measurement of the hysteresis curves in the frequency domain.
5. Measurement of the dynamic load using DC magnetic field.
In the simplest measuring system, only one magnetic characteristic of steel is used. In more sophisticated measuring systems, a whole network of hysteresis curves is measured both in the time and frequency domain. Most recently, spectral analysis of the voltage induced in the sensing coil has been used to give valuable information about residual stress and fatigue, and further research is proceeding.
The measuring system feeds the primary winding of the sensor and processes the voltage induced in the sensor coil or coils. Values of stress or force and temperature in

the measured cross-section are output directly. It is possible to connect more sophisticated measuring systems to the control computer through the serial interface RS232 to allow remote control.

5.3. ACCURACY AND RELIABILITY

Uncertainty of the measurement by magnetoelastic method is influenced by:
1. Changes in the measuring system parameters with temperature and time.
2. Change of measured element temperature.
3. Scatter of magnetoelastic characteristics of the measured element and their change with time.

Before every measurement the change of the measuring parameters is excluded by autocalibration (built-in or external).

The temperature effect may be excluded mathematically, using the known temperature dependence of magnetoelastic characteristics and the measured element temperature, or by using the differential sensor.

The scatter of magnetoelastic characteristics is obviated by sensor calibration with a sample of the real element used. Changes of magnetoelastic characteristics with time are under laboratory testing , it seems that long-time stability of the magnetoelastic characteristics is high (change under 1%).

Reliability of the magnetoelastic method depends only on the time-changes of the measured steel magnetic characteristics. In case of a differential sensor, this effect is partially compensated as both the measured and the compensating element are under the same conditions and the magnetoelastic characteristics of the used steel specimen can be measured after a long time, to exclude the change of the magnetoelastic sensitivity. The magnetoelastic sensor characteristics (geometry of windings) are stable in time. Expected lifetime of the sensor is more than 50 years (mainly the lifetime of the connecting cables and casting resin).

6. Applications of the Magnetoelastic Method

1. Prestressed concrete structures (pre-tensioned and post-tensioned) were the first applications of the magnetoelastic method in Slovakia.
2. Cable and suspension bridges. The first cable bridges in the Czech Republic were measured by this method.
3. Prestressed concrete bridges. During last 10 years, more than ten new bridges and several bridges under reconstruction were monitored by the magnetoelastic method. Measured were the forces in both the grouted and external tendons, distribution of the force between the strands in the cable, losses of prestressing force due to friction, long time stress behaviour of grouted and external tendons (relaxation losses).
4. External and grouted tendons prestressed strand by strand. In many cases (especially reconstruction) it is impossible to use heavy prestressing jack for prestressing tendons containing from 6 to 18 strands 0,6". Prestressing these

tendons by light mono-jack strand by strand is there the only acceptable technical solution. Measurements on four new bridges and two bridges under reconstruction in Slovakia during period 1990-95 confirmed the reliability of that prestressing technology (stress losses caused by group effect being acceptable).

5. Unbounded tendons. The magnetoelastic method is probably the only method that enables the measurement of force distribution along MONOSTRAND (strand coated with PE tube) tendon without destroying the PE cover. Such measurements have been made on decks, prestressed by unbounded parabolic tendons, waste water tank prestressed by circular MONOSTRAND tendons or unbounded tendons in the envelope of the nuclear reactor, prestressed to 10MN.

6. Rock and soil anchors. Magnetoelastic sensors have also been used in soil and rock anchors for instantaneous force measurements during prestressing (consolidation of the anchor) and long-term force monitoring. Simple of installation (small sensor on each strand) and high reliability are the main advantages. It is possible to measure real force in the anchor at any time.

7. Force measurement in rigid steel structures. It is a tedious task to measure force in rigid steel constructions, where high internal stress without measurable axial deformation may occur (e.g. armoblocks connected together by welded or pressed connections). Since magnetoelastic sensors are sensitive only to stress and but not to strain, they represent the only choice for this situation.

8. Small stress change measurement. The high sensitivity of the differential sensor, enables e.g. to measure low relaxation in the wires and strands.

9. Investigation of corrosion in stay cables. Magnetoelastic method enables also under certain conditions (constant force and temperature) to measure changes of the cross-section of a stay cable protected by PE tube.

7. Conclusion

Our experience over the past 12 years confirms that the magnetoelastic method is reliable, accurate and generally applicable to many structural monitoring situations, when other methods of measuring the force or stress in steel are inapplicable. But keep in mind that:

1. Magnetoelastic sensor is not a real sensor, it is only the input (magnetising winding) and output (sensing coil) part of the sensor - real sensor is the "smart" steel structure itself. Magnetic properties of a steel are sensitive to stress (also internal stress).

2. It is impossible to calibrate a magnetoelastic sensor - we can calibrate only the certain steel specimens with certain type of magnetising and sensing coil.

3. Scatter of construction steel magnetic properties is not given by the manufacturer (like working diagram, breaking stress etc.). For successful application of the magnetoelastic method knowledge of magnetic characteristics and its scatter is essential. It is a tedious work to measure these characteristic for different steel grades, manufacturers and batches, but it is the only way how to avoid calibration

114

of the magnetoelastic sensor with steel specimen. Our experience with low-carbon prestressing strands 0,5" and 0,6"/1800MPa produced in Slovakia, Spain, UK and USA shows, that the scatter of their magnetoelastic characteristics is within ±5%.

4. Magnetic properties of the steel change also with temperature and the temperature error is up to ±10MPa/deg. Precise measurement of the temperature of steel for elimination of the temperature error is necessary. For very precise measurement of small stress changes (e.g. relaxation measurement), the differential sensor (like dummy gauge) and the pulse method is the best solution.

5. Magnetoelastic sensor uses for transfer of the stress from the structure to the sensing coil the change of the magnetic flux. There are no problems with insulation resistance but ferromagnetic surrounding of the sensor affects the distribution of the magnetic flux. In cases, when ferromagnetic surrounding of the sensor is not stable, the magnetic shielding of the sensor is necessary.

References

1. Jarosevic, A., Fabo, P., Kyska, R., Hatala, M. (1992) Vorspannungmessungen an Baukonstruktionen, Braunschweiger Bauseminar, Braunschweig, Germany November 1992, Heft **97**, 71-82

2. Jarosevic, A., Fabo, P., Chandoga, M., Begg, D.W. and Heinecke O. (1996) The Elastomagnetic Method of Force Measurement in Prestressing Steel, Structural Assesment - The Role of Large and Full Scale Testing, Joint Institution of Structural Engineers/City University International Seminar, London, UK July 1-3 1996, 64.1-64.8

3. Kvasnica B., Fabo P. (1996) Highly Precise Noncontact Instrumentation for Magnetic Measurement of Mechanical Stress in Low-Carbon Steel Wires, *Measurement Science and Technology* 7, 763-767

4. Jiles, D.C., Atherton, D.L.(1986) *Journal of Magnetism and Magnetic Materials*, **61**, 48

5. Jarosevic, A. and Chandoga, M. (1994) Force Measuring of Prestressing Steel, Slovak Report on Prestressed Concrete, XII. International Congress of the FIP, Washington 1994, in *Inzinierske stavby* **42**, 2-3, 56-63

6. Jarosevic, A. and Kvasnica, B. (1997) Mechanical Relaxation of Steel Cable Measured by Incremental Permeability, 4-th Japanese-Czech-Slovak Joint Seminar on Applied Electromagnetics, Stara Lesna, Slovakia September 18-20 1997 In: *Journal of Electrical Engineering*, **48**, 8/s, 170-173

7. Kvasnica, B. and Kundracik, F. (1996) Fitting experimental anhysteretic curves of ferromagnetic materials and investigation of the effect of temperature and tensile stress, *Journal of Magnetism and Magnetic Materials* **162** , 43-49

8. Chandoga, M., Halvonik, J., Jarosevic, A. and Begg D.W. (1997) Relationship between design, modelling and in-situ measurement of pre-tensioned bridge beams, *Proceedings of the International Conference "Computational Methods and Experimental Measurements"*, Rhodes, Greece May 21-23 1997

DAMAGE ASSESSMENT, RELIABILITY ESTIMATION AND RENOVATION OF STRUCTURAL SYSTEMS

S. JENDO* and J. NICZYJ**

*Institute of Fundamental Technological Research, Polish Academy of Sciences, Warsaw, Poland

**Dept. of Civil Engineering and Architecture, Technical University of Szczecin, Szczecin, Poland

1. Introduction

Structures are composed of many members. The reliability of a structural system will be a function of the reliability of its members. Any real structural system usually requires a combination of both serial and parallel descriptions of its subsystems. The paper defines the damage state of a structure in terms of fuzzy sets. The use of interval arithmetic and α-cuts gives considerable advantage for evaluation of fuzzy systems. In this study we have shown that by describing uncertainty of damage state of elements with fuzzy sets, we may estimate the reliability of existing construction in case of incomplete or uncertain knowledge of its parameters. The proposed method may be used in expert systems.

In the classical theory of structural reliability, the theory of probability and statistics is applied to interpret the degree of safety in terms of the failure probability for a given type of structure. For many existing constructions, gathering of statistical data concerning the loads and resistance of these constructions are difficult. The major problem addressed in this paper is the presentation of the fuzzy system failure probability by fuzzy linguistic variables and by a fuzzy number over an interval of confidence. The examples of bar structures demonstrate the applicability and usefulness of the fuzzy reliability approach (in conjunction with the theory of system reliability) for the failure analysis of structural systems.

2. Element Damage Assessment by Fuzzy Sets

Most of the structural systems that are composed of multiple components can be classified as serial-connected, parallel-connected or combined systems. Structural systems which are modelled as series systems mean that the failure of any one or more of these components implies the failure of the entire system. For example, statically determinate truss can be modelled by a series system.

115

J. Holnicki-Szulc and J. Rodellar (eds.), Smart Structures, 115–123.

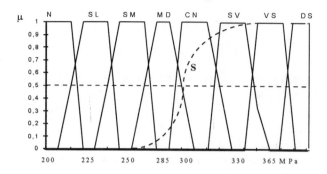

Figure 1. The membership functions $\mu(DE_i)$

For the reliability analysis, various types of uncertainty must be taken into account [1, 2, 7]. Some uncertainties can be described in terms of probability density functions. Identification of uncertainties for complex systems may be difficult. If the probability density function (PDF) is known, we can transform this function into a corresponding membership function [4, 5, 6, 9, 11]. Usually, to obtain PDF, enough data to model the statistical distribution is required. A fuzzy damage set is obtained by the fuzzy set theory and arithmetic operations. Then, by identifying a fuzzy limit state function and using probability of fuzzy events, the failure probability is calculated.

We assumed eight linguistic terms for damage: *none (N), slight (SL) small (SM), moderate (MD), considerable (CN), severe (SV), very severe (VS)* and *destructive (DS)*. The membership functions for them, $\mu(DE_i)$, for steel elements are shown in Fig. 1. Fuzzy variables with membership functions are estimated for a given attribute by experiments based on different assessments of the same real quantity by several experts.

A fuzzy limit state function $S(x;\delta,\beta,\gamma)$ is assumed to have an S-shaped form as shown in Figs. 1 and 2. The function $S(x;\delta,\beta,\gamma)$ is monotonically increasing in the whole interval $[\delta,\gamma]$, $\mu_S(\delta)=0$ and $\mu_S(\gamma)=1$, with the shape function defined as:

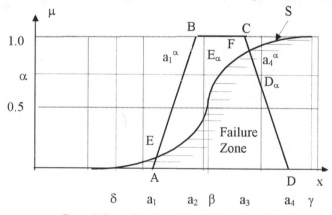

Figure 2. Fuzzy damage (ABCD) and fuzzy limit state function S

$$
S(x,\ \delta,\beta,\gamma) = \begin{cases} 0, & \text{if} \quad x \le \delta, \\ 2\left(\dfrac{x-\delta}{\gamma-\delta}\right)^2, & \text{if} \quad \delta \le x \le \beta, \\ 1-2\left(\dfrac{x-\gamma}{\gamma-\delta}\right)^2, & \text{if} \quad \beta \le x \le \gamma, \\ 1, & \text{if} \quad x \ge \gamma, \end{cases}
$$

where δ, $\beta=(\delta+\gamma)/2$ and γ are shape parameters of the S-function, and x represents damage. The values of parameters δ, β and γ depend on the type of stresses and materials of the elements. For example, some statistical parameters of the steel elements used here are: mean value $\beta = R_e = 285$ MPa, standard deviation $\sigma_{Re} = 22.5$ MPa, reliability index $t_{Re} = 2$, such that $\delta = R_e - t_{Re}\sigma_{Re} = 285-2\times22.5 = 240$ MPa, $\gamma = R_e + t_{Re}\sigma_{Re} = 330$ MPa.

An appropriate definition of fuzzy limit state functions by fuzzy sets is of importance here. A fuzzy limit state function, based on expertise and experience, is defined to determine whether events fail or not. By taking human subjective uncertainty into account, the resultant possibility-based reliability is more flexible and realistic.

The possibility distribution of fuzzy damage set DE_i and limit state function S are shown in Fig. 2. The failure probability for $\alpha = 0$ and for the other α-cut levels are defined as ratio of the area of overlap of fuzzy damage and limit state function, to the area of fuzzy damage, i.e.,

$$
p_f(DE_i) = \frac{\displaystyle\int_{x=A}^{x=D} \mu_S(x)dx}{\displaystyle\int_{x=A}^{x=D} \mu(DE_i)dx}, \qquad p_f^{\alpha}(DE_i) = \frac{\displaystyle\int_{x=E_\alpha}^{x=D_\alpha} \mu_S(x)dx}{\displaystyle\int_{x=a_1^\alpha}^{x=a_4^\alpha} \mu(DE_i)dx}. \tag{1}
$$

Using Eqs. (1), numerical results of the failure probability, $p_f^{\alpha}(DE_i)$, of the fuzzy linguistic variables can be calculated as shown in Table 1. Using the probability method, limit states are defined crisply. For fuzzy sets analysis, when cumulative damage is a little greater than or equal to zero, systems are denoted as safe, whereas failure occurs when damage is a little less than or equal to unity.

There are many different classes of fuzzy numbers. We have used a special class of fuzzy numbers called trapezoidal numbers. The numbers were obtained from estimating the state of the elements [8]. A trapezoidal fuzzy number can be defined by a set $\prod$ (a_1, a_2, a_3, a_4) shown in Fig. 2.

The membership function is defined as

$$\mu(DE_i)=\begin{cases} 0, & x < a_{i1}, \\ \dfrac{x-a_{i1}}{a_{i2}-a_{i1}}, & a_{i1} \le x \le a_{i2}, \\ 1, & a_{i2} \le x \le a_{i3}, \\ \dfrac{a_{i4}-x}{a_{i4}-a_{i3}}, & a_{i3} \le x \le a_{i4}, \\ 0, & x > a_{i4}, \end{cases}$$

Defining the interval of confidence at α-cut level parallel to the horizontal axis, we represent the trapezoidal fuzzy number by

$$DE_i^{\alpha} = [a_{i1}{}^{\alpha}, a_{i4}{}^{\alpha}] = [(a_{i2}-a_{i1})\,\alpha + a_{i1}, \; -(a_{i4}-a_{i3})\,\alpha + a_{i4}].$$

TABLE 1. Element failure probability estimated by fuzzy linguistic variables for various α-cut levels

α	N	SL	SM	MD	CN	SV	VS	DS
1.0	0.0	0.0	0.0	0.0	0.0	0.0	0.833333	1.000000
0.9	0.0	0.0	0.0	0.0	0.0	0.420881	0.999173	1.000000
0.8	0.0	0.0	0.0	0.0	0.0	0.705742	0.999603	1.000000
0.7	0.0	0.0	0.0	0.0	0.102660	0.812357	0.999745	1.000000
0.6	0.0	0.0	0.0	0.0	0.237522	0.865132	0.999815	1.000000
0.5	0.0	0.0	0.0	0.0	0.367955	0.896421	0.999857	1.000000
0.4	0.0	0.0	0.0	0.021400	0.488121	0.917004	0.999885	1.000000
0.3	0.0	0.0	0.0	0.071228	0.582824	0.931496	0.999905	1.000000
0.2	0.0	0.0	0.0	0.137331	0.652082	0.942199	0.999919	1.000000
0.1	0.0	0.0	0.0	0.215628	0.704587	0.949116	0.999930	1.000000
0.0	0.0	0.0	0.033652	0.312405	0.745528	0.956842	0.999939	1.000000

3. Structural System Reliability Estimation

For the static scheme of an examplary construction (Fig. 3a), the theory of system reliability produces a corresponding complex system with elements comprising both series and parallel systems, Fig. 3b. Consider a system (Fig. 3b) composed of n elements E_i, i=1,2,...,n. We associate with each element E_i a single linguistic variable representing the damage of the element $DE_i \in \{N, SL, SM, MD, CN, SV, VS, DS\}$.

For the whole structure, we define the vector DS_i whose elements are linguistic terms that describe the level of damage for each structural element, i.e., $DS_i= (DE_1,...,DE_{12})$. For example, for four independent states, the vectors DS_i are listed in Table 2.

Consider a system of n=12 components with fuzzy reliabilities $p_f(DE_i)$ with statistical independence of components assumed. For statically determinate trusses it is clear that the whole structural system fails as soon as any structural element fails. For a statically indeterminate truss structure the whole structural system will not always fail when one of the structural element fails.

For the vectors DS_2 and DS_3, the change in the 6-th and 12-th elements are introduced, so we can observe the influence this change has on the reliability of the truss and the complex system. Thus, the fuzzy probability of failure of the series system (truss) is

$$p_f^\alpha(T) = 1 - \prod_{i=1}^{7} [1 - p_f^\alpha(DE_i)]. \tag{2}$$

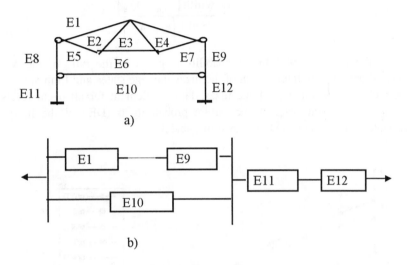

a)

b)

Figure 3. A complex structural system

TABLE 2. Damage levels for structural elements

DE_i	DS_1	DS_2	DS_3	DS_4
DE_1	N	SL	SL	SL
DE_2	SM	SM	SM	SM
DE_3	N	N	N	N
DE_4	SL	MD	MD	MD
DE_5	N	SM	SM	SM
DE_6	SL	CN	MD	MD
DE_7	SM	SM	SM	SM
DE_8	SM	SM	SM	SM
DE_9	SL	N	N	N
DE_{10}	SL	SM	SM	SM
DE_{11}	SL	SL	SL	SL
DE_{12}	SL	SM	CN	MD

The fuzzy probability of failure of a complex system is

$$p_f^\alpha(C)=1-\left\{1-\left[1-\prod_{i=1}^{9}\left(1-p_f^\alpha\left(DE_i\right)\right)\right]p_f^\alpha\left(DE_{10}\right)\right\}[1-p_f^\alpha(DE_{11})][1-p_f^\alpha(DE_{12})]. \tag{3}$$

For all levels α (the grade of membership function), using Eqs. (2) and (3) we obtain the fuzzy probability of failure as shown in Fig. 4.

Next, we introduce a range of values (intervals) of the fuzzy probability of failure [1, 2, 3, 10, 12]. An upper bound of the interval is defined as

$$p_{i,U}{}^{\alpha} = \frac{\text{width}\left[E_{i,\alpha},\ D_{i,\alpha}\right]}{\text{width}\left[a_{i,1}^{\alpha},\ a_{i,4}^{\alpha}\right]},$$ (4)

where the measure of failure probability is given as the ratio of the width of the interval obtained by intersection of the fuzzy damage states and limit state function, to the width of the interval of the fuzzy damage element for all α-cut levels (Fig. 2). Using Eq. (4), intervals of the failure probability $p_f^{\alpha}(DE_i)$ of the fuzzy linguistic variables can be calculated as shown in Table 3.

p_f

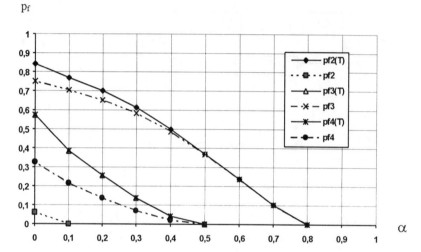

Figure 4. The probability of failure for all α-cut levels

TABLE 3. The intervals of the system failure probability estimated by fuzzy linguistic variables for α-cut levels

α	N	SL	SM	MD	CN	SV	VS	DS
1.0	[0, 0]	[0, 0]	[0, 0]	[0, 0]	[0, 0]	[0, 0]	[0, 0.83333]	[1, 1]
0.9	[0, 0]	[0, 0]	[0, 0]	[0, 0]	[0, 0]	[0, 0.81061]	[1, 1]	[1, 1]
0.8	[0, 0]	[0, 0]	[0, 0]	[0, 0]	[0, 0]	[1, 1]	[1, 1]	[1, 1]
0.7	[0, 0]	[0, 0]	[0, 0]	[0, 0]	[0, 0.43021]	[1, 1]	[1, 1]	[1, 1]
0.6	[0, 0]	[0, 0]	[0, 0]	[0, 0]	[0, 0.67708]	[1, 1]	[1, 1]	[1, 1]
0.5	[0, 0]	[0, 0]	[0, 0]	[0, 0]	[0, 0.85714]	[1, 1]	[1, 1]	[1, 1]
0.4	[0, 0]	[0, 0]	[0, 0]	[0, 0.18942]	[1, 1]	[1, 1]	[1, 1]	[1, 1]
0.3	[0, 0]	[0, 0]	[0, 0]	[0, 0.36509]	[1, 1]	[1, 1]	[1, 1]	[1, 1]
0.2	[0, 0]	[0, 0]	[0, 0]	[0, 0.53947]	[1, 1]	[1, 1]	[1, 1]	[1, 1]
0.1	[0, 0]	[0, 0]	[0, 0]	[0, 0.73512]	[1, 1]	[1, 1]	[1, 1]	[1, 1]
0.0	[0, 0]	[0, 0]	[0, 0.6]	[1, 1]	[1, 1]	[1, 1]	[1, 1]	[1, 1]

Using (4) as a measure of uncertainty of the interval of failure probability, we obtain the system failure probabilities as follows:

$$p_f^\alpha(T) = \left[1 - \prod_{i=1}^{7}\left\{1 - p_{i,L}^\alpha\right\}, \ 1 - \prod_{i=1}^{7}\left\{1 - p_{i,U}^\alpha\right\}\right], \tag{5}$$

and

$$p_f^\alpha(C) = \left[1 - \left(1 - \left\{1 - \prod_{i=1}^{9}\left(1 - p_{i,L}^\alpha\right)\right\}p_{10,L}^\alpha\right)\left(1 - p_{11,L}^\alpha\right)\left(1 - p_{12,L}^\alpha\right),\right.$$

$$\left.1 - \left(1 - \left\{1 - \prod_{i=1}^{9}\left(1 - p_{i,L}^\alpha\right)\right\}p_{10,L}^\alpha\right)\left(1 - p_{11,U}^\alpha\right)\left(1 - p_{12,U}^\alpha\right)\right], \tag{6}$$

where

$$[p_{i,L}{}^\alpha, \ p_{i,U}{}^\alpha] = \begin{cases} [0,0], & if \ \ [E_{i,\alpha}, D_{i,\alpha}] \not\subset [a_{i,1}^\alpha, a_{i,4}^\alpha], \\ \left[0, \dfrac{width\,[E_{i,\alpha}, D_{i,\alpha}]}{width\,[a_{i,1}^\alpha, a_{i,4}^\alpha]}\right], & if \ \ [E_{i,\alpha}, D_{i,\alpha}] \subset [a_{i,1}^\alpha, a_{i,4}^\alpha], \\ [1,1], & if \ \ [E_{i,\alpha}, D_{i,\alpha}] = [a_{i,1}^\alpha, a_{i,4}^\alpha], \end{cases}$$

and $p_{i,L}{}^\alpha$ and $p_{i,U}{}^\alpha$ represent lower and upper bounds for each α of the i-th element. Using Eqs. (5) and (6) for each level α we obtain the intervals of the fuzzy probability of system failure for all α-cut levels as shown in Fig. 5.

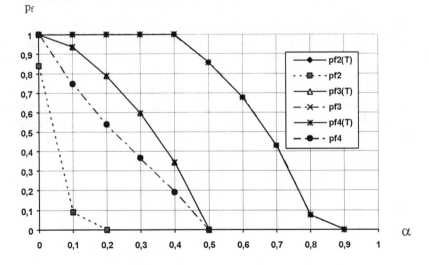

Figure 5. Confidence intervals of the probability of system failure for all α-cut levels

4. Construction State Assessment and Renovation

The methods created in the last years are more efficient in maintenance, functional quality (usability) and repair of existing constructions. The presented method may be used to estimate an actual state of construction and to establish what kind of actions should be undertaken. The system modelling of construction (Fig. 3b) and the probability of system failure (Eq. 3) allow to get information about an element damage influence on the state of the whole construction. For the simplest estimate of the state of the construction, the civil engineer, who makes an estimation, was asked two questions conserning the damage of each element (DE_i), and influence of one for the behaviour of construction [8]. The following linguistic terms for damage assessment of structural elements: *none (N), slight (SL) small (SM), moderate (MD), considerable (CN), severe (SV), very severe (VS)* and *destructive (DS)* have been assumed. Remedial action as a result of the current damage state of elements and their importance in reliability model of construction descibed as *none (NM), small (SM), medium (MM), large (LM)* and *major (RM)* are shown in Table 4. For the action we assume: Enhanced Monitoring (Enh.Monit.), Repair and Replacement (Replac.).

TABLE 4. Remedial action as a result of current state of elements and their importance in rerliability model

	N	SL	SM	MD	CN	SV	VS	DS
RM	-	Enh.Monit.	Repair	Repair	Replac.	Replac.	Replac.	Replac.
LM	-	Enh.Monit.	Repair	Repair	Repair	Replac.	Replac.	Replac.
MM	-	Enh.Monit.	Enh.Monit.	Repair	Repair	Repair	Replac.	Replac.
SM	-	-	Enh.Monit.	Enh.Monit.	Repair	Repair	Replac.	Replac.
NM	-	-	-	Enh.Monit.	Enh.Monit.	Repair	Repair	Replac.

The probability of structural failure shown in Fig. 5 is helpful in determination of the membership function which should be calculated for an actual state of construction. By comparison of the results, the estimation of state of the whole construction contains not only the structural state described by linguistic terms but also the actions that should be undertaken. The six linguistic terms of the type: *very good (VS), good (GS), medium (MS), severe (SS), emergency (ES)* and *collapse (CS)* state have been used. Depending on the state of the whole construction, the possible exemplary actions are presented in Figure 6.

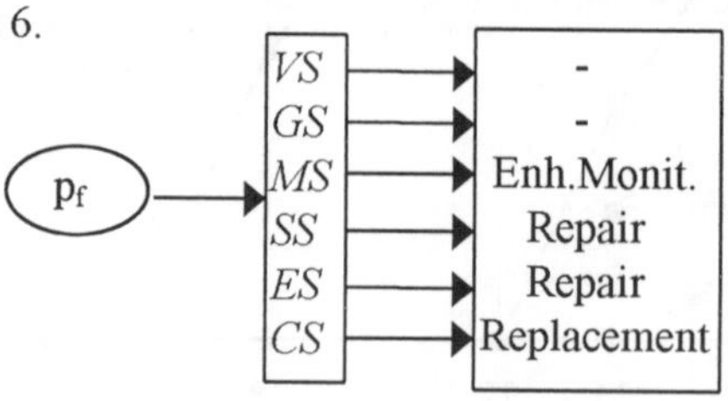

Figure 6. Possible exemplary actions

5. Conclusions

The simple calculation based on interval formulation of the problem can be easily used for qualitative decision making. The examples demonstrate the applicability and usefulness of the fuzzy reliability approach (in conjunction with the theory of reliability) for failure analysis of complex structures.

The major problem addressed in this paper is the presentation of the fuzzy system failure probability by fuzzy linguistic variables and by a fuzzy number over an interval of confidence. The use of interval arithmetic and α-cuts gives considerable advantage for evaluation of fuzzy systems.

In this study we have shown that by describing uncertainty of the damage state of elements with fuzzy sets, one may estimate the reliability of the existing construction in case of incomplete or uncertain knowledge of its parameters. The proposed method may be used in expert systems.

The method of α-cut levels allows to estimate the probability of failure of the construction for different degrees of uncertainty (fuzziness) of data.

Visual estimation of the construction state through periodic monitoring allows to qualify the actual state of construction and to undertake the necessary actions for elements and/or for the whole construction.

References

1. Cheng, C. H. and Mon, D. L. (1993) Fuzzy system reliability analysis by interval of confidence. *Fuzzy Sets and Systems*. **56**, 29-35.
2. Cox, E. (1994) *The Fuzzy Systems Handbook*, Academic Press, London.
3. Dubois, D. and Prade, H. (1980) *Fuzzy Sets and Systems: Theory and Applications*, Academic Press, San Diego.
4. Jager, R. (1995) *Fuzzy Logic in Control*, Thesis, Technische Universiteit Delft, Delft.
5. Lee, C.C. (1990) *Fuzzy Logic in Control Systems: Fuzzy Logic Controller - Part I*, IEEE Transactions on System, Man an Cybernetics, **20**, 2, 404-418.
6. Lee, C.C. (1990) *Fuzzy Logic in Control Systems: Fuzzy Logic Controller - Part II*, IEEE Transactions on System, Man an Cybernetics, **20**, 2, 419-435.
7. Melchers, R. E. (1987) *Structural Reliability: Analysis and Prediction*. Ellis Horwood Limited, Southampton, Great Britain..
8. Niczyj, J. (1995) The fuzzy linguistic variables defining a state of members, in J.B. Obrêbski (ed.), *Proc. of the Int. Conf. on Lightweight Structures in Civil Engineering*, Warsaw Univ. of Technology, Warsaw, Poland. 546-551.
9. Pedrycz, W. (1993) *Fuzzy control and fuzzy systems*. New York, John Wiley.
10. Terano, T., Asai, K. and Sugeno, M. (1992) *Fuzzy Systems Theory and its Applications*, Academic Press, London.
11. Wang, L. X., (1994) *Adaptive Fuzzy Systems and Control - Design and Stability Analysis*, Prentice Hall, Englewood Cliffs.
12. Zimmermann, H. J. (1994) *Fuzzy Set Theory*, Kluwer Academic Publishers, Boston/Dordrecht/London.

CONTROLLING PANTOGRAPH DYNAMICS USING SMART TECHNOLOGY

H.W. JIANG
Dept. of Mechanical Engineering, University of Sheffield, UK
F. SCHMID
Advanced Railway Research Center, University of Sheffield, UK
W. BRAND
Dept. of Mechanical Engineering, University of Sheffield, UK
G.R. TOMLINSON
Dept. of Mechanical Engineering, University of Sheffield, UK

1.   Introduction

Most modern high-speed railway systems use electric power for their operation. The current for the traction power is transferred from an overhead catenary structure to the train via a sliding contact between the contact wire and the mechanical structure known as the pantograph, which is mounted on the roof of the train. The contact wire is

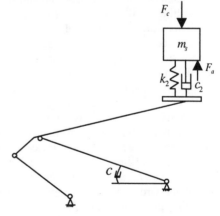

Figure 1. DSA-350 S high Performance Pantograph   *Figure 2.* Schematic Model of DSA 350S Pantograph

mechanically tensioned and is suspended from the catenary using dropper wires. An example of a modern pantograph is shown in *Figure 1*. In this paper a research programme aimed at improving the pantograph dynamics through the active control using a PZT smart material actuator is described, investigated and simulated. The PZT actuation system is designed to produce and amplify linearly the displacements and actuation forces to maintain the contact force within a given bandwidth at the contact point between the contact wire and pantograph head. The analysis of the actuation

J. Holnicki-Szulc and J. Rodellar (eds.), Smart Structures, 125–132.
© 1999 *Kluwer Academic Publishers. Printed in the Netherlands.*

properties for the piezoelectric material is presented. Classical control theory is applied for this SISO, time-invariant system design.

2. Background

For high-speed operation of trains the power collection performance is highly important. The Shinkansen high-speed train running at 330km/h requires 16,000kW of power; the Eurostar for the English Channel running at 300km/h requires 14,000kW of power. For these high speed trains the current collection must be assured at 600~1000A at 25kV AC. It is obvious that a good contact based on maintaining the contact force between the pantograph head and overhead contact wire within a given bandwidth is a crucial issue for high-speed trains.

The dynamic interaction between the pantograph and contact wire has a considerable effect on the power collection performance. Ideally the pantograph collector strips should slide on the contact wire with sufficient force while ensuring that the contact wire is never significantly displaced. Increasing the up-lifting force in order to maintain contact does not ensure better electrical performance because it increases the likelihood of exciting the response of the overhead structure, resulting in to loss of contact and an increase in the bending stress and the wear of the contact wire. If the applied up-lifting force is not large enough, any disturbance would cause the loss of contact easily. When loss of contact occurs, not only is the power to the train interrupted, but also electrical arcing occurs. Arcing can severely damage a contact wire and greatly decrease its operating life. A key factor affecting the contact force is the uneven elasticity of the overhead wires. The contact wire is relatively stiff at the support tower locations and comparatively soft in the middle of a span. Increasing the up-lifting force causes larger variations in contact wire displacement, which can lead to greater dynamic excitation. The resulting oscillations may cause loss of contact in the stiffest regions of the contact wire. This indicates that the ideal contact wire has a uniform stiffness over its length. In order to achieve good contact performance for a high-speed train, considerable effort has been spent on improving these aspects of passive pantographs.

Several configurations of active pantograph designs have been investigated in an attempt to maintain the contact force within a given force bandwidth. Hydraulic actuators were incorporated by Sikorsky Aircraft into the frame of an August-Stemman pantograph [8]. Simulations showed that active elements have the potential to improve significantly the dynamic performance of the pantograph [1]. In this paper a piezoelectric material actuation device is presented and described. The PZT actuator device is an amplifier mechanism which amplifies the expansion and contraction of two PZT stacks and transfers the forces generated to the contact point between the collectors and contact wire. It is located between the pantograph collector strips and the central bar of the pantograph head (*Figure 3.*).

3. The Dynamic Model of the Pantograph Head and the Catenary Contact Wire

In the present study a simplified model of the pantograph head is used, based on the schematic show in *Figure 2*.

The pantograph head and the contact wire are modelled as a two degrees of freedom system mirroring the set-up in the laboratory test. In this model the pantograph frame is modelled as the ground reference due to the low cut-off point of its frequency response. The pantograph collector strips and contact wire are modelled as two lumped masses connected by the simulated stiffness and damping of the contact wire with the external disturbance force pressing on the lumped mass of the contact wire (*Figure 4.*).

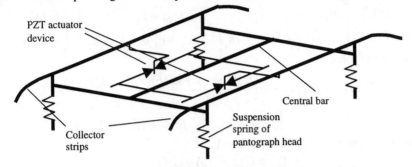

Figure 3. The Location of the PZT Actuator Device in the Pantograph Head

The dynamic equation of this model can be obtained as:

$$m_c \ddot{x}_1 + c_1(\dot{x}_1 - \dot{x}_2) + k_1(x_1 - x_2) = F_e$$
$$m_s \ddot{x}_2 - c_1 \dot{x}_1 + (c_1 + c_2)\dot{x}_2 - k_1 x_1 + (k_1 + k_2)x_2 = -F_a$$

(1)

The contact force can be expressed by the deflection force stored in k_1 and the damping force of c_1, as below:

$$F_c = k_1(x_1 - x_2) + c_1(\dot{x}_1 - \dot{x}_2)$$

(2)

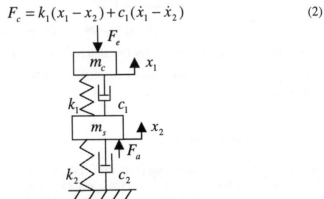

Figure 4. The Dynamic Model of the Pantograph Head

Where, F_c : The contact force between the contact wire and pantograph head;

F_e : Simulated disturbance force of contact wire;

F_a : Active control force generated by PZT actuator;

k_1 : Simulated stiffness of contact wire;

k_2 : Stiffness of pantograph head suspension;

c_1 : Simulated damping coefficient of contact wire;

c_2 : Damping coefficient of pantograph head suspension;

c : Simulated damping coefficient of lifting power system;

m_s : Mass of pantograph collector strips; m_c : Mass of simulated contact wire;

Using Laplace Transformation the solution of the above equation is:

$$\begin{pmatrix} x_1(s) \\ x_2(s) \end{pmatrix} = \begin{bmatrix} m_c s^2 + c_1 s + k_1 & -(c_1 s + k_1) \\ -(c_1 s + k_1) & m_s s^2 + (c_1 + c_2)s + k_1 + k_2 \end{bmatrix}^{-1} \begin{pmatrix} F_e(s) \\ -F_a(s) \end{pmatrix} \quad (3)$$

and the contact force is:

$$F_c(s) = (k_1 + c_1 s) \dfrac{(m_s s^2 + c_2 s + k_2)F_e(s) + m_c s^2 F_a(s)}{\begin{vmatrix} m_c s^2 + c_1 s + k_1 & -(c_1 s + k_1) \\ -(c_1 s + k_1) & m_s s^2 + (c_1 + c_2)s + k_1 + k_2 \end{vmatrix}} \quad (4)$$

Further, the contact force can be expressed as:

$$F_c(s) = G(s)F_a(s) + G_e(s)F_e(s) \quad (5)$$

with:

$$G(s) = \dfrac{m_c s^2 (c_1 s + k_1)}{\begin{vmatrix} m_c s^2 + c_1 s + k_1 & -(c_1 s + k_1) \\ -(c_1 s + k_1) & m_s s^2 + (c_1 + c_2)s + k_1 + k_2 \end{vmatrix}} \quad (6)$$

and:

$$G_e(s) = \dfrac{(m_s s^2 + c_2 s + k_2)(c_1 s + k_1)}{\begin{vmatrix} m_c s^2 + c_1 s + k_1 & -(c_1 s + k_1) \\ -(c_1 s + k_1) & m_s s^2 + (c_1 + c_2)s + k_1 + k_2 \end{vmatrix}} \quad (7)$$

Where $G(s)$ represents the transfer function between the actuation force and the contact force, while $G_e(s)$ establishes the relationship between the disturbance force and the contact force.

4. The Control System Design

In this section the linear actuation properties of the piezoelectric material is analyzed. The control system to be designed for the contact force here is a SISO, linear time-invariant system. Therefore the classical control approach can be used in this system design. Two controllers are designed. One controller is necessary for the cancellation of the undesired complex-conjugate poles resulting from the low damping of the pantograph head suspension. The other, a phase-lead controller, is designed using the root locus method. It ensures the stability of the feedback loop, satisfies the transient response performance and the steady state error requirement.

4.1. ANALYSIS OF THE ACTUATING PROPERTIES OF THE PIEZOELECTRIC ACTUATOR

The piezoelectric actuators convert electrical voltage signals into mechanical displacements or forces. Expansion is produced when the piezoelectric stack is placed in an electrical field which has the same polarity as the piezoelectric stack. The equation of the free expansion of a PZT stack is:

$$\Delta l_0 = E d_{33} L_0 = \frac{d_{33}}{d_s} L_0 U \qquad (8)$$

Where, U : Operating voltage of the electric field; E : Strength of the electric field;

Δl_0 : Free motion of PZT stack; $\qquad$ L_0 : Length of PZT stack;

d_s : Single PZT disc thickness; $\qquad$ d_{33} : PZT coefficient;

Generally, the actuator will be working against the external load, therefore the real extension of the PZT is less than the free expansion. An actuation force equivalent to the force in a spring is generated during the non-free expansion, as shown below.

$$F = k_s \Delta l \qquad (9)$$

Where, k_s : The stiffness equivalent to the external load;

and thus the force equation of the PZT:

$$F_a = k_p (\Delta l_0 - \Delta l) \qquad (10)$$

Where, k_p : The stiffness of the PZT stack;

The force equilibrium is:

$$k_s \Delta l = k_p (\Delta l_0 - \Delta l) \qquad (11)$$

Therefore, the actuation force generated by the PZT stack can be represented as:

$$F_a = k_p \frac{d_{33}}{d_s} L_0 \frac{k_s}{k_s + k_p} U \qquad (12)$$

This equation can be understood from *Figure 5*.
The transfer function of the PZT actuator is:

$$G_a(s) = \frac{F_a(s)}{U(s)} = k_p \frac{d_{33}}{d_s} L_0 \frac{k_s}{k_s + k_p} \qquad (13)$$

In the actuator arrangement schematically shown (*Figure 3*), two PZT stacks are preloaded against a common actuating level. They are excited with opposite polarity voltages which causing one PZT stack to extend and the other to contract. This displacement is amplified to vary the force at the contact wire interface. A prototype actuator was manufactured as part of the present project and is currently being refined.

4.2. THE FEEDBACK CONTROL LOOP DESIGN

4.2.1. *Design of the Controller for Cancellation of Complex- Conjugate Poles*

The evaluation of the parameters of the open loop transfer function $G(s)$, from (6)

$$G(s) = \frac{s^2(s+60666.7)}{(s^2+0.0024s+98.67)(s^2+0.0156s+862)} \tag{14}$$

shows that there are two pairs of complex-conjugate poles in the $G(s)$ resulting from the low damping of the pantograph head suspension. One pair of the complex-conjugate poles is very close to the imaginary axis, in the left half of the s-plan. Also, we noticed that there are two zero zeros in the numerator of $G(s)$ which cause very big

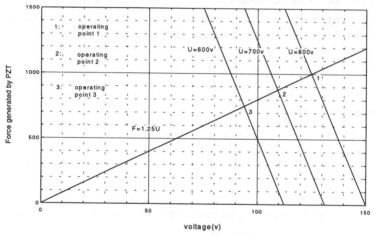

Figure 5. The Relationship between Force, Expansion and Operation Voltage in a PZT Stack

steady-state error. In order to ensure the stability and improving the steady-state error the first controller's transfer function is designed as below:

$$G_{c1}(s) = \frac{s^2+0.0024s+98.67}{s^2} \tag{15}$$

The numerator and denominator of $G_{c1}(s)$ are designed to cancel the undesired poles and two zero zeros of $G(s)$ respectively.

4.2.2. *Design of the Phase-Lead Controller*
A phase-lead controller was also designed to work together with the first one. It is intended to have the effect of pulling the root locus to the left, making the system more stable and speed up the settling of the transient response.
The transfer function of the phase-lead controller is defined as below:

$$G_{c2}(s) = \frac{k(s+a)}{(s+b)} \tag{16}$$

Based on the specification of the control system performance, ($\xi = 0.45$, $\omega_n = 200 rad/s$), the desired complex-conjugate poles of the closed loop are:

$$s_{1,2} = -\xi\omega_n \pm \sqrt{1-\xi^2}\omega_n j = -90 \pm 178.6j$$

The compensated open loop transfer function is:

$$G_o(s) = G(s)G_{c1}(s)G_{c2}(s) = \frac{k(s+a)(s+60666.7)}{(s+b)(s^2 + 0.0156s + 862)} \qquad (17)$$

Supposing the system is a single-loop system, then the phase-lead controller must satisfy the conditions of:

$$|G_o(s)|\angle G_o(s) = -1 + 0j \qquad (18)$$

Therefore,
$$G_{c2}(s) = \frac{10(s+9)}{(s+338.7)} \qquad (19)$$

The block diagram of the feedback control system is shown in *Figure 6*. The whole

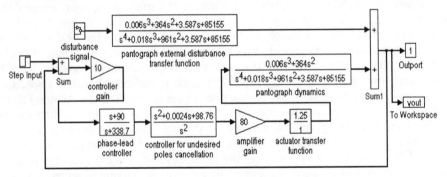

Figure 6. Block Diagram of the Control System

system was simulated using the Matlab SIMULINK software. Some of the simulation results are shown below.

5. The Simulation Results from Matlab SIMULINK

Once the simulation model has been built in Matlab SIMULINK as shown in *Figure 6*. The time domain step responses are obtained with the help of SIMULINK toolbox. *Figure 7* shows the step response of the control loop. *Figure 8* compares the step responses of the active contact force with disturbance coming into the control loop and the contact force of the passive pantograph. The input of the disturbance signal is set to be a random function.

132

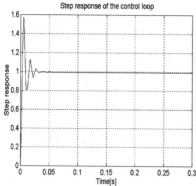

Figure 7. Step Response of the Control Loop

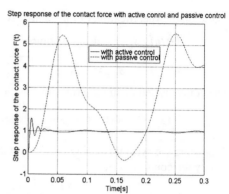

Figure 8. Step Response of the contact force with
Active Control and Passive Control

6. Conclusion

In this paper the authors have presented the results of a study using smart materials to control the contact force between the pantograph head and contact wire. They have shown that the PZT smart actuator has the potential of enhancing the swiftness of the control response for high-speed train by a head actuated active control system. It can be seen from the simulation results that the step response of the control loop is significantly improved, and the active contact force is maintained within a specified bandwidth after the disturbance has been applied to the system. The low damping of the pantograph head suspension is compensated by the controller design to increase the relative stability of the system. In addition, the dynamic model in this paper is not the complete model of the pantograph and the catenary system. The lumped mass of the contact wire in the model only mirrors the laboratory test. From a practical point of view the mechanical amplifier necessary to achieve the required pantograph head displacement from the small expansion of PZT stacks also needs further development. Controlling the pantograph dynamics using smart material technology should thus be considered as a preliminary study in this research field.

References

1. D. N. O'Connor, S. D. Eppinger, W. P. Seering, and D. N. Wormley, (1997) Active Control of High-Speed Pantograph, *Journal of Dynamic Systems, Measurements and Control*, Vol. 119/1
2. David O'Conner, (1984) Modelling and Simulation of Pantograph-Catenary System *Master's thesis, Massachusetts Institute of Technology*
3. Kurt Armbruster, (1983) Modelling and Dynamics of Pantograph/Catenary, *Master's thesis, Massachusetts Institute of Technology*
4. Richard C. Dorf and Robert H. Bishop, (1995) *Modern Control System, 7th edition*, Addison-Wesley Publishing Company, Inc., United States
5. Katsuhiko Ogata, (1990) Modern Control Engineering, 2nd edition, Prentice-Hall Inc.
6. J. W. Waanders, (1991) *Piezoelectric Ceramics, Properties and Applications, 1st eddition*, Philips Components and Eindhoven
7. Pierre Boissonnade, (1975) Catenary design for high speeds, Report, Rail International Chef de la Division des Indtallations Fixes de Traction Electrique, Direction de J'Equipement, SNCF
8. Sikorsky Aicraft, (1970) Design and Development of a Servo-Operated Pantograph for High-Speed Trains, Final Technical Report, U.S. Department of Transportation

PIEZOGENERATED ELASTIC WAVES FOR STRUCTURAL HEALTH MONITORING

GRZEGORZ KAWIECKI
Department of Mechanical and Aerospace Engineering and Engineering Science, The University of Tennessee Knoxville, TN 37996-2210

1. Abstract

The purpose of the presented study is to demonstrate experimentally the feasibility of nondestructive damage detection by an array of piezotransducers embedded or bonded to the surface of a monitored structure. This paper shows the results of experiments with transmission of elastic waves among systems of piezotransducers surface-bonded to aluminum and concrete plates, and embedded in a concrete block. Structural damages were simulated by masses of various magnitudes placed at the surface of tested specimens, and by drilled holes. Experimental results obtained so far indicate that there is a strong correlation between the size and location of a simulated damage and the character of signals transmitted among piezotransducers. Anomalies in those signals are repeatable and have distinct characters when they are caused by a simulated damage and when they are due to other causes. Piezotransducers used in this study were able to produce and receive signals well above the noise level both when surface-bonded to aluminum specimens and when embedded in a concrete block. The anomalies caused by simulated damages were significant enough to be used for training a neural network.

2. Introduction

Major structures such as bridges, buildings, airport runways, passenger and cargo aircraft, etc., constitute a significant portion of national wealth. Proper structural maintenance costs are huge and that is why even a small percentage cost reduction amounts to a significant saving. One of the more cost effective maintenance methods is structural condition monitoring. Early detection of problems such as cracks, delaminations, decay of cement matrix, concrete spalls and corrosion losses can prevent a catastrophic failure or structural deterioration beyond repair. Structural defect detection and analysis methods can be divided into destructive and nondestructive. Those latter are in most cases more efficient, cheaper, easier to apply

J. Holnicki-Szulc and J. Rodellar (eds.), Smart Structures, 133–142.

and as such, continue to attract a considerable attention of civil engineering community. Some of the most popular methods used for nondestructive evaluation are: visual inspection, magnetic field, eddy current, x-ray, ultrasound, acoustic emission, strain gage, thermal contours, laser interferometry and neutron radiography. Unfortunately, most of those methods have significant disadvantages. In general, they work best for assessing the condition of portions of a structure or individual members and are quite impractical for comprehensive monitoring of large structures with complex geometry. Also, each of those methods has serious limitations associated with their principles of operation. For instance, x-ray analysis poses a health hazard and requires a good access to the evaluated element [1]. Eddy currents do not penetrate deeply into the inspected material and can be used mainly for surface inspection [2]. Pulse echo method works well only for concrete and requires very experienced operators for proper collected data interpretation [3].

A distinct, and a very promising approach is damage detection based on vibration testing. That approach is based on the fact that structural faults do change stiffness, and in some cases also damping and inertial characteristics of damaged structures. In theory, it is sufficient to monitor mode shapes, natural frequencies and modal damping coefficients in order to detect a defect, localize it and even estimate its size. However, the results obtained to-date are quite controversial and to some extent, contradictory. Both successes and failures were reported in the literature. Most work was concentrated on damage detection in beams [4, 5], frames [6, 7], and plates [8, 9].

A related family of methods is based on the application of elastic waves for structural health monitoring [10]. Those methods offer a promise of identifying local damages or impacts. Methods based on global natural frequencies or modal shape changes do not perform well in local damage detection. In fact, methods based on stress waves propagation, and ultrasonic inspection in particular, are already well established in the engineering community [11]. Ultrasonic waves are usually generated and received by piezotransducers. However, those transducers are bulky and require good coupling with the monitored structure. These difficulties have been addressed recently by several researchers proposing the use of piezotransducers surface-bonded to the structure or embedded in the walls of the structure.

Moetakef [12] analyzed experimentally and numerically the capability of piezoceramic patches to generate elastic waves in beams and plates. Sun et al. [13] used piezoelements bonded to the monitored structure and investigated the effect of damages on the electric impedance of piezotransducers. Choi and Chang [14] proposed a system of distributed piezosensors for impact detection. They demonstrated the system feasibility by applying it to monitor the magnitude and location of impacts on a beam. A computer code based on a theoretical model of the tested specimen was used to determine impact magnitude and location. Jones et al. [15] applied neural networks to find the magnitude and location of an impact on isotropic plates. An array of piezotransducers surface-bonded to the plate was used for detecting elastic waves generated by the impact. Lakshmanan and Pines [16] have shown the application of piezogenerated elastic waves to damage detection in helicopter rotor flexbeams. Their method was based on measuring resonant frequencies and wave scattering properties.

The studies described above indicate the feasibility of a comprehensive system for structural health monitoring based on an array of piezotransducers capable of generating and receiving elastic waves. Anomalies in signals transferred between two elements of that array contain information about a possible damage in that area. However, to the best of the knowledge of the author, only three published studies explored such systems. Castanien and Liang [17] tested the performance of an array of piezoelements bonded to the surface of a portion of an aircraft fuselage. Their method was based on measuring the electromechanical impedance. Statistical methods were used for signal processing. Shen et al. [8] showed the feasibility of using an actuator/receiver pair embedded in a composite plate for monitoring the cure condition and impact location. Lamb waves generated and received by the elements of the piezotransducer network were used for structural inspection. Keilers [9] presented a concept of using flexural waves transferred between elements of an array of piezotransducers surface-bonded to a composite plate to detect delaminations. He showed theoretical results predicting the effect of a delamination on signal frequency response.

The intention of the present paper is to demonstrate experimentally the feasibility of a similar system generating and receiving simple frequency sweeps. It will be shown that anomalies in transmitted signals carry information strongly associated with the location and size of structural damage.

3. Concept Description

The proposed concept is presented in Fig. 1, where 1 is a plate, 2 denotes damage, and 3 represents a pair of piezoelectric transducers. The plate is instrumented with four pairs of piezoelements (1, 5, 6, 8) bonded to both faces of the plate. Each of these pairs can be used as a transmitter or as a receiver. Assume that the piezotransducer 1 generates elastic waves with a profile appropriate for a given task (burst chirp, harmonic signal, etc.). This signal will be picked up by piezotransducers 5, 6, and 8. After recording signals received by transducers 5, 6, and 8, the transducer 5 will become a sender, and signals received by transducers 6, 8, and 1 will be recorded. This procedure will be repeated until a set of twelve transfer functions, with three functions generated by each of the piezotransducers, is recorded. Any change in the structure's characteristics, such as a local change in stiffness or inertia or a change in mechanical admittance caused by a material defect, will create anomalies in propagated elastic waves and modify the received signals. These anomalies will be analyzed and interpreted by a neural network, which will use this information to estimate the location and size of the defect. The same concept will be used to monitor, e.g., the condition of a typical civil engineering structure. See Fig. 2. The feasibility of both components of the proposed concept: the transducer array and the neural network

136

based signal processing system have been already successfully demonstrated. Sample results are presented in following sections.

4. Approach

The experimental feasibility verification of the proposed concept had three objectives: (1) to show that anomalies in signals transferred between elements of the piezotransducer network are consistent with the size and extent of simulated damages, (2) to show that the results are repeatable, and (3) to show that signal anomalies associated with simulated damages are sufficiently large to be used by an automatic damage identification system, e.g., a neural network. Frequency ranges in which transfer functions seemed to show greatest sensitivity to damages were selected by trial and error for each group of experiments. All of the presented experimental results were produced using pairs of piezoelements actuated in-phase, i.e., producing compressive waves.

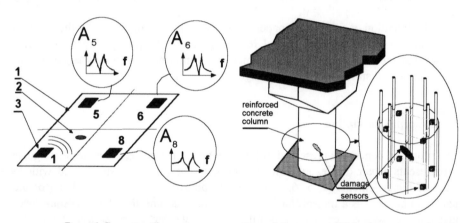

Figure 1 Concept outline. *Figure 2* Instrumented highway overpass column.

Experiments were performed using an experimental set-up consisting of an HP3562A dynamic signal analyzer, and a piezoamplifier (Model 60.102 from Piezo Systems, Inc., Cambridge, MA). The excitation signal generated by one channel of the dynamic signal analyzer was fed to the piezoamplifier. The amplified signal was then sent to the pair of piezoelements used as a sender. The generated elastic wave was received by the piezotransducer set up as a sensor, and sent to the other channel of the dynamic signal analyzer. Both the input and output signals were recorded on a diskette, and then translated from the proprietary HP data format to MATLAB format. Four different specimens were tested: an aluminum beam, an aluminum plate, a concrete beam, and a concrete block. Tested specimens are shown in Figure 3, where 1 indicates piezotransducers, 2 depicts the aluminum beam, and 3 is a weight simulating damage, in Figure 4, where 1 is the concrete beam, 2 indicates a piezotransducer, and 3 is a weight simulating structural damage, in Figure 5, where 1 is a piezoamplifier, 2 is the aluminum plate, 3 is a weight simulating damage, and 4 represents piezotransducers bonded symmetrically at each corner to both faces of the plate, and in Figure 6, where 1 is the concrete block, 2 represents a weight simulating damage, 3 is a surface-bonded piezoelement, and 4 indicates the filed-off edges of embedded piezoelements. Both the concrete beam and the block were fabricated from Quickrete No 1240, which is a mixture of portland cement and silica sand. Piezotransducers were surface-bonded to tested specimens using ECCOBOND 45LV epoxy resin and CATALYST 15LV epoxy resin hardener (Emersons & Cuming, Inc., Woburn, MA). The concrete block was also instrumented with two pairs of piezotransducers embedded as shown in Figure 6. All piezoelements were cut from 0.25 mm thick PSI-5A-S2 plates purchased from Piezo Systems, Inc., Cambridge, MA.

The concrete and aluminum beams were tested mainly to evaluate the effect of a simulated damage on the character of transmitted elastic waves. The concrete block was tested to determine whether piezotransducers embedded in that type of material are capable of generating and receiving elastic waves. The aluminum plate was used to test the effect of simulated defect location and size on signal transmission along various paths.

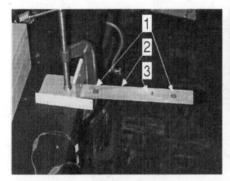

Figure 3. Tested aluminum beam.

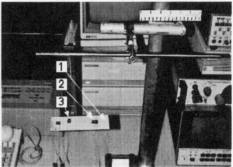

Figure 4. Tested concrete beam.

Figure 5. Tested aluminum plate.

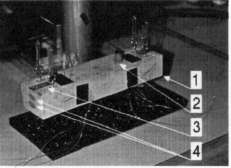

Figure 6. Tested concrete block.

138

5. Results

5.1. DAMAGE DETECTION IN BEAMS

Testing of the aluminum beam (see Fig. 3) was conducted using burst chirps with a duration of 0.8 s and frequency range 100 Hz to 1.1 kHz.

Figure 7 shows the effect of weight size on the signal transmitted between transducers. Transfer function obtained for a pristine beam is compared against signals recorded with 10 G, 4 G, and 2 G placed at the location shown in Fig. 3. It is clear that even a 2 G weight affects significantly the transfer function, particularly around 250 Hz, and 650 Hz.

The effect of a simulated damage on signal transmission in the concrete beam (see Fig. 4) is shown in Figure 8. We can note that the frequencies for which we have

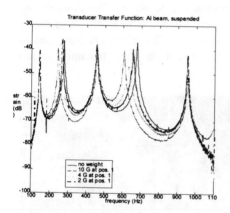

Figure 7. Effect of a simulated damage in Al beam.

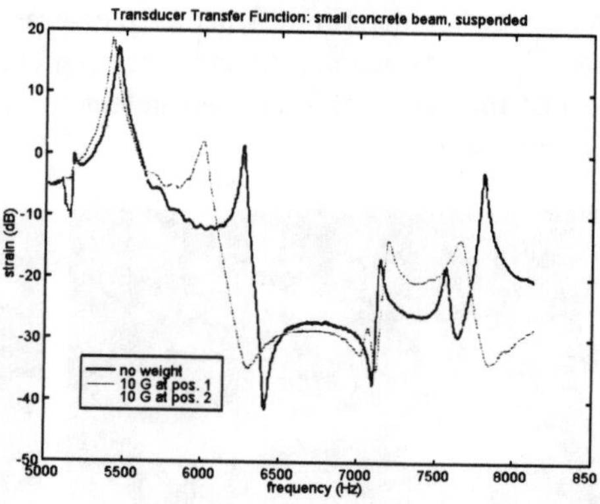

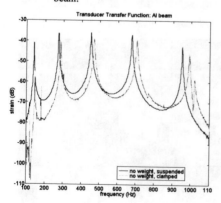

Figure 9. Boundary conditions effect for a pristine beam.

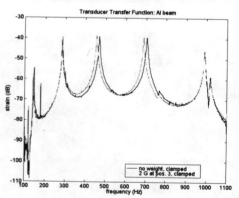

Figure 8. Effect of a simulated damage in concrete beam.

Figure 10. The effect of a 2 G weight for a clamped beam.

high sensitivity to a simulated damage are higher than for metal specimens. The differences between frequency response function characteristics recorded for a pristine beam and for a beam with a simulated damage are sufficiently large to be used for a neural network training.

Another set of experiments demonstrated that the effect of a simulated damage differs from the effect of a change in boundary conditions. Such a difference is very important to avoid triggering the damage detection system by a change in boundary conditions or other environmental factors.

Figures 9 and 10 show the differences between transfer functions recorded for suspended and clamped beams with no damage, and with 2 G weight located as shown in Fig. 3. Fig. 9 indicates that the most pronounced effect of a boundary condition change is a natural frequencies drift, the higher the frequency, the larger the difference between the equivalent natural frequencies measured for the two types of support. On the other hand, Figure 10 indicates that the effect of damage seems to have a different character than that due to a change in boundary conditions. In this case frequency shifts do not follow a regular pattern. This observation was confirmed by other results, not shown. Those differences are very important for the damage detection system, so that it can learn to distinguish anomalies generated by real damages from spurious information.

5.2. DAMAGE DETECTION IN THE ALUMINUM PLATE

Plate testing was conducted using periodic chirps with a duration of 1.0 s and frequency range 300 Hz to 800 Hz. Only bottom transducers (No. 5, 6, and 8, see Fig. 1) were used as receivers.

Figure 11 shows the effect of receiver location on the transfer function recorded for a 10 G weight placed at location shown in Figs. 1 and 5. We can note that signals received by transducers 5 and 8 are virtually identical, as they should be, because the simulated damage is located on the diagonal of the plate. Then, the consistency of the

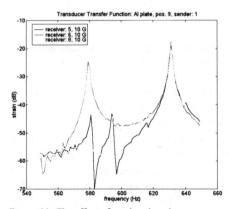

Figure 11. The effect of receiver location.

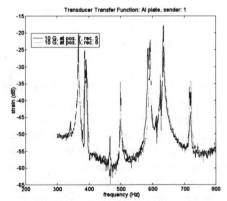

Figure 12. Damages and receivers symmetrical w.r.t. the diagonal of the plate.

proposed method was evaluated. Figure 12 shows transfer functions received by two transducers located symmetrically with respect to the diagonal of the plate (i.e., by piezoelements 5 and 8, see Figure 1) registering damages located symmetrically with respect to the diagonal. Even though these plots have been obtained for a large weight the frequency transfer functions have exactly the same character.

5.3. DAMAGE DETECTION IN THE CONCRETE BLOCK

Concrete block testing was conducted using burst chirps with a duration of 0.16 s and frequency range 20 kHz to 25 kHz. The main objective of this test was to determine the performance of piezotransducers embedded in concrete. The effect of a simulated damage is shown in Fig. 13. The damage was simulated by placing a 10 G weight at a location shown in Fig. 6. The effect of an actual damage has been demonstrated in Fig. 14. The damage was introduced by drilling a transverse 3/16" (4.76 mm) diameter hole, and then by drilling a 1/4" (6.35 mm) hole at the same location. In both cases the effect of damage on transmitted signals was strong. Also, the obtained results seem to indicate a very good adhesion of embedded piezotransducers to the material of the block.

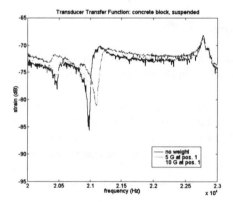 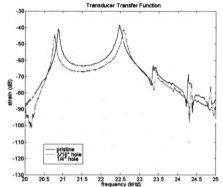

Figure 13. The effect of simulated block damage.   *Figure 14.* The effect of a transverse hole in the block.

Fig. 15 shows the effect of concrete curing time on the character of transfer functions. Measurements were made on the 5th, the 6th, and the 7th day after pouring the concrete block. The process of concrete curing was not complete until approximately three weeks after fabricating the block. Shifted transfer functions indicate that the proposed method also can be used to monitor the fabrication process. Although no experiments have been performed for larger specimens, similar effects can be expected. Fig. 16 demonstrates transfer functions obtained for the same block on several consecutive days after the curing process was complete. We can see that the results are almost identical. This experiment confirmed the repeatability of the proposed method.

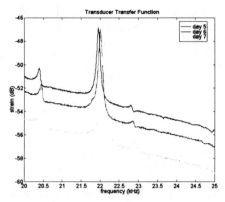
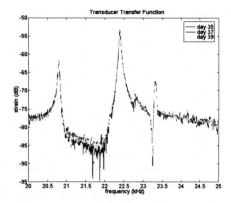

Figure 15. The effect of material curing.   *Figure 16.* Repeatability of the proposed method.

6. Conclusions

Results presented in this study indicate that elastic waves transferred between elements of an array of piezotransducers are strongly affected by damages resulting in local inertia changes. Such damages produce anomalies in piezogenerated elastic waves propagating through the structure. Those anomalies have a repeatable character. The changes in transferred signals have a different character when the anomalies are caused by a simulated damage and when they result from boundary conditions alteration. That indicates that it should be possible to design a data processing system capable of distinguishing changes in structural behavior due to a damage from changes due to other causes. Indeed, the development of a neuro-fuzzy system capable to associate anomalies in transferred signals with damage size and location will be a next step in the presented study.

Acknowledgment

This project was partially supported by the National Science Foundation under grant CMS-9402802 with Dr. S. C. Liu as a the technical monitor. His guidance and advices are gratefully acknowledged.

References

1. Richardson, M. H. (1993) Are Modes a Useful Diagnostic in Structural Fault Detection ?, in *Proceedings of the 47th Meeting of the Mechanical Failure Prevention Group*, Virginia Beach, VA, pp. 37-50.
2. Thompson, D. O. and Schmerr Jr., L. W. (1994) The Role of Modeling in the Determination of Probability of Detection, in X. P. V. Maldague (ed.), *Advances in Signal Processing for Nondestructive Evaluation of Materials*, NATA ASI Series, Series E: Applied Sciences - Vol. 262, Kluwer Academic Publishers, pp. 285-301.

3. Muenow, R. A. (1990) Field Proven Nondestructive Testing Methods for Evaluating Damage and Repair of Concrete, in H. L. M. dos Reis (ed.), *Nondestructive Testing and Evaluation for Manufacturing and Construction*, Hemisphere Publishing Corporation, pp. 279-284.

4. Banks, H. T., Inman, D. J., Leo, D. J. and Wang, Y. (1996) An Experimentally Validated Damage Detection Theory in Smart Structures, *Journal of Sound and Vibration* **191**, 859-880.

5. Stubbs, N. and Kim, J.-T. (1996) Damage Localization in Structures Without Baseline Modal Parameters, *AIAA Journal* **34**, 1645-1649.

6. Hearn, G. and Testa, R. B. (1991) Modal Analysis for Damage Detection in Structures, *Journal of Structural Engineering* **117**, 3042-3063.

7. Lyon, R. H. (1995) Structural Diagnostics Using Vibration Transfer Functions, *Sound & Vibration*, January, pp. 28-31.

8. Shen, B. S., Tracy, M., Roh, Y.-S. and Chang, F.-K. (1996) Built-In Piezoelectrics for Processing and Health Monitoring of Composite Structures, in *AIAA/ASME/AHS Adaptive Structures Forum*, Salt Lake City, UT, pp. 390-397.

9. Keilers, C. H. (1997) Search Strategies for Identifying a Composite Plate Delamination Using Built-In Transducers, in F.-K. Chang (ed.), *Proceedings of the International Workshop on Structural Health Monitoring*, Technomic Publishing Co., pp. 466-477.

10. Cawley, P., 1997, "Quick Inspection of Large Structures Using Low Frequency Ultrasound," Proceedings of the International Workshop on Structural Health Monitoring, Stanford, CA, September 18-20, Edited by Fu-Kuo Chang, Technomic Publishing Co., pp. 529-540.

11. Sansalone, M., and Carino, N. J. (1991), in V. M. Malhotra and N. J. Carino (eds.), *Stress Wave Propagation Methods - CRC Handbook on Nondestructive Testing of Concrete*, CRC Press, pp. 275-304.

12. Moetakef, M. A., Joshi, S. P., and Lawrence, K. L. (1996) Elastic wave generation by piezoceramic patches, *AIAA Journal*, **34**, 2110-2117.

13. Sun, F. P., Chaudhry, Z., Liang, C., and Rogers, C. A. (1994) Truss Structure Integrity Identification Using PZT Sensor-Actuator, in *Proceedings of the 2nd International Conference on Intelligent Materials*, VA Technomic Pub. Co., Inc., pp. 1210-1222.

14. Choi K. and Chang, F.-K. (1996) Identification of Impact Force and Location Using Distributed Sensors, *AIAA Journal*, **34**, 136-142.

15. Jones, R. T., Sirkis, J. S., and Friebele, E. J. (1997) Detection of Impact Location and Magnitude for Isotropic Plates Using Neural Networks, *Journal of Intelligent Material Systems and Structures*, **7**, pp. 90-99.

16. Lakshmanan, K. A., Pines, D. J. (1997) Modeling damage in rotorcraft flexbeams using wave mechanics, *Smart Materials and Structures*, **6**, 383-392.

17. Castanien, K. E. and Liang, C. (1996) Application of Active Structural Monitoring Technique to Aircraft Fuselage Structures, in *Proceedings of the SPIE Smart Structures and Materials Conference*, San Diego, California, Vol. 2721.

DESIGN OF ADAPTIVE STRUCTURES FOR IMPROVED
CRASHWORTHINESS

L. KNAP[1,2], J. HOLNICKI-SZULC[1]
1. Institute of Fundamental Technological Research,PAS
Swietokrzyska 21, 00-049 Warsaw, Poland
2. Warsaw University of Technology,
Institute of Machine Design Fundamentals
Narbutta 84, 02-524 Warsaw, Poland

1. Introduction

The problem of safe, self-adaptive structures with maximum dissipation of impact energy is discussed. The main point is incorporation of the largest portion of the structure to the failure process, e.g. in order to protect passengers in a train accident. The overall effect desired in safe design is analogous with application of structural members with elasto-plastic properties, built into the structure designed for the most expected impact. In our proposition the effect of *active adaptation* (during transient dynamic analysis) of yield limits in several chosen members to detected the impact will be applied, which gives an important increase of impact energy dissipation.

Modifications of the yield limits in chosen members and their adaptation to detected (in real time) impacts can be possible by application of special devices (so-called *adaptive energy dissipaters,* see Fig. 1a). On the other hand, the structure should be equipped: i) with a sensor system (e.g. accelerometers) detecting and identifying an impact, ii) with the energy dissipating devices able to provoke the generation of local distortions in a controlled way, and iii) with a controller realising in real time the pre-designed control strategy of yield stress modifications in dissipaters. An important feature of the proposed concept is dissipative character of actuation. It means that always $\sigma_i \Delta\varepsilon_i^o \geq 0$, what means that no external energy sources are required to „move" the actuators.

In truss structures, the energy-dissipating device can be realised as a piston with controlled valves opening the flow of fluid (Fig. 1a). Assume that characteristic of an element (Fig. 1b) defines the desired maximum upper and lower limit values σ^u, $\sigma^l = -\sigma^u$ beyond which a plastic behaviour occurs. The dissipater device, controlled (by opening, if it is necessary, valve A_1 or A_2) to keep pressures in the cylinder (Fig.1a) $p_1 \leq \sigma^l$ and $p_2 \leq \sigma^u$, provokes plastic-like overall behaviour (simulated by *plastic distortions* ε^0) of the member. In this way, the original structure equipped with properly located dissipaters can behave like an ideal elasto-plastic structure with

143

J. Holnicki-Szulc and J. Rodellar (eds.), Smart Structures, 143–152.

144

controllable limit stress level. Optionally, devices with easy controllable MRF (magneto rheological fluid) (Ref. 5) can be applied as adaptive energy absorber.

The above concept of adaptive energy absorption will be verified experimentally (Ref. 3) with use of the truss cantilever model equipped with hydraulic dissipaters with controllable valves.

a)   b)   c)

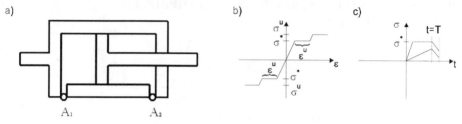

Figure 1. Adaptive Energy Dissipater

The VDM based approach (Refs. 1, 2) allows the development of an efficient computational method for numerical dynamic analysis of elasto-plastic structures modelling the behaviour of elastic structures equipped with controlled (due to characteristics from Fig.1b) actuators. In each time step of analysis (the transient problem), the stress redistribution due to plastic zone's development can be done through a sequence of corrections to the deformation field. These corrections are realised through *virtual distortions* (free of any geometrical and statical constraints) and corresponding states of compatible deformations and self-equilibrated forces. Constraining our considerations to small deformations, the global non-linear behaviour of the structure is obtained from superposition of linear elastic and residual states, and thus reformation of the global stiffness matrix is not required. The proposed VDM technique can be simultaneously applied also for simulation of material redistribution (Ref. 1) and solution of the complex smart structure design.

2. Accumulation/Release Strategy of Impact Energy Dissipation

Let us now formulate the problem of adaptive strategy for impact energy dissipation. Applying a discretized time description, this problem can be expressed as follows:

$$\max U = \max \sum_{t=0}^{T} \sum_{i} A_i l_i \sigma_i(t) \Delta \varepsilon_i^o(t) \tag{1}$$

subject to the condition:

$$\left|\sigma_i(t)\right| \le \sigma^u , \left|\varepsilon_i^o(t)\right| \le \varepsilon^u , \sigma_i(t)\Delta\varepsilon_i^o(t) \ge 0 \text{ for } 0 \le t \le T , \tag{2}$$

where: $\Delta\varepsilon_i^o(t) = \dot{\varepsilon}_i^o(t)\Delta t$, i runs through all members of truss structure and

$$\sigma_i(t) = \sigma_i^L(t) + \sigma_i^R(t) = \sigma_i^L(t) + \sum_{\tau=\tau_o}^{t} \sum_{j} D_{ij}(t-\tau)\Delta\varepsilon_j^o(\tau) . \tag{3}$$

So-called *dynamic influence matrix* $D_{ij}(t-\tau)$ describes the stress caused in the truss member i and in the time instant $t \geq \tau$, due to the unit *virtual distortion* impulse $\Delta\varepsilon_j^o(\tau)=1$ generated in the member j in the time instant τ. The vector σ_i^L denotes the stress distribution due to the external load applied. The matrix D stores information about the entire structure properties (including boundary conditions) in calculation of structural response.

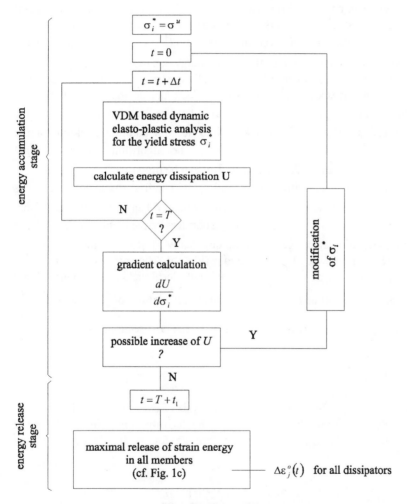

Figure 2. The Flow-Chart of the Impact Energy Absorption Algorithm

Let us postulate the maximal dissipation of impact energy in the time interval (0,T), where T is chosen arbitrarily, however, before the average movement of structural nodes change the direction. (before the object turns back). The number of control parameters $\Delta\varepsilon_i^o(t)$ for the above problem (1)-(3) is $n = n_t n_k$ where n_t denotes the number of time steps Δt (in the interval $(0,T)$) and n_k denotes the number of dissipaters. However, realisation of the control strategy with a free scenario of $\Delta\varepsilon_i^o(t)$ variation is technically unrealistic in crash problems. Therefore, the following accumulation/release strategy of impact energy dissipation (tested in the static case (Ref.6)) is proposed. Assuming the yield stresses ($\sigma_i^* \leq \sigma^u$ constant in time) defined for each dissipater separately as control parameters, let us postulate that the scenario of dissipater's activity can be divided into two stages. In the first one (for $t<T$), maximal accumulation of the strain energy is performed, taking into account the control parameters ε_i^o only in the set A of overloaded members ($|\sigma_i| \geq \sigma_i^*$). On the other hand, in the second stage, maximal release of the strain energy is performed due to the action of all dissipaters for $t>T$. The overall behaviour of the structure in the first stage of energy accumulation corresponds to the process of elasto-plastic adaptation (except for the unloading cases) to yield stresses σ_i^* and can be determined through the Shitkowski's procedure (Ref.4). Then, taking into account additional search for the best distribution of σ_i^* among active members, the algorithm shown in Fig.2 determining the best strategy of dissipater's control can be proposed.

The outputs from the above algorithm are σ_i^* and $\varepsilon_i^o(t)$. The first parameter determines the level of stresses in each active member when dissipation has to be activated. Then, $\varepsilon_i^o(t)$ determines its plastic-like distortions (e.g. shortness of the element due to piston movement in a hydraulic device) generated automatically in active members equipped with actuators which mimic the plastic behaviour (Fig.1b). The analysis corresponds to elasto-plastic dynamic analysis of a truss structure with local yield stresses defined and modelled by plastic distortions $\varepsilon_j^o(t)$. Analytical calculation of derivatives $\dfrac{dU}{d\sigma_i^*}$ required by the algorithm (Fig.2) will be discussed in the next section.

3. Sensitivity Analysis

Taking advantage of the VDM based dynamic analysis of elasto-plastic behaviour, let us derive an analytical formula for calculation of the derivatives required by the above algorithm. From Eqs. (1), (3) the following formula for gradient's calculation is obtained:

$$\frac{dU}{d\sigma_i^*} = \sum_{t=0}^{T} \frac{\partial U}{\partial\Delta\varepsilon_i^o(t)} \frac{\partial\Delta\varepsilon_i^o(t)}{\partial\sigma_i^*} + \frac{\partial U}{\partial\sigma_i^*}, \tag{4}$$

where (cf.(1) for $\sigma_i(t) \equiv \sigma_i^*$ in plastified members)

$$\frac{\partial U}{\partial \Delta \varepsilon_i^o(t)} = A_i l_i \sigma_i^* ; \quad \frac{\partial U}{\partial \sigma_i^*} = \sum_{t=0}^{T} A_i l_i \Delta \varepsilon_i^o(t) \tag{5}$$

and $\dfrac{\partial \Delta \varepsilon_i^o(t)}{\partial \sigma_i^*}$ can be determined from (3), assuming $\sigma_i(t) = \sigma_i^*$ in active members. Then:

$$\frac{\partial \Delta \varepsilon_i^o(t)}{\partial \sigma_i^*} = \frac{1}{D_{ii}(0)}. \tag{6}$$

Substituting (5), (6) to (4), the final analytical formula for the sensitivity can be derived:

$$\frac{dU}{d\sigma_i^*} = \sum_{t=0}^{T} A_i l_i \left[\Delta \varepsilon_i^o(t) + \frac{\sigma_i^*}{D_{ii}(0)} \right]. \tag{7}$$

The superposition in the above formula covers all active members in the sequence of time steps.

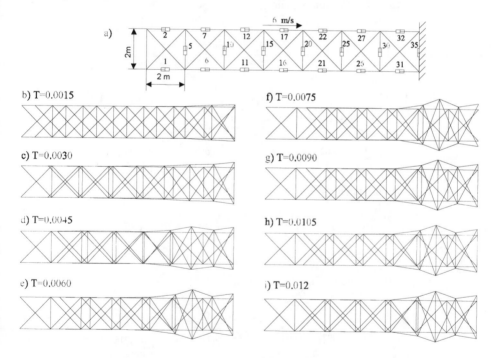

Figure 3. Impact scenario for the One-Stage Control Strategy
a) adaptive structure, b), c), d), e), f), g), h), i) structure response after T sec.

148

4. Numerical Simulation

Simulation of the dynamic response of adaptive structure (Fig.3a) with total mass 1550 kg hitting the rigid wall with initial velocity $v = 6$ m/s is demonstrated in the sequence of time steps in Figs. 3b,c,d,e,f,g,h,i. Almost 98 % of the initial kinetic energy 25,5 kJ has been dissipated after T=0.012 seconds.

a) dissipated

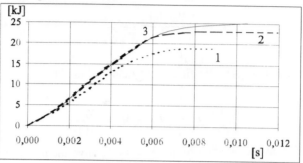

b) strain recoverable energy

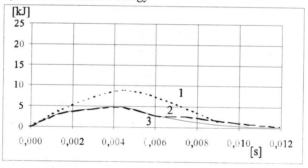

c) kinetic energy

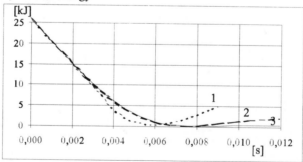

Figure 4. Energy dissipation effect for: 1 – original adaptive structure with fixed stress yield limit σ^u,

2 – structure with modified stress yield limit (σ_i^* unchanged after t>T period),

3 – structure with accumulation/release strategy of energy dissipation (cf. Fig.1c)

TABLE 1 Level of stress limits and generated distortions.

Element No.	Original Adaptive Structure		Modified Structures with:			
			stress yield limits modified			accumulation /releasestrategy
	σ_i^* [MPa]	ε_i^o	σ_i^* (in time t=T) [MPa]	ε_i^o	$\dfrac{dU}{d\sigma_i^*}$ [J/Pa]	ε_i^o
1,2	200	-	200	-	-	-
5	200	-	150	-	-	0,00004
6,7	200	-	200	-	-	- 0,00010
10	200	-	150	-	-	0.00011
11,12	200	- 0.00004	200	-	-	- 0.00020
15	200	-	150	-	-	0.00012
16,17	200	- 0.00083	155	-0.00080	0.234e-11	- 0.00126
20	200	0.00016	76	0.00230	0.509e-12	0.00247
21,22	200	- 0.00137	156	-0.00422	-0.259e-11	- 0.00444
25	200	0.00045	64	0.00619	0.544e-12	0.00613
26,27	200	- 0.00279	200	-0.00278	-0.752e-06	- 0.00341
30	200	0.00307	150	0.00117	-0.421e-05	0.00142
31,32	200	- 0.00262	200	-0.00026	-0.377e-05	- 0.00295
35	200	-	82	0.00316	0.377e-12	0.00346
Dissipated energy (percent of initial value U=25,5kJ)	75%		90%			98%

Numerical simulations were performed for three types of energy dissipation processes for: original structure, structure with modified stress yield limit in time period <0,T>, and structure with modified stress yield limit during the whole crash time period. Comparison of the energy distribution for those different approaches is presented in Fig. 4.

Figs. 5 and 6 show fluctuations of the stress during the whole dissipation process in active members of the structure with modified stress yield limits in the time period <0,T>. The corresponding plastic-like virtual distortions are presented in Fig. 7.

Additionally, comparison of the energy dissipation distribution over the structure sections for different approaches was performed and is presented in Figs.8a, 8b and 8c.

150

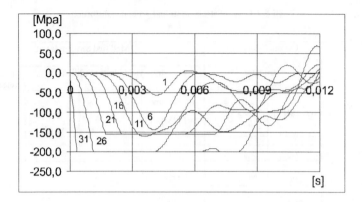

Figure 5. Fluctuation of stress in compressed horizontal active members.

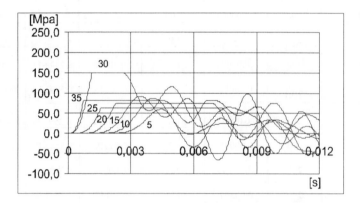

Figure 6. Fluctuation of stress in tensioned vertical active members.

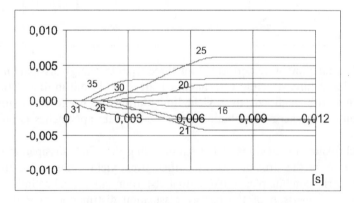

Figure 7. Evaluation of plastic-like virtual distortions in active members.

a)

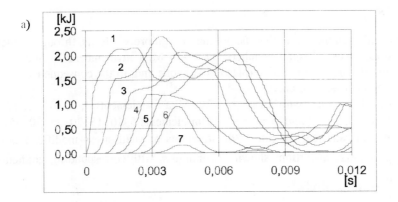

b)

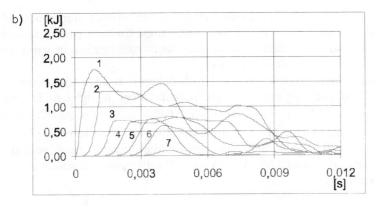

c)

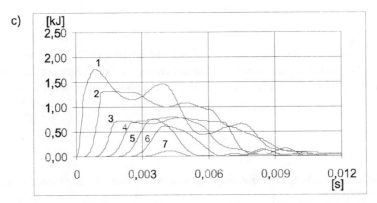

Figure 8. Strain recoverable energy distribution in sections (section No. 1 contains members: 31,32,33,34,35) of the a) original adaptive structure, b) structure with modified stress yield limit (σ_i^* unchanged after t>T period), c) structure with accumulation/release strategy of energy dissipation.

5. Conclusions

The numerical example shows that the optimal design process leads, through reduction of yield stress levels in some members (to 64-76 MPa in members 20 and 25, see Table 1), to extension of the active (plastic-like) zone and, in the consequence, to 23% increment of the dissipated energy.

For a modified structure (with initial $\bar{\sigma} = 200$ MPa for elements 1, 2, 67, 11, 12, 16, 17, 21, 22, 26, 27, 31, 32 and $\bar{\sigma} = 150$ MPa for elements 5, 10, 15, 20, 25, 30, 35), the optimal solution was reached (after 300 iterations of yield stress modifications) for adaptive plastic-like distortions shown in column ε_i^o with corresponding gradients in column $\dfrac{\mathrm{d}U}{\mathrm{d}\sigma_i^*}$, respectively.

It has been checked that maximal initial velocity for the above testing model allowing still for a safe crash scenario is v=8 m/s. Then, stress limits in passive members are reached (220 Mpa). Also, effectiveness of stress limits modifications in active members decrease when the initial velocity grows up.

The natural extension of the proposed methodology is an additional modification of material redistribution to maximise the impact energy dissipation. In this way, searching for the best proportions between the additional control parameters A_i, the safe crash velocity (with the structure still adapted) can be effectively increased.

Numerical verification of the VDM based elasto-plastic dynamic analysis codes versus the classical approach (ABAQUS) shows the divergences reaching up to 1% (in the examples analyzed).

Acknowledgement

This work was supported by the grant No. KBN.8T11F00196C/2778. from the Institute of Fundamental Technological Research, funded by the National Research Committee and presents a part of the Ph.D. thesis of the first author, supervised by the second author.

References

1. J.Holnicki-Szulc, J.T.Gierlinski *Structural Analysis, Design and Control by the VDM Method,* J.Wiley & Sons, Chichester, 95
2. L.Knap and J.Holnicki-Szulc, *Dynamic Analysis of Adaptive, Visco-PlasticStructures,* Proc. Complas 5, Barcelona, 17-19 March 97
3. *Adaptive Impact Absorption,* Rep.of the KBN.Project,IFTR, Patent licence applied for.
4. K.Shitkowski, QLD procedure, 1985
5. B.F.Spencer,Jr., J.David Carlson, M.K.Sain and G.Yang, *On the Current Status of Magnethoreological Dampers: Seismic Protection of Full-Scale Structures,* Proc. Of the 1997 American Control Conference, Albuquerque New Mexico, June 4-6, 1997
6. J.Holnicki-Szulc, A.Mackiewicz and P.Kolakowski, *Design of Adaptive Structures for Improved Load Capacity,* AIAA Journal, vol.36, No.3, March 1998

MODELING AND EXPERIMENTS OF THE HYSTERETIC RESPONSE OF AN ACTIVE HYDROFOIL ACTUATED BY SMA LINE ACTUATORS

D.C.LAGOUDAS, L.J.GARNER, O.K.REDINIOTIS. AND N.WILSON
Aerospace Engineering Department, Texas A&M University
College Station, Texas 77843-3141

1. Introduction

All movable control surfaces in surface ships and submarines, to date, involve a fixed stock (pivot axis) about which the appendage rotates. However, the location of the center of pressure on the appendage varies relative to the fixed stock over the range of angles of attack. The fundamental problem is that a parasitic and harmful hydrodynamic moment is created about the stock by the force-center being located some distance away from the stock centerline. Moreover, this force-center moves as the angle of attack changes. Preliminary research [1],[2] indicates that a control surface with a small fraction (order of 10%) of its aft section being continuously deformed can move the center of pressure to just about any desired position on (or off) the surface, for a given angle of attack, thus controlling the torque about the primary stock. Early studies on Shape Memory Alloy (SMA) actuated control surfaces also indicated that large increases in the lift generated by the control surface could be attained via smooth deflection of the trailing edge [3]. It is possible to position the center of pressure such that there is zero parasitic torque on the primary stock while getting the equivalent lift as a traditional appendage. The use of active materials to effect the shape control can result in an all-electric actuation system without the noise associated with hydraulically moving mechanical components. Furthermore, optimal shape control will result in a hydrodynamically more efficient control surface, delaying or avoiding boundary layer separation, a phenomenon associated with shear layer shedding in the wake, a process which is acoustically harmful.

In actuator technology, active or "smart" materials have opened up new horizons in terms of actuation simplicity, compactness and miniaturization potential. Piezoelectric materials, Shape Memory Alloys, Magnetostrictive materials, Electrorheological Fluids (ERF) and Magnetorheological Fluids (MRF) [4] are the most often recognized types with the first three being the most common. These materials most commonly develop strains or, more applicably, displacements when exposed to electric, thermal and magnetic fields. Our experience with or knowledge of these material types indicates that SMAs are the most appropriate candidates for the present application where large forces and strains are required and the aquatic environment offers an ideal heat sink that will dramatically accelerate SMA actuation and increase actuation bandwidth. A few arguments to substantiate the appropriateness of SMAs are given below.

J. Holnicki-Szulc and J. Rodellar (eds.), Smart Structures, 153–162.

For the quasi-static or low-bandwidth applications of interest here (less than 2Hz), the high-frequency-response capabilities of piezoceramic and magnetostrictive materials (in the kHz range), are unnecessary and superfluous. Moreover, these materials are only capable of producing small strains/displacements (measured in a few hundred microstrain) compared to those attained by SMA materials (as high as 8% for one-way trained and 4% for two-way trained SMAs).[5],[6] SMAs provide the highest energy densities among the three active material types, as well as the highest stresses (up to 200 MPa). Additional SMA advantages include: (a) Simplicity of actuation mechanism. SMAs are all-electric devices and can be used as "Direct Drive Linear Actuators" with little or no additional gear reduction or motion amplification hardware. These merits permit the realization of small or even miniature actuation systems in order to overcome space limitation restrictions. Reduced production cost and improved reliability are other side benefits. (b) Silent actuation. Since no acoustic signature is associated with such an actuation system, acoustic detectability will be drastically reduced. (c) Low driving voltages. Nitinol (NiTi) SMAs can be actuated with very low voltages (5V to 12V), thus requiring very simple power driving hardware. In contrast, piezoceramic materials typically require voltages on the order of 100V for their actuation.

The tailoring and implementation of the accumulated knowledge into submersible vehicles is a task of multidisciplinary nature with two of the dominant fields being actuation and hydrodynamic control. Motivated by the ideas described above, an active hydrofoil with SMA actuators is being designed, modeled with FEA, and built for experimental validation. The active hydrofoil is modeled in the present work using a thermomechanical constitutive model for SMA coupled with a CFD code to model the hydrodynamic load. FEA has been performed and the model predictions have been compared with experimental measurements. The second section will give a more complete description of the active hydrofoil and its mode of operation. In the third section, the modeling of the hydrofoil in a fully-coupled hydro-mechanical environment is also discussed. Finally, a comparison of experimental results is drawn and the direction of ongoing work is discussed in the fourth section.

2. Description of the Active Hydrofoil

SMA actuator technology is presently applied to control the hydrodynamic forces and moments on a hydrofoil. The actuation elements are two sets of thin SMA wires attached to an elastomeric element that provides the main structural support and a hinge-less bend joint in the trailing edge of the hydrofoil. Controlled heating of the two wire sets generates bi-directional bending of the elastomer, which in turn deflects the trailing edge of the hydrofoil. The aquatic environment of the hydrofoil lends itself to cooling schemes that utilize the excellent heat transfer properties of water.

A schematic of the active hydrofoil is shown in Figure 1 and a photograph of the hydrofoil model can be seen in Figure 2. The model is comprised of three sections. The leading edge section encompasses the front of the model and is constructed of foam and fiberglass. This section is hollow to allow the mounting of the pitch sensor within the model. This section also has a shaft passing through it upon which the model is mounted

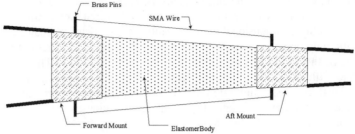

Figure 1. Schematic of hydrofoil center section

in the water tunnel as shown in Figure 3. The bearings and the exit point on the shaft are waterproofed before the model is submerged. The wires for the on-board inclinometers and thermocouples also pass through the leading edge section and then out of the model through the mounting shaft. Attached to the rear of the leading edge section is the SMA-actuator. The elastomer is mounted in a Plexiglas frame that is fastened to the leading-edge section. The trailing edge, which is also constructed out of fiberglass, is attached to the elastomer in a similar fashion. In order for the model to balanced, fore and aft, about the shaft, lead weights are added to the hollow area in the leading-edge section of the model. Balancing weights are also added to make the model neutrally buoyant.

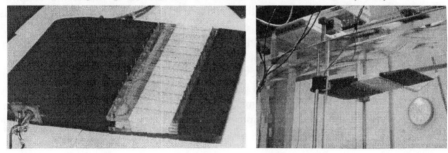

Figure 2. Photograph of the active hydrofoil showing placement of SMA wires [7].   *Figure 3.* Water tunnel setup with hydrofoil.

The SMA actuator consists of a SMA wire that has been conditioned to acquire a two-way shape memory effect, so that the wire can change as much as 3.5% in length during a temperature cycle. The positioning of the SMA wires is such that when the wires on the upper surface are heated by electrical current, the trailing edge is deflected up. By having the wire exposed directly to the water flow, the maximum possible cooling rate can be achieved.

3. Finite Element Modeling of an Active Hydrofoil

A constitutive model for SMA, which will be discussed later, has been developed and has been successfully integrated into the commercial Finite Element program ABAQUS. Using this constitutive model and ABAQUS, it is possible to model fully coupled thermomechanical problems or the mechanical response to a specified temperature variation. In the current analysis, the temperature of the SMA is taken from experimental measurements in order to reduce computational time and simplify the model. This three-dimensional model exploits the spanwise periodicity of the design by modeling only one cell (section between two adjacent SMA wires) of the hydrofoil and consists of the center and aft sections of the hydrofoil only. This model was used to correlate with in air tests of the hydrofoil. The SMA wires are represented by one-dimensional truss elements rigidly fixed to planar elements representing the pins on the ends. The elastomer is represented by 20-node brick elements.

In order to model the hydro-mechanical coupling the FEM model mesh need to be expanded to include the entire airfoil surface. Also, results showed the variation in stress along the span caused by periodic arrangement of the SMA actuators is negligible. Therefore, the mesh was reduced to a two-dimension constructed of plane strain elements with the same truss elements representing the SMA.

3.1 MODELING OF HYDRO-MECHANICAL COUPLING

The nature of the aquatic environment introduces substantial hydrodynamic loads to the analysis. These loads are dependent of the deflection of the hydrofoil trailing edge. This dependency creates a strong coupling between the hydrodynamic loads and the mechanical response. An iterative procedure was developed to model the coupling between the hydrodynamic forces and the mechanical response of the hydrofoil. In this procedure, shown in
Figure *4*, the SMA wires are actuated in the absence of hydrodynamic forces, resulting in the maximum possible deflection of the hydrofoil trailing edge. For this deflected shape, the hydrodynamic forces are evaluated and are used to modify the loading boundary conditions for the mechanical problem. A new deflected shape is evaluated by the modified FEA and the hydrodynamic forces are computed again for the updated shape. This iterative procedure is complete when the difference in the lift coefficient between consecutive iterations becomes smaller than a given tolerance.

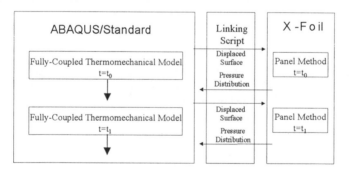

Figure 4. Schematic of hydro-mechanical coupling procedure.

Since the FEM model can only track displacements at nodes, this is the only data that can be supplied to the panel method to describe the airfoil shape. In order to produce good results the panel method requires fine resolution of the leading and trailing edges. This requires the placement of many nodes on the leading and trailing edges in the FEM model. Since these areas have very low stress gradients and are not of structural interest, this is a great penalty. While it is true that leading and trailing edges are essentially rigid bodies, their presence in the FEM model is required in order to accurately calculate hydrodynamic forces and moments. Also, the nodes comprising the upper and lower surface will not remain vertically aligned as the airfoil deforms the panel methods. This problem leads to more inaccuracy and creates a blunt trailing edge. The use of Xfoil (a 2-D panel method CFD code with boundary layer capabilities [8]) eliminates both of these problems by creating its own model of the hydrofoil based on coordinates supplied. It is therefore only required to provide enough nodal coordinates to accurately describe the airfoil shape. Xfoil is also able to accurately calculate the pressures for airfoils with blunt trailing edges.

Figure 5 shows the predicted deflection of the trailing edge for a freestream velocity of 1ft/sec. It should be noted that while ignoring the hydrodynamic load produces only small errors in displacement, the error in the predicted lift is substantial. Also shown in the figure is the convergence of the lift coefficient.

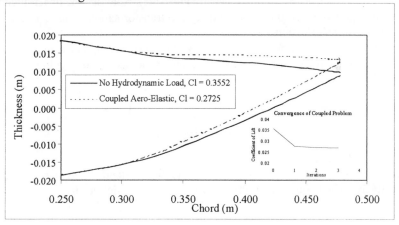

Figure 5. Hydrofoil trailing edge deflection showing the effect of hydrodynamic forces and rate of convergence of the iterative coupling scheme.

3.2. SMA CONSTITUTIVE MODEL

For the numerical simulation of this problem, the 3-D incremental formulation of the SMA constitutive model, developed by Boyd and Lagoudas [9], is used to predict the thermomechanical response of SMA. The model consists of three sets of equations: The constitutive equations, which describe the increment of strain, in terms of the increments of stress, temperature, and the volume fraction of martensite, $\dot{\sigma}_{ij}, \dot{T}, \dot{\xi}$ respectively. i.e.,

$$\dot{\varepsilon}_{ij} = S_{ijkl}\dot{\sigma}_{kl} + \alpha_{ij}\dot{T} + Q_{ij}\dot{\xi} \qquad (1)$$

the transformation equation, which relates the increment of martensite volume fraction to the transformation strain, i.e,

$$\dot{\varepsilon}_{ij}^{t} = \Lambda_{ij}\dot{\xi} \tag{2}$$

and the transformation surface equation, which controls the onset of the forward and reverse phase transformations, i.e.,

$$\pi = \sigma_{ij}^{eff}\Lambda_{ij} + \frac{1}{2}\Delta S_{ijkl}\sigma_{ij}\sigma_{kl} + \Delta\alpha_{ij}\sigma_{ij}\Delta T + \rho\,\Delta c\left[\Delta T - T\ln\left(\frac{T}{T_o}\right)\right]$$

$$+ \rho\,\Delta s_o T - b\xi - \mu_1 \pm \frac{1}{4}\Delta\rho\,b - \rho\,\Delta u_o = \pm Y^{*} \tag{3}$$

The plus sign on the right hand side in Eq. (3) should be used for the forward phase transformation (austenite to martensite), while the minus sign should be used for the reverse phase transformation (martensite to austenite). Note that the material constant Y^{*} can be interpreted as the threshold value of the transformation surface, π, for the onset of the phase transformation.

In the above equations $S_{ijkl} = (C_{ijkl})^{-1}$ is the elastic compliance tensor, α_{ij}, is the thermal expansion coefficient tensor, given in terms of the volume fraction of martensite by

$$C_{ijkl} = C_{ijkl}^{A} + \xi\left(C_{ijkl}^{M} - C_{ijkl}^{A}\right)$$

$$\alpha_{ij} = \alpha_{ij}^{A} + \xi\left(\alpha_{ij}^{M} - \alpha_{ij}^{A}\right) \tag{4}$$

where the superscripts "A" and "M" denote the austenitic and martensitic phases, respectively. The various other terms in Eqs. (1)- (3) are defined by

$$Q_{ij} = \Delta S_{ijkl}\sigma_{kl} + \Delta\alpha_{ij}\left(T - T_o\right) + \Lambda_{ij} ,$$

$$\Lambda_{ij} = \begin{cases} \dfrac{3}{2}H\left(\overline{\sigma}^{eff}\right)^{-1}\sigma_{ij}^{eff}, & \dot{\xi} > 0 , \\[2mm] H\left(\overline{\varepsilon}^{t}\right)^{-1}\varepsilon_{ij}^{t}, & \dot{\xi} < 0 , \end{cases}$$

$$\overline{\sigma}^{eff} = \left(\frac{3}{2}\sigma_{ij}^{eff\,'}\sigma_{ij}^{eff\,'}\right)^{\frac{1}{2}}, \quad \sigma_{ij}^{eff\,'} = \sigma_{ij}^{eff} - \frac{1}{3}\sigma_{kk}^{eff}\delta_{ij} ,$$

$$\overline{\varepsilon}^{t} = \left(\frac{3}{2}\varepsilon_{ij}^{t}\varepsilon_{ij}^{t}\right). \tag{5}$$

Note that $H = \varepsilon^{t,max}$ corresponds to the maximum uniaxial transformation strain and is found from uniaxial stress-strain curve for detwinning of martensite at temperatures below M_0^{S}. The remaining terms that are defined with the prefix "Δ" indicate the difference of a quantity between the austenitic and martensitic phase , and are given by

$$\Delta S_{ijkl} = \frac{\partial S_{ijkl}}{\partial \xi} = -\left(S_{ijpq}\right)\Delta C_{pqmn}\left(S_{mnkl}\right),$$

$$\Delta C_{ijkl} = C_{ijkl}^{M} - C_{ijkl}^{A}, \quad \Delta \alpha_{ij} = \alpha_{ij}^{M} - \alpha_{ij}^{A},$$

$$\Delta c = c^{M} - c^{A}, \quad \Delta s_{o} = s_{o}^{M} - s_{o}^{A}, \tag{6}$$

$$\Delta u_{o} = u_{o}^{M} - u_{o}^{A}, \quad \Delta T = T - T_{o},$$

$$\Delta \rho b = \rho b^{A} - \rho b^{M}.$$

where ρ, c, s_{o}, u, b are the mass density, specific heat, specific entropy, specific internal energy at the reference state, and the transformation strain hardening constant, respectively.

The hardening function, $f(\xi)$, describes transformation induced strain hardening in the SMA material and is given by

$$f(\xi) = \begin{cases} f^{M0}(\xi) + \dfrac{1-\xi}{1-\xi^{R}}\left[\left(f^{A}(\xi^{R}) - f^{M0}(\xi^{R})\right)\right] \\ f^{M0}(\xi) + \dfrac{\xi}{\xi^{R}}\left[\left(f^{M}(\xi^{R}) - f^{A0}(\xi^{R})\right)\right] \end{cases} \tag{7}$$

where

$$f^{M0}(\xi) = \frac{1}{2}\rho b\xi^{2} + \left(\mu_{1} + \mu_{2}\right)\xi$$

$$f^{A0}(\xi) = \frac{1}{2}\rho b\xi^{2} + \left(\mu_{1} - \mu_{2}\right)\xi \tag{8}$$

and ξ^{R} is the martensitic volume fraction at a point during the phase transformation characterized by a change in the sign of $\dot{\xi}$. The material constants μ_{1}, μ_{2}, are used to describe the accumulation of the elastic strain energy at the onset of the forward phase transformation and to enforce the continuity of $f(\xi)$ at $\xi = 1$, respectively.

The implementation of the incremental SMA constitutive model is performed using incremental Newton-Raphson iteration method based on displacement formulation. The scheme requires the tangent stiffness tensor and the stress tensor at each integration point to be updated in each iteration for given increments of strain and temperature. In order to obtain the stress tensor and tangent stiffness, the transformation function Φ is defined using Eq. (1) as follows [10]

$$\Phi = \begin{cases} \pi - Y^{*}, & \dot{\xi} > 0 \\ -\pi - Y^{*}, & \dot{\xi} < 0 \end{cases}. \tag{9}$$

The forward phase transformation (austenite to martensite) is characterized by $\Phi = 0$ and $\dot{\xi} > 0$, while the reverse phase transformation (martensite to austenite) is characterized by $\Phi = 0$ and $\dot{\xi} < 0$. Using the consistency condition $\dot{\Phi} = 0$, the evolution of the martensitic volume fraction can be derived. Substituting this result into the stress rate form of Eq.(1), the tangent stiffness tensor L_{ijkl} and tangent thermal moduli l_{ij} are given by

$$L_{ijkl} = C_{ijkl} + \frac{1}{B} C_{ijmn} Q_{mn} R_{kl} \; , \quad l_{ij} = C_{ijkl} \left(Q_{kl} \frac{S}{B} - \alpha_{kl} \right),$$

$$R_{kl} = C_{ijkl} \frac{\partial \Phi}{\partial \sigma_{ij}} \; , \quad B = \frac{\partial \Phi}{\partial \xi} - \frac{\partial \Phi}{\partial \sigma_{pq}} C_{pqrs} Q_{rs} \; , \tag{10}$$

$$S = \frac{\partial \Phi}{\partial T} - \frac{\partial \Phi}{\partial \sigma_{ij}} C_{ijkl} Q_{kl} \; .$$

To calculate the increment of stress for given strain and temperature increments according to SMA constitutive model, a return mapping integration algorithm proposed by Ortiz and Simo [11] has been used. The work by Lagoudas *et. al.* [10] can be consulted for details of the implementation.

4. Comparison of Experimental and FEM Model Results

The SMA wires used in the model were previously trained using thermal cycles under constant applied stress. The material constants used in the model were obtained is a series of test performed by cycling the temperature under constant load. These tests were used to determine the transformation temperatures, (M_0^F, M_0^S, A_0^S, A_0^F) and the slopes of the transformation boundaries as seen on a stress temperature plot. These parameters as well as the amount of pre-strain (training) were also used as input to the model. In order to obtain properties for the elastomer, the hydrofoil was subjected to a known load on the tip with the hydrofoil body being rigidly fixed and the SMA wires being removed. The deflection of the tip was measured and used to calibrate the FEM model of the hydrofoil structure.

The results of a low frequency test in air are shown in Figure 6. In-air testing is done to verify the FEM model without hydrodynamic forces being present. For this test, the hydrofoil is again clamped and the upper set of SMA wires are heated by electrical current. The wire temperature and the deflection of the trailing edge are measured and shown in the figure. The test time was protracted to allow the thermocouples sufficient time to assure accurate measurement of temperature. Prior to the application of the temperature profile the hydrofoil was cooled to −20°C to assure an all martensite initial state. The temperature profile shown (experimentally measured) was used as an input to the analysis. The tip deflection predicted by the FEA using the temperature profile from the experiment is also shown in the figure. As is seen in the figure, reasonable agreement between the model and experiment was achieved.

Tests of the hydrofoil in the water tunnel were also preformed. Figure 7 shows the results of water tunnel tests at two flow velocities and the results from the in-air test described above. The data shown are the tip deflection plotted against the SMA wire

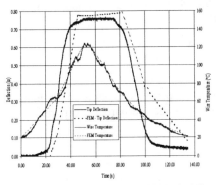

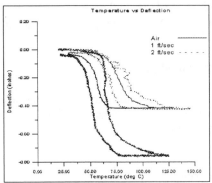

Figure 6. Experimental and numerical hydrofoil (trailing edge) tip deflection and wire temperature versus time for quasi-static actuation in still air.

Figure 7. Hydrofoil experimental test results comparing trailing edge deflection vs., wire temperature of hydrofoil in air and in water tunnel at two fluid velocities.

temperature. As can be seen in the figure there is a large change in the tip deflection when the hydrofoil is submerged. This change can be attributed to the combination of much larger heat transfer which prevents the wires from being able to be heated sufficiently to cause complete transformation and the shift in the transition temperatures due to the increased load of the hydrodynamic forces. The change in the transition temperatures due to increased load can be seen in the figure in the shift of the hystersis loops to the right as the flow speed increases. The figure also shows that the temperature measured by the thermocouples becomes unreliable as the velocity of the flow increases.

The results of these hydrofoil tests in the water tunnel identified several difficulties in modeling the response of the hydrofoil in this complex environment. (1) Since the martensitic finish temperature of the wire was below the temperature of the water tunnel, full reverse transformations are not possible. This creates uncertainty in the initial state of the martensitic volume fraction of subsequent temperature cycles. (2) The thermocouples attached to the wires are exposed to the flow and register a temperature, which is influenced by the temperature of the water. This renders the experimental measurement of the wire temperature unreliable. This difficulty will require using the fully-coupled thermomechanical model. (3) The control surface needs to be able to deflect through a full range of positions. The ability to model minor loops (partial transformations) is desired. (4) The amount of load on the SMA in this configuration is dependent on the amount of deflection. A more advanced material model capable of modeling non-proportional loading is needed.

Work is currently underway to resolve these difficulties. The solutions to these problems are as follows. (1)New Ni-Ti-Cu wire have been obtained which have martensitic finish temperatures above 30°C. (2)Both the SMA constitutive model and the FEM program are capable of modeling fully-coupled thermomechanical problems. The hydrofoil response will be modeled as a function of the electrical current supplied to the SMA wire actuators. (3)A new SMA constitutive model which is capable of modeling minor loops has been developed by Bo and Lagoudas and implemented in the

162

FEM code [12]. (4)The new constitutive model by Bo and Lagoudas can accurately model SMA under non-proportional loading.

Acknowledgements

The authors wish to acknowledge the financial support of this research by the Office of Naval Research, Contract No. N00014-97-1-0840 and Aeroprobe Corporation, Project No. 96-386.

References

1. HE&D, Inc. 1990, "FACS: Flap Assisted Control Surface," Progress Report, Arlington, Virginia.
2. Traub, L.W., Singh, K., Aguilar, V., Kim, K. and Rediniotis, O.K., 1996, "Conceptual Analysis of a Flap Assisted Control Surface," Texas A&M University, College Station, Texas.
3. Beauchamp, C., Nadolink, R, and Dean, L., 1992, "Shape Memory Alloy Articulated (SMAART) Control Surfaces), *SPIE Conf, ActiveMaterials and Adaptive Structures – Session 25.*
4. Giurgiutiu, V., Chaudhry, Z., Rogers, C.A., 1994, "The Analysis of Power Delivery Capability of Induced-Strain Actuators for Dynamic Applications", Proc. Second International Conf. On Intelligent Materials, ICIM '94, Colonial Williamsburg, VA, Technomic Pub. Co.
5. Jackson, C.M., Wagner, H.J., and Wasilewski, R.J., 1972, "55-Nitinol – The Alloy with a Memory: Its Physical Metallurgy, Properties, and Applications, Technical Report", NASA-SP 5110.
6. Hodgson, D. , 1988, *Using Shape Memory Alloys*, Shape Memory Alloy Applications, Sunnyvale, CA
7. Rediniotis, O.K., Lagoudas, D.C., Garner, L. and Wilson, N., 1998, "Experiments and Analysis of an Active Hydrofoil with SMA Actuators," AIAA Paper No. 98-0102, 36th AIAA Aerospace Sciences Meeting, Reno, Nevada.
8. Drela, M., 1996, "Xfoil 6.8 User Primer", MIT Aero &Astro.
9. Boyd, J.G., and Lagoudas, D.C., 1996, "A Thermodynamical Constitutive Model for Shape Memory Materials. Part I. The Monolithic Shape Memory Alloy," *International Journal of Plasticity* (in print).
10. Lagoudas, D.C., Bo, Z., and Qidwai, M.A., 1996, "A Unified Thermodynamic Constitutive Model for SMA and Finite Element Analysis of Active Metal Matrix Composites," *Mechanics of Composite Materials and Structures,* Vol.4, pp. 153-179.
11. Ortiz, M. and Sumo, J.C., 1986, "An Analysis of a New Class of Integration Algorithms for Elastoplastic Constitute Relations", *International Journal for Numerical Methods in Engineering*, Vol. 23 pp. 353-366
12. Bo, Z., Lagoudas, D., and Miller, D., 1998, "Material Characterization of SMA Actuators under Non-Proportional Thermomechanical Loading", submitted to JEMT Special Edition on SMA.

VIBRATION AND STABILITY CONTROL OF GYROELASTIC THIN-WALLED BEAMS VIA SMART MATERIALS TECHNOLOGY

LIVIU LIBRESCU
Engineering Science and Mechanics Department
Virginia Polytechnic Institute and State University
Blacksburg, VA 24061, USA

and

OHSEOP SONG AND HYUCK-DONG KWON
Mechanical Engineering Department
Chungnam National University, Taejon 305-764, South Korea

1. Introduction

The problems of the mathematical modeling, eigenvibration response and stability behavior of cantilevered thin-walled beams carrying a rotor at its tip, and incorporating adaptive capabilities are investigated. The structure modelled as a thin-walled beam, encompasses non-classical features such as anisotropy, transverse shear and secondary warping and in this context, a special ply-angle configuration inducing a structural coupling between flapping-lagging-transverse shear is implemented. The adaptive capabilities are provided by a system of piezoactuators bonded or embedded into the master structure. Based on the converse piezoelectric effect, the piezoactuators produce a localized strain field in response to an applied voltage, and as a result, an adaptive change of the dynamic response characteristics is obtained. Two feedback control laws relating the piezoelectrically induced bending moments at the beam tip with the kinematical response quantities appropriately selected are used, and their beneficial effects upon the closed-loop eigenvibration are highlighted.

2. Preliminaries

A great deal of interest for the study of gyroscopic systems has been manifested in the last decade. This interest was stimulated by the need of a better understanding of the behavior of a number of important technological devices such as drive shafts of gas turbines, rotor systems used in helicopter and tilt rotor aircraft, as well as of robotic manipulators and gyro-devices used in space applications. The extensive list of references supplied in the papers [1,2] reflects in full the interest afforded

J. Holnicki-Szulc and J. Rodellar (eds.), Smart Structures, 163–172.
© 1999 *Kluwer Academic Publishers. Printed in the Netherlands.*

to the study of this problem. However, in the last years, (see [3,4]), a more encompassing concept of gyroscopic systems (referred to as gyroelastic ones) was developed. According to this concept, a gyroelastic system is constituted of a body considered to be continuous in mass, stiffness, and in gyricity as well. Based on this concept, an efficient possibility of modeling elastic structures that are equipped with a large number of small spinning rotors spread over the structure can be devised. Moreover, within this concept, also the case of the continuum equipped with discrete spinning rotors can be accommodated. In several recent papers this concept was exemplified for the cases of a Bernoulli-Euler solid beam containing a 1-D distribution of gyricity along the neutral axis of the beam and subjected to a conservative external load, [see e.g. Ref. 5], whereas in [Ref. 6] this concept was applied to the case of a solid beam carrying a spinning rotor located at the beam tip. Herein, an encompassing structural model equipped with a gyro-device is considered. In this sense, an anisotropic thin-walled beam equipped, at its free end, with a small spinning rotor is considered. The beam model as considered in this paper was developed in Refs. 6 through 9. In this context, one of the goals of this paper is to model and analyze the vibrational behavior of the gyroelastic system as a function of the spin rate Ω_z of the rotor and on the conservative compressive force acting along the longitudinal z-axis. Another goal of this research is to show that the vibrational and the instabilities, static (divergence) and dynamic (flutter), as featured by these systems can be enhanced by using the adaptive capability provided by a system of piezoactuator devices bonded to the surface of the beam. However, the results related with the latter issue will not be displayed here. In order to control the flapping-lagging coupled motions, one considers that the piezoactuators are mounted symmetrically on the upper and bottom surfaces of the beam, and on the lateral surfaces of the beam as well. Since, we consider the piezoactuators spread over the entire span of the beam, by applying out-of-phase voltages, bending moments at the beam tip, $\hat{M}_X$ and $\hat{M}_Y$, are piezoelectrically induced. Implementation of a feedback control law relating the piezoelectrically induced bending moments with the appropriately selected kinematical response quantities of the structural system, a closed-loop eigenvalue problem has to be solved. The solution methodology based upon the Extended Galerkin Method (EGM) (see Refs. 6 through 10) will enable one to solve the closed-loop eigenvalue problem and consequently, a comprehensive picture of the efficiency of the control methodology used to enhance both the vibrational response and the instability behavior of the structure will be provided. Moreover, in order to induce specific structural couplings beneficial for the structure, a *circumferentially uniform stiffness configuration* is implemented. Within this ply-angle configuration the entire system of governing equations splits exactly into two independent systems, one of them featuring the flapping- lagging-transverse shear coupling, and the other one, the twist- extension coupling. Herein, only the motion induced by the flapping - lagging - transerverse shear couplings will be considered.

3. Coordinate Systems. Basic Assumptions

The case of a straight flexible thin-walled beam of length L equipped with a spinning rotor of mass m located at the beam tip is considered. (see Fig. 1) The offset between the beam tip and the centroid of the rotor (assumed to be rigid), is denoted as r_m. The rotor is assumed to have an axis of inertial symmetry which is coincident with the longitudinal axis of the beam.

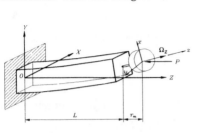

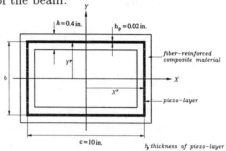

Figure 1. Composite beam with a tip rotor

Figure 2. Cross-section of the beam with piezoactuator

Two sets of coordinates are considered, an inertial frame OXYZ and a body fixed frame Gxyz attached to the rotor, where G is the rotor mass center. The external force P acting at the beam tip is assumed to remain parallel to the Z axis of the inertial frame, whereas the constant spin rate vector direction of the rotor Ω_z about the Gz axis, is assumed to coincide with the longitudinal OZ axis of the beam at any time t. Toward its modeling the following assumptions are adopted: i) the original cross-section of the beam is preserved, ii) both primary and secondary warping effects are included, iii) transverse shear effects are incorporated and finally, iv) the constituent material of the structure features anisotropic properties, and, in this context, a special layup inducing flapping-lagging-transverse shear coupling is implemented.

4. Kinematics

In light of the previously mentioned assumptions and in order to reduce the 3-D elasticity problem to an equivalent 1-D one, the components of the displacement vector are represented as in Refs 1, 2 and 6 through 9.

$$u(X,Y,Z;t) = u_0(Z;t) - Y\phi(Z;t), \quad v(X,Y,Z;t) = v_0(Z;t) + X\phi(Z;t),$$

$$w(X,Y,Z;t) = w_0(Z;t) + \theta_x(Z;t)\left[Y(s) - n\frac{dX}{ds}\right] + \theta_Y(Z;t)\left[X(s) + n\frac{dY}{ds}\right]$$

$$- \phi'(Z;t)[F_\omega(s) + na(s)].$$

$$(1a-c)$$

In these equations $u_0(Z;t)$, $v_o(Z;t)$, $w_0(Z;t)$ denote the rigid body translations along the X, Y and Z axes respectively, while $\phi(Z;t)$, $\theta_X(Z;t)$, $\theta_Y(Z;t)$ denote the twist about the Z-axis and rotations about the X and Y-axis respectively, $F_\omega(s)$ and $na(s)$ play the role of primary and secondary warping functions, respectively. In these equations, as well as in the forthcoming ones, the primes denote differentiation with respect to the longitudinal Z-coordinate. The beam carries a rotor of mass m at its tip. The rotational displacement of the beam tip can be expressed as $\Theta = [\theta_X, -\theta_Y, \phi]^T$, while the angular velocity Ω of the rotor elements due to the elasticity of the beam and represented through the symbolism of Hughes, (see Ref. 3) as:

$$
\Omega = \dot{\Theta} - \frac{1}{2}\Theta^\times\dot{\Theta} = [\dot{\theta}_X + \frac{1}{2}(\theta_Y\dot{\phi} - \dot{\theta}_Y\phi); -\dot{\theta}_Y + \frac{1}{2}(\theta_X\dot{\phi} - \dot{\theta}_X\phi);
$$
$$
\dot{\phi} + \frac{1}{2}(\theta_X\dot{\theta}_Y - \dot{\theta}_X\theta_Y)]^T.
$$
(2)

In Eq. (2), $\Theta^\times$ is the skew-symmetric matrix based upon the elements of the column matrix Θ, considered in the sense of Hughes (Ref. 3) as

$$
\Theta^\times = \begin{vmatrix} 0 & -\phi & \theta_Y \\ \phi & 0 & -\theta_X \\ -\theta_Y & \theta_X & 0 \end{vmatrix}.
$$
(3)

In addition, the supperposed dots and superscript T denote time derivatives and transpose of a matrix, respectively. Needless to say, Ω as expressed by Eq. (2) has to be evaluated at the beam tip. As a result, the terms associated with its effect should appear in the boundary conditions at $Z = L$.

5. Governing Equations

Toward the goal of deriving the equations of motion of beams equipped with a spinning rotor at its tip, and the associated boundary conditions, Hamilton's variational principle is used. It may be stated as:

$$
\delta J = \int_{t_0}^{t_1} \left[\int_\tau \sigma_{ij}\delta\epsilon_{ij}d\tau - \delta K - \int_{\Omega_\sigma} s_i\delta v_i d\Omega - \int_\tau \rho H_i\delta v_i d\tau \right] dt = 0,
$$
(4)

where

$$
U = \frac{1}{2}\int_\tau \sigma_{ij}\epsilon_{ij}d\tau \text{ and } K = \frac{1}{2}\int_\tau \rho(\dot{\mathbf{R}} \cdot \dot{\mathbf{R}})d\tau,
$$
(5a, b)

denotes the strain energy functional and the kinetic energy, respectively. In these equations, t_0 and t_1 denote two arbitrary instants of time; $d\tau(\equiv dndsdz)$ denotes the differential volume element; $s_i(\equiv \sigma_j n_j)$ denote the prescribed components of the stress vector on a surface element of the undeformed body characterized by the outward normal components n_i; H_i denote the components of the body forces; Ω_σ denotes the external area of the body over which the stresses are prescribed; ρ denotes the mass density; an undertilde sign identifies a prescribed quantity while δ denotes the variation operator. In Eqs. (4) and (5) the Einstein summation convention applies to repeated indices where Latin indices, range from 1 to 3. In the same equations, $(v_1, v_2, v_3) \equiv (u, v, w), (x_1, x_2, x_3) \equiv (X, Y, Z)$. For the present structural system it is necessary to consider the energies pertaining to the different parts of the system, i.e. to the beam and rotor. Considering in Eq. (4) the various energies related to the beam and the rotor, and taking into account Hamilton's condition stating that $\delta v_i = 0$ at $t = t_0, t_1$, one can obtain an explicit form of the variational equation (4). In order to induce the elastic coupling between flapwise bending and chordwise bending, a special ply angle distribution referred to as *circumferentially uniform stiffness* (CUS) configuration (see Refs. 6 and 7) achieved by skewing angle plies with respect to the beam axis according to the law $\theta(Y) = \theta(-Y)$ and $\theta(X) = \theta(-X)$ is implemented. Angle θ denotes the dominant ply orientation in the top and bottom as well as in the lateral walls of the beam. In this case, from the variational equation (4) expressed in full, the equations of motion and the boundary conditions featuring this type of coupling are obtained. Employment of constitutive equations and strain-displacement relationships, the equations of motion of the system about a static equilibrium position, expressed in terms of displacement quantities are obtained. These equations are:

$$\delta u_0 : \ a_{43}\theta_X'' + a_{44}(u_0'' + \theta_Y') - b_1\ddot{u}_0 - Pu_0'' = 0,$$
$$\delta v_0 : \ a_{52}\theta_Y'' + a_{55}(v_0'' + \theta_X') - b_1\ddot{v}_0 - Pv_0'' = 0,$$
$$\delta\theta_Y : \ a_{22}\theta_Y'' + a_{25}(v_0'' + \theta_X') - a_{44}(u_0' + \theta_Y) - a_{43}\theta_X' - (b_5 + b_{15})\ddot{\theta}_Y = 0,$$
$$\delta\theta_X : \ a_{33}\theta_X'' + a_{34}(u_0'' + \theta_Y') - a_{55}(v_0' + \theta_X) - a_{52}\theta_Y' - (b_4 + b_{14})\ddot{\theta}_X = 0.$$
$$(6a - d)$$

For cantilevered-beams, the *boundary conditions* at $Z = 0$ are:

$$u_0 = 0, \ v_0 = 0, \ \theta_Y = 0, \ \text{and} \ \theta_X = 0, \qquad (7a - d)$$

while those at $Z = L$ are:

$$\delta u_0 : \ a_{43}\theta_X' + a_{44}(u_0' + \theta_Y) - Pu_0' + m(\ddot{u}_0 - r_m\ddot{\theta}_Y) = 0,$$
$$\delta v_0 : \ a_{52}\theta_Y' + a_{55}(v_0' + \theta_X) - Pv_0' + m(\ddot{v}_0 - r_m\ddot{\theta}_X) = 0,$$

$$\delta\theta_X : \ a_{22}\theta_Y' + a_{25}(v_0' + \theta_X) + J_{xx}\ddot{\theta}_X - \underline{J_{zz}\Omega_z\dot{\theta}_Y} - mr_m\ddot{v}_0 - \hat{M}_X = 0,$$

$$\delta\theta_Y : \ a_{33}\theta_X' + a_{34}(u_0' + \theta_Y) + J_{yy}\ddot{\theta}_Y + \underline{J_{zz}\Omega_z\dot{\theta}_X} - mr_m\ddot{u}_0 - \hat{M}_Y = 0.$$

$$(8a - d)$$

The coefficients $a_{ij} = a_{ji}$ and b_i appearing in these equations as well as in the forthcoming ones denote stiffness and reduced mass terms, respectively. Their expressions can be found in Refs. 6 through 10. The underscored terms appearing in the boundary conditions at $z = L$ are associated with the gyroscopic effects. It should be remarked that for $\theta = 0°$ and $90°$, the coupling stiffness quantities a_{43} and a_{52} become immaterial, and when also $\Omega_z = 0$, the governing system and boundary conditions split exactly into two independent groups associated with flapping and lagging motions. However, even for $\theta = 0°, 90°$, in which case the decoupling appears in the governing equations, when $\Omega_z \neq 0$, the coupling is still manifested in the boundary conditions at $Z = L$. It can readily be shown that in the context of above considered ply angle configuration, in addition to the already mentioned elastic couplings, the flapwise transverse shear is also coupled with lagwise bending, and lagwise transverse shear is couple with flapwise bending. The quantities $\hat{M}_X$ and $\hat{M}_Y$ are the piezoelectrically induced bending moments about the axis X and Y, respectively, appearing as non-homogeneous terms in the boundary conditions at $Z = L$, Eqs. (8). These induced bending moments are different from zero only if external voltages of opposite signs (out-of-phase activation) are applied on the top and bottom, and on the lateral piezoactuator layers, respectively. Their expressions are not displayed here.

6. The Control Law

In the previously displayed equations, due to the distribution of piezoactuators along the entire beam span, the piezoelectrically induced bending moments $\hat{M}_X$ and $\hat{M}_Y$ intervene solely in the boundary conditions prescribed at the beam tip and, hence, they play the role of the *boundary moment controls* (See Refs. 8 through 10). The adaptive nature of the structural system is introduced by requiring the applied electric field $\mathcal{E}_3$ to be related to one of the mechanical quantitites characterizing its dynamic response. As a result, a number of control laws can be implemented. Herein, the displacement and acceleration control laws are used to control the free vibration and instability of the system. For the sake of identification, these control laws are labelled as (CL1) for Control Law 1 and (CL2), for Control Law 2. These are expressed as:

and
$$\text{CL1}: \quad \hat{M}_X = K_d v_0(L, t), \quad \hat{M}_Y = K_d u_0(L, t) \qquad (9a, b)$$

$$\text{CL2}: \quad \hat{M}_X = K_a \ddot{\theta}_X(L, t), \quad \hat{M}_Y = K_a \ddot{\theta}_Y(L, t) \qquad (10c, d)$$

where K_d and K_a stand for the displacement and acceleration feedback gains, respectively. Within the CL1, the piezoelectrically induced bending moments at the beam tip, $\hat{M}_X$ and $\hat{M}_Y$ are proportional to the displacements $v_0(L, t)$ and $u_0(L, t)$, respectively, while within the CL2, the induced moments should be proportional to the angular accelerations $\ddot{\theta}_X(L, t)$ and $\ddot{\theta}_Y(L, t)$, respectively.

7. Vibration and Stability of Gyroelastic System

Toward the goal of solving the closed-loop eigenvalue problem of the gyroelastic system loaded by a compressive edge load applied at its tip, the following steps will be implemented. The first step consists of the representation of displacement functions in the form

$$u_0(Z, t) = \mathbf{U}^T(Z)\mathbf{q}_u(t), \quad v_0(Z, t) = \mathbf{V}^T(Z)\mathbf{q}_v(t),$$

$$\theta_{\mathbf{X}}(Z, t) = \boldsymbol{\phi}^T(Z)\mathbf{q}_\phi(t), \quad \theta_{\mathbf{Y}}(Z, t) = \hat{\boldsymbol{\phi}}^T(Z)\mathbf{q}_{\hat{\phi}}(t), \qquad (11a - d)$$

where $\mathbf{U}$, $\mathbf{V}$, ϕ, and $\hat{\phi}$ are the vectors of trial functions, $\mathbf{q}_u$, $\mathbf{q}_v$, $\mathbf{q}_\phi$ and $\mathbf{q}_{\hat{\phi}}$ are the vectors of generalized coordinates, while superscript T denotes the transpose operation. Replacement of representations (11) in the variational integral, Eq. (4), and of the energy quantities not displayed in this paper, carrying out the indicated variations and the required integrations one obtains the equation

$$\mathbf{M}\ddot{\mathbf{q}}(t) + \mathbf{G}\dot{\mathbf{q}}(t) + \mathbf{K}\mathbf{q}(t) = 0 \tag{12}$$

where $\mathbf{M}$ is the real symmetric postive definite mass matrix, $\mathbf{K}$ is the symmetric stiffness matrix including in its elements the external load P, $\mathbf{G}$ is the skew symmetric gyroscopic matrix, while $\mathbf{q}(\equiv [\mathbf{q}_u^T, \mathbf{q}_V^T, \mathbf{q}_\phi^T, \mathbf{q}_{\hat{\phi}}^T]^T$ is the overall vector of generalized coordinates. Notice that in the case of the non-spinning rotor i.e., when $\Omega_z = 0$, the matrix $\mathbf{G}$ becomes immaterial.

For synchronous motion, upon expressing in Eqs. (12), $\mathbf{q}(t) = \mathbf{Z}\, exp\,(i\omega t)$, where $\mathbf{Z}$ is a constant vector and ω a constant valued quantity, both generally complex, and following the usual steps one obtains the eigenvalue problem

$$-\omega^2 \mathbf{M}\mathbf{Z} + i\omega\mathbf{G}\mathbf{Z} + \mathbf{K}\mathbf{Z} = 0 \tag{13}$$

where ω must satisfy the characteristic equation:

$$\Delta(\omega^2) \equiv det\,|- \omega^2\mathbf{M} + i\omega\,\mathbf{G} + \mathbf{K}| = 0 \tag{14}$$

Since Δ contains only even powers of ω, it is expedient to use the notation $\omega^2 \equiv \Omega$ Since $\mathbf{K}$ includes the contribution of the external load which can be tensile or compressive, we can have $\mathbf{K} > 0, \mathbf{K} = 0$ or $\mathbf{K} < 0$. As a result, one can distinguish three cases.

1) pure oscilliatory motion, i.e. stable motion, occuring when K is positive definite, for which case the eigenvalues occur in pairs of pure imaginary complex conjugates,$\pm i\omega_r$, $(r = \overline{1,\, n})$, where ω_r are the natural frequencies,

2) instability by divergence is found from $\Delta(\Omega) = 0$ when $\Omega = 0$,

3) $\mathbf{K}$ is negative definite. In this case it is still possible to have stable motion. This implies that in this case the gyroscopic effects stabilize an unstable conservative system. However for the same case, when $\mathbf{K}$ is negative definite, the eigevalues can be complex conjugate with at least one of them having a negative imaginary part, which results in unstable motion of the flutter type. As a result, the divergence boundary can be found by satisfying the equation, $\Delta(\Omega) = 0$ in conjunction with $\Omega = 0$. As clearly emerges, the instability by divergence depends on the external load P, only. As the rotational speed of the rotor Ω_z and the compressive load P increase, two consecutive eigenvalues $\Omega_p^2, \Omega_p^2 + 1$ may coalesce, and beyond that point of coalescence, the eigenfrequencies become complex conjugate. Ω and Ω_z corresponding to the coalescence point are referred to as the flutter frequency Ω_{fl}, and the flutter rotational speed $(\Omega_z)_{fl}$, respectively.

8. Numerical Illustrations

The numerical illustrations are carried out for the case a composite box-beam. It features the following dimensions: c = 10 in., h = 0.4in., L = 80in., and is of prescribed cross-section aspect ratio $R(\equiv b/c)$. It is also assumed that the material of the host structure is of a graphite- epoxy composite whose elastic properties can be found in Refs. 1 and 2. As concerns the piezoactuators, these are of the PZT-4 piezoceramic whose properties can be found e.g. in Refs. 8 and 9. In addition, $h_{piezo} = 0.02$in. (See Figure 2).

Figures 3 and 4 depict the variation of the decoupled first and second dimensionless frequencies in flapping and lagging as a function of the dimensionless displacement and acceleration feedback gains $\overline{K}_d(\equiv K_d L^2/a_{33})$ and $\overline{K}_a(\equiv K_a L\overline{\omega}^2/a_{33})$ in the case of the non-rotating rotor. The system is characterized by R = 0.5, $\overline{M}(\equiv m/b_1 L) = 0.2$, $\theta = 90°$. Herein $\overline{\omega}_i = \omega_i/\overline{\omega}$, where $\overline{\omega} = 221.59$ rad/s is the first eigenfrequency of the unactivated system. This remains valid throughout this paper. The results emerging from these plots reveal that the displacement feedback control (CL1) is more efficient in enhancing the lower mode eigenfrequencies, whereas the acceleration feedback control (CL2), has an opposite influence in the sense of enhancing the higher mode eigenfrequencies.

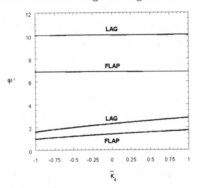 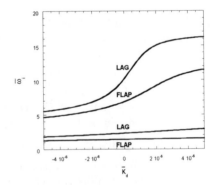

Figure 3. Influence of the displacement feedback control (CL1) on the first decoupled eigenfrequencies of the system when the rotor is non-rotating $(\Omega_z = 0)$.

Figure 4. Influence of the acceleration feedback control (CL2) on the first decoupled eigenfrequencies of the system when the rotor is non-rotating.

In Figs. 5 and 6 there are depictions of the variation of first two and next two successive coupled eigenfrequencies, respectively, as a function of the rotor spin rate for selected values of the acceleration feedback gain. The results reveal that: i) the coupled odd mode eigenfrequencies diminish with the increase of the rotor spin rate, whereas the even mode frequencies feature the opposite trend, ii) the acceleration feedback control plays, in general, a beneficial effect specially on the higher mode eigenfrequencies, and iii) with the increase of the rotor spin rate, the efficiency of the feedback control tends to decay.

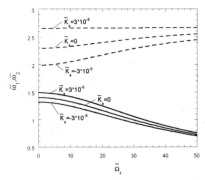

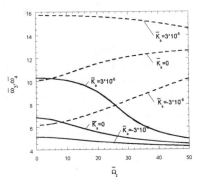

Figure 5. Influence of the angular velocity of the rotor $\bar{\Omega}_z(\equiv \Omega_z/\bar{\omega})$ on the first two eigenfrequencies of the activated and unactivated system. Herein the acceleration feedback control (CL2) is used, $(\underline{\hspace{1cm}} \bar{\omega}_1, ---- \bar{\omega}_2)$

Figure 6. The counterpart of Fig. 5 for the third and fourth eigenfrequencies, $(\underline{\hspace{1cm}} \bar{\omega}_3, ---- \bar{\omega}_4)$

As concerns the effect of the ply-angle coupled with that of the strain actuation control methodology, Figs. 7 and 8 depict the variation of the first two and of next two successive coupled eigenfrequencies, respectively, as a function of the displacement feedback gain $\overline{K}_d$ for two ply angles, $\theta = 0°$ and $90°$.

Whereas the effect of the implementation of the smart materials technology reveals a similar overall trend for each of the considered ply-angles, $\theta = 90°$ yields a substantial increase in the eigenfrequencies as compared to $\theta = 0°$. This trend is perfectly explainable on the ground that in contrast to $\theta = 0°$, at $\theta = 90°$, the maximum flexural stiffness is experienced.

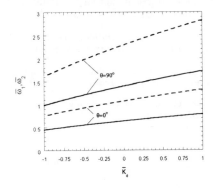

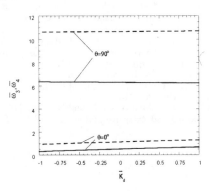

Figure 7. Influence of the displacement feedback control and of the ply-angles $\theta = 0°$ and $90°$ upon the first two eigenfrequencies of the system whose rotor is rotating $(\bar{\Omega}_z = 10)$, $(\underline{\hspace{1cm}} \bar{\omega}_1, ---- \bar{\omega}_2)$

Figure 8. Influence of the displacement feedback control and ply-angles $(\theta = 0°, 90°)$ upon the third and fourth eigenfrequencies of the system whose rotor is rotating $(\bar{\Omega}_z = 10)$, $(\underline{\hspace{1cm}} \bar{\omega}_3, ---- \bar{\omega}_4)$.

9. Conclusions

The displayed results restricted to only the eigenfrequencies occuring in the case of the rotating and non-rotating rotor, reveal that via this control capability is possible to tune conveniently the eigenfrequencies of the system. Moreover, the results also reveal that the directionality property featured by the material of the host structure can be used as to enhance the response of the structure. The results revealing the efficiency of this control technology should encourage its application to more encompassing gyroscopic systems.

References

1. Song, O. and Librescu, L., (1997) "Anisotropy and Structural Coupling on Vibration and Instability of Spinning Thin-Walled Beams," *Journal of Sound and Vibration*, **204 (3)**, 477-494.
2. Song, O. and Librescu, L., (1997) "Modeling and Vibration of Pretwisted Spinning Composite Thin-Walled Beams," *Proceedings of the 38th AIAA/ASME/ASCE/AHS/ASC, Structures, Structural Dynamics, Materials Conference and Exhibition and AIAA/ASME/AHS/Adaptive Structures Forum, Part I, AIAA 97-1091*, pp. 312-322, Kissimmee, FL, April 7-10.
3. D'Eleuterio, G.M.T. and Hughes, P.C., (1987) "Dynamics of Gyroelastic Spacecraft," *Journal of Guidance, Control and Dynamics*, Vol. **10**, 4, 401-405.
4. Yamanaka, K., Heppler, G.R. and Huseyin, K., (1996) "Stability of Gyroelastic Beams," *AIAA Journal*, **34**, 6, 1270-1278.
5. Yamanaka, K. Heppler., G.R. and Huseyin, K., (1994) "On the Dynamics and Stability of a Beam with a Tip Rotor," *Proceedings of the 34th AIAA/ASME/ASCE/AHS Structures, Structural Dynamics and Materials Conference and Exhibit, Hilton Head, SC.*
6. Song, O. and Librescu, L., (1993) "Free Vibration of Anisotropic Composite Thin-Walled Beams of Closed Cross-Section Contour," *Journal of Sound and Vibration*, 167(1), 129-147.
7. Librescu, L., Meirovitch, L. and Song, O., (1996) "Refined Structural Modeling for Enhancing Vibrational and Aéroelastic Characteristics of Composite Aircraft Wings," *La Recherche Aerospatiale*, 1, pp. 23-35.
8. Librescu, L., Song, O. and Rogers, C.A., (1993) "Adaptive Vibrational Behavior of Cantilevered Structures Modeled as Composite Thin-Walled Beams," *International Journal of Engineering Science*, 31, No. 5, 775-792.
9. Librescu, L., Meirovitch, L. and Song, O., (1996) "Integrated Structural Tailoring and Adaptive Materials Control for Advanced Flight Vehicle Structural Vibration," *Journal of Aircraft*, 33, 1, Jan.-Feb. 203-213.
10. Librescu, L., Meirovitch, L. and Na, S.S., (1997) "Control of Cantilever Vibration via Structural Tailoring and Adaptive Materials," *AIAA Journal*, **35**, 8, 1309-1315.

A CRITICAL OVERVIEW OF THE PUBLISHED RESEARCH ON CONTROL OF VIBRATIONS OF CIVIL ENGINEERING STRUCTURES

F. LÓPEZ-ALMANSA
Technical University of Catalonia
Avda. Diagonal, 649; 08028 Barcelona; Spain

1. Introduction

The objective of this paper is to present a critical state-of-the-art review of the literature about control of vibrations of civil engineering structures. Some global conclusions and remarks are discussed to highlight the feasibility of the proposed technologies.

2. Vibrations of structures

Some examples where vibrations are important and some solution based on control has been suggested (not a comprehensive list):

- Flexible structures (slender) or structures housing sensitive equipment (semiconductor facilities, ultra-precision machining, measuring instruments).
- Civil engineering: tall buildings, long-span bridges, high speed train bridges, offshore structures, slender towers and chimneys, nuclear power plants, telescopes, electric energy transportation lines, perforation elements, cables (at cable-stayed bridges), lifelines (pipelines), cable-ways and high velocity lifts (elevators).
- Aerospace engineering: plane wings, panels, lightweight space structures (free vibrations).
- Automotive engineering: panels, vehicles body, pantographs of high speed trains.
- Mechanical engineering: coolers, washing machines, pressing machines, forging machines, flexible robots, pipelines and industrial facilities with dangerous products, shaking tables.
- Ships, submarines.

Problems generated by the vibrations of structures:

- *Relative displacements* (between different points of the structure): structural safety (strength and fatigue).

173

J. Holnicki-Szulc and J. Rodellar (eds.), Smart Structures, 173–182.
© 1999 *Kluwer Academic Publishers. Printed in the Netherlands.*

- *Amplitude* (absolute displacements): noise. Noise control is not dealt with in this paper.
- *Absolute accelerations*: safety of equipment and/or human comfort conditions. Third temporal derivative of the displacement $\overset{...}{x}$ ("jerk") influences the human comfort as well.
- Traditionally these parameters have been reduced by the proper design of the vibrating structure; *control* is an innovative a approach that can be defined as *any mean to reduce vibrations without altering the main structure*. The simplest example can be adding external mass to the structure to change their natural frequencies.

Scope of this paper:

- Innovative systems for reduction of vibrations of structures.
- Focus in civil engineering structures.
- Moreover: notions on smart structures.

Close fields (quasi-static problems): adaptive structures, active stress control (crush energy dissipation), satellite positioning, shape control (telescopes, air foils), etc. Other close dynamic problems: noise control.

3. Systems of control of vibrations of structures

Classification:

- *Passive systems*. Some "inert" devices (here inert means that their behaviour can not be modified "on-line") are added to the structure to dissipate and/or to deflect energy (this last means that the dynamic parameters of the structure are modified as to input excites mostly non harmful modes). Passive systems are not able to adapt to the unexpected characteristics of the excitation.
- *Active systems*. The devices (actuators) are active (they are able to push the structure) and are governed by a controller (typically a computer) that tries to minimise the response of the structure following some strategy (control algorithm). The system operates as a closed loop where the response of the structure is measured on-line by sensors and this information is used by the controller to decide which amount of forces have to be applied by the actuators in order to control the vibrations. These systems are not very feasible because: (i) to move the massive civil engineering constructions big forces are required compared to lighter structures at mechanical and aeronautic engineering (consequently, a huge energy source is needed) and (ii) reliability is low since, for example, energy supply can be interrupted during a strong input (earthquake, wind, etc.). In building structures the feasibility of active control is higher for wind than for seismic excitation because the required control forces are smaller (roughly hundreds of tons compared to thousands).

- *Semi-active systems.* Their operation is similar to the one of the active systems but the actuators are only able to restrain the structure instead of pushing it. The devices are much smaller and only a minor amount of energy is required (typically about 25 W, some batteries can supply it). These systems are more feasible than the active ones. The performance is better than in the passive case but it is worst than in the active one.
- *Hybrid systems.* Combination (series of parallel) of active (or semi-active) and passive systems. The main idea is that the passive systems provides the biggest reduction of response (by absorbing or deflecting energy) and the active one is used for further lowering (for protection of sensitive equipment, for example) of displacement of accelerations as no big control forces are required and so the system is more feasible.

It is remarkable that none of these devices (active, semi-active or passive) participates in the main load-carrying system, so they can be temporarily removed and easily replaced.

4. Passive systems

Classification of main passive systems for civil engineering structures:

- *Seismic isolation* (base isolation). The structure is partially uncoupled from the foundation by using flexible bearings instead of traditional (rigid) connections. Vibration to be controlled can be transmitted from the ground to the structure or, conversely, from the structure to the ground (vibrating elements isolation).
- *Energy dissipators* (passive). Energy is dissipated by external devices fixed to the structure. Input: earthquakes, wind, etc.
- *Mass dampers* (passive). Energy is transformed into kinetic energy (translation or rotation) in massive devices (possessing big mass or big moment of inertia). Input: earthquakes, wind, etc. It is remarkable that, in spite of these devices are known usually as dampers, sometimes their main effect is to deflect energy instead of dissipating it.

Appendages used for reducing wind vibration (helical strakes, fins, slats, splitter plates, etc.) can not be included in any of these groups.

4.1. SEISMIC (BASE) ISOLATION

Two (opposite) situations:

- *Seismic (or base) isolation of structures*: to isolate one structure from the seismic accelerations (or other sources of vibration as trains, vehicles, etc.) transmitted by the foundation (flexible connection); since the structure becomes partly uncoupled from the ground, these systems can be effective even if the isolators behave linearly (no energy dissipation). Applications: buildings (isolators are placed in between the foundation and the lower floor slab), floors, nuclear power plants, sensitive industries, hospitals, important buildings, pipe lines, bridges (isolators are placed in between the top of columns and piers and the deck), radar and offshore platforms. Vertical components of the excitation usually are not affected (stiff connection) but in some cases are also filtered (3D floor isolation). It is a well-known technique (more than 2000 years old in several cultures). There are much more than 1000 isolated structures, mostly buildings or bridges; a number of companies commercialise isolators. In the USA there exist specific rules at the UBC 97 about seismic isolation; these rules have been suggested by the SEAONC. Main countries where this technology has been implemented for buildings or bridges: Armenia, Azerbaijan, Canada, Chile, China, Ecuador, England, France, Greece, Italy, Japan, Korea, Macedonia, Mexico, New Zealand, Portugal, Rumania, Russia, South Africa and USA. Design parameters: fundamental period of the isolated structure (must be far from the peak of the response spectrum and from the fundamental period of the fixed-base structure), maximum displacement in the isolators (gap) and maximum base shear. Uplift has to be avoided. Limitations: stiff soil and stiff structure (about <30 floors); these constraints deal mostly with non-friction isolators. Nonlinearities are concentrated in the isolators and the structure is considered to keep in the elastic domain; general non-linear computer codes usually fail and ad-hoc programs can be used for design (3D-BASIS, DRAIN-2D, DRAIN-3D, SADSAP among others). Usually isolation systems are designed for the maximum credible earthquake; in light to moderate seismic events, notably in aftershocks, such systems may provide little decrease in structural response; as well, effectiveness with near-field earthquakes has been contested. In general, isolation is effective for reducing the resonance peak but can increase the response in the high frequency range.

• *Isolation of vibrating elements*: to isolate a facility or environment from the vibrations transmitted for some elements. Applications: vibrating machinery, underground (subway), trains, pipelines, etc.

These technologies are useful for both new and existing constructions (retrofit). Potential market: facilities dealing with dangerous products.

Types of isolators (these devices can be combined both in series and in parallel):

• Flexible lower floor (can be considered as a forerunner of the modern isolators). This technology is still used at Russia.
• Isolators with elastomeric materials: "Rubber bearings" (RB) (internal reinforcement steel laminae can be planar or "V-shaped"), "Lead-Rubber Bearings" (LRB), "High Damping Rubber Bearings" (HDRB) (ferrite rubber), Dowel Type Rubber Bearings, "Rubber Layer Roller Bearing" (two rubber layers + an intermediate layer of steel balls), Multi-stage rubber bearings (to increase stability for floor isolation or mass damper support), "Pendulum Rubber Bearings" (PRB). The horizontal stiffness of these devices can be non-isotropic to fit the characteristic of the superstructure. To increase the stability of the isolator for high horizontal

displacement, bigger lower base surfaces have been suggested (i.e. truncated-cone-shaped). Rubber isolators are very spread.

• Isolators based on friction: "Friction Pendulum Systems" (FPS), "TAISEI Shake Suppression System" (TASS), "Improved Friction Damper", friction dampers (cylinder + inner wedge), "Wire Friction Damper". The isolators that support the weight of the building use teflon + steel in the contact surface (low friction coefficient + high strength).

• Hydraulic and viscous dampers: "Viscous Shear Damper" (VSD), "Oleo-dynamic Damper" (+ spring), visco-elastic materials (polymers). The performance of visco-elastic materials depends on its temperature, it can be important in long lasting inputs because of the heating generated during the energy dissipation.

• Isolators with rolling devices (+ damper + re-centering device): "Ball Bearings", "Eccentric Roller Bearings".

• "Mushroom Rocking Support". Reinforced concrete columns with lower spherical ends. It is extremely simple and cheap. Has been used at China and Russia.

• Isolators based in plastification of metals (steel). According to the shape of the steel member can be classified as: PC steel rods (+ viscoelastic body connected in series = Composite Damper), Type "E", "T", "U" and "Y", Helical (coil) springs, Rings, etc.

• "Lead Extrusion Damper".

• Electro Inductive Energy Dissipators (the mechanical energy is converted into electrical energy that is dissipated in a circuit as heat; the behaviour is similar to the viscous one).

• Pneumatic cushions (air springs). For floor isolation only (low vertical load).

• Hangers. For floor isolation only (low vertical load).

• "Shock Transmission Units" (STU). During major earthquakes, they are useful to reduce horizontal pounding forces between adjacent decks in bridges.

Some of these devices can be used as well as dissipators and/or as bearings for mass dampers.

4.2. ENERGY DISSIPATORS

External devices are connected to the structure as when it vibrates they deform and dissipate energy; after major earthquakes (or other strong inputs), such mechanisms can be easily replaced. In this way, no ductility in the main carrying-load system is required since damage (permanent strains) is concentrated in dissipators. Applications: buildings (up to 150 floors or more), slender chimneys, piping, equipment, dams (the dissipators are placed in expansion or construction joints), etc. Places to be installed (in buildings): in braces (diagonals), between adjacent buildings (taking profit of pounding effect), parallel to an isolation system, in the connection between columns and beams (as in the "World Trade Center" at NY), between stiff inner cores and outer flexible structures or, in general, in any place where the (horizontal) oscillations of the building generate strains. In suspended buildings (those where floors are hanged from the core), dissipators can be placed in the connections between both systems. Unless base isolation, dissipators are useful to absorb the energy from any source: earthquake, wind, vibrating machinery, fluids flow, sea waves, vehicles, humans, etc. This technology is useful for both new and existing constructions (retrofit). This approach is cheaper than base isolation because the devices do not need to withstand the weight of the building (they are not a part of the main load-carrying system).

About design, the same computer codes as in base isolation can be used (non-linearities concentrate in the dissipators). The energy approach has been extensively used. A crucial point in the design is the number of cycles before cracking (as in isolators).

Classification of energy dissipators (most of these devices can be combined both in series and in parallel). Some of the devices used in isolation systems can be used as dissipators as well. Only specific mechanisms are mentioned here:

• Plastification of metals (steel, lead) "Added Damping and Stiffness" (ADAS), TADAS, "Triangular-ADAS" (Tri-ADAS), "Reinforced ADAS" (RADAS), "Honeycomb Damper", "Steel Slit Damper", "Bell Damper", "Shear Link", pre-casted concrete walls with unbonded steel braces, etc. Steels with low yield point (about 100 MPa) are often used. These devices do not start dissipating energy until motion is significant and the maximum force they can apply on the structure is bounded.
• Friction based: Pall damper, Oiles damper, etc.
• Visco-elastic materials (dissipators VED and walls VD-walls). These devices start dissipating energy for low displacements and the force they can transmit to the structure is very big.
• Shape memory alloys (super-elasticity) (Ni-Ti, Cu-Zn-Al, Cu-Al-Ni).
• "Wire Energy Absorbing Rope" (WEAR).
• "Repulsive clathrats".
• Mechanical snubbers (the energy is deflected to rotate a mass).
• Mechatro damper (the energy is dissipated in an energy generator).
• DREAMY (Damping REsistance Amplifier by Means of Yoke).

The friction dissipators and those based in hysteretic cycles with sharp corners (force-displacement diagram) introduce energy in the higher modes of the structure (high frequencies). It affects mostly to accelerations (important for human comfort and equipment safety). This is also true for friction-based isolators.

4.3. MASS DAMPERS

They are used with tall buildings and similar structures (towers of suspension bridges, industrial chimneys, ventilation ducts, microwave radio link supports, radio and television towers, rigs, industrial towers, telescope towers, tower cranes, offshore platforms, etc.) and are placed usually in the top floors to control mainly the first mode (if higher modes are to be controlled additional dampers installed in lower levels are required, typically one damper per mode) in each direction; as well, first torsional mode can be controlled as well. The mass of the damper is about 2-5% of the total mass of the building (necessary items as ice storage can be used). The relative displacements between the damper and the upper floor are big. Other application is the reduction of vertical vibrations of floors (aerobic studios, ball rooms, event halls, foot bridges, etc.).

These devices are also referred as DVA (Dynamic Vibration Absorbers). Mass dampers are a special type of energy dissipators and so are useful for the same kind of inputs (earthquake, wind, machinery, traffic, human generated and others).

Other applications: vibrations of train coaches and other vehicles.

Mass dampers (passive) devices. Examples:

- "Tuned Mass Damper" (TMD), Two-Mass TMD, Connected Mass TMD.
- "Tuned Roller-Pendulum Damper" (TRD).
- "Multi-Stage Pendulum" (height is dramatically reduced compared to ordinary TMD).
- "Tuned Liquid Damper" (TLD); U-shaped ("Tuned Liquid Column Damper" TLCD), V-shaped and LCVA ("Liquid Column Vibration Absorber"). Liquid sloshing is used as damping dynamic forces. If water is used (low performance), since its natural damping is small, baffles can be installed inside the container to generate turbulence. As well, other viscous fluids can be used: magnetic fluid, colloidal solutions, heavy mud, etc. These devices can be installed in pendulum systems.
- Kinetic energy (rotational). STREAM (STabilizer for REsponse Abating Method). These devices can be considered as dissipators as well.
- (DMSTMD) "Distributed Multi-Stage Tuned Mass Dampers". A light mass dissipator is placed on a floor of a building.

5. Active and semi-active systems

The *smart systems* are those that can know its state (normally by means of distributed sensors) or report about it; even can control it. Example: concrete able of detecting its cracks (by embedded optics fibres, for example) and healing them (by hollow fibres filled with crack-sealing material which would be released if the fibres are broken, for example). Controlled properties: cracking, corrosion, pollution, fire, strain (stress), shape, position, noise or vibration (displacement or acceleration). The knowledge of the state ("Remote Sensing", "Health Monitoring", "Damage Detection") arises from sensors (some times distributed, "Distributed Sensing").

The active, semi-active, hybrid and passive systems can be viewed as particular cases of smart materials where the vibration is the controlled property. In the active systems vibrations are measured and can be controlled, in the semi-active ones the vibrations are also measured but the control ability is limited (small actuators).

Semi-active systems. Advantages and drawbacks (compared to active and passive):

- Feasibility (technological simplicity, moderate size and cost and independence of the energy source).
- Since they are intrinsically stables and their time delay is short (the inertia is small) stringent control algorithms can be implemented.
- Performances comparable to those of active systems and bigger to those of passive ones.

Active actuators in civil engineering (big forces required):

- "Active Mass Damper", at mechanical engineering, similar (yet smaller) devices are known as "Reaction-Mass Actuators". These devices are used usually as passive TMD but during the construction of a long span cable-stayed bridge, vertical vibrations due to wind gusts were controlled by an AMD.
- "Active Tendon System" and ABS ("Active Bracing System"). Active braces are used to apply horizontal control forces on building structures. In the control loop, the control signal can be converted into mechanical forces by a hydraulic cylinder governed by a servo-valve.
- Moving bearings (in cable structures). Tension force is controlled.
- Active sloshing. Sloshing in a TLD is generated by a rotating axis.
- Propeller thruster.
- Active bridges. Active connections among adjacent (tall) buildings.
- Piezoelectric actuators. Pads stuck to both sides of columns to control bending.
- Active Air-bag.
- Air-jet. Forces are generated as pulses.
- Existing cables (in cable-stayed and suspension bridges) can be used as actuators.

Active actuators in Aeronautics and Astronautics (low forces required):

- Mechanical actuators.
- Piezoelectric materials (PZT, PVDF), Piezoactive actuators.
- Magneto-strictive.

Active actuators in mechanical engineering:

- Engines for Robot arms.
- Piezoelectric materials ("Segmented Active Constrained Layer").

Semi-active actuators in civil engineering:

- Dampers with variable orifices ("Active Variable Damping", variable resistance damper).
- Dampers with electro-rheological and magneto-rheological fluids.
- Variable friction devices.
- VFB ("Variable Friction Bearing").
- Variable tuned liquid dampers.
- "Active Variable Stiffness" (AVS). "Innervated structures".
- Gravity actuators.
- Active fins, flaps or appendages. "Aerodynamic Flap System" (AFS). Useful for wind-generated vibrations.

Sensors:

- Accelerometers (absolute or relative acceleration).
- Velocity sensors.
- LVDT or other displacement transducers (relative displacement).
- Strain gauges.
- Optical fibres (strain).
- Piezo-electric (can be used as actuators at the same time).

6. Hybrid systems

It consists of the combination (series of parallel) of an active (or semi-active) system and a passive one trying to combine the advantages of both. Usually the active system does not require a big source of energy (it is provided by the passive system) so it can be considered as semi-active. Examples:

- Active (intelligent) isolation.
- "DUOX" (TMD + AMD = APTMD or HMD).
- Base isolation + AMD.

7. Conclusions and remarks

- The main conclusion of this study is that *the control of vibrations of civil engineering structures has reached an important degree of maturity*. These assert relies on: there is a huge number of papers published both in scientific journals and in conferences proceedings, several journals devoted to this field exist, a number of international and local associations have been created, two international conferences (WCSC and ECSC) are organised regularly (every 4 years) with a growing number of attendants, some workshops are celebrated (like this one) and many constructions all over the world incorporate control devices.

182

- Some benchmark control problems (building under wind and seismic loads) have been proposed to test the efficiency of the proposed techniques; for exemple, information on a particular proposal can be obtained at the following Internet address: *http://www.eng.uci.edu/~anil/benchmark/html*.
- Energy dissipators for buildings are not necessarily hi-tech devices; simple, cheap yet effective and reliable systems have been proposed.
- Further research in design guidelines of systems of dissipators is needed.
- Seismic isolators for buildings are more expensive than dissipators (because they need to withstand the weight of the building among other factors), but simple and cheap solutions have been proposed (Mushroom Rocking Support).
- Research on the definition of seismic input for isolated structures is currently on progress.
- Semi-active devices are more feasible than the active ones and look promising.
- Application of active control for seismic or wind protection of big structures is questionable.
- For the development of these technologies, codes, standard and practices are of crucial importance.

Acknowledgements

This work has been partially funded by NATO under collaborative research grant SA.-5-2-04 OUTREACH (CRG 950316) 470(95)472.

References

Because of the restrained space and the big number of papers consulted to write this paper, no individual references are quoted. Interested reader can address to:

1. "Seismic Isolation and Response Control for Nuclear and Non-Nuclear Structures", SMiRT 11 Conference, Tokyo, Japan 1991.
2. Proceedings of "Seminar on Seismic Isolation, Passive Energy Dissipation and Active Control", ATC (Applied Technology Council), San Francisco, CA 1993.
3. Kelly J.M. "Earthquake-Resistant Design with Rubber" Springer-Verlag 1993.
4. Korenev B.G., Reznikov L.M. "Dynamic Vibration Absorbers. Theory and Technical Applications" John Wiley & Sons 1993.
5. Skinner R.I., Robinson W.H., McVerry G.H. "An Introduction to Seismic Isolation" John Wiley & Sons 1993.
6. Proceedings of 1WCSC (First World Conference on Structural Control), Los Angeles, CA 1994.
7. Proceedings of International Symposium MV2 ("Active Control in Mechanical Engineering"), Lyon, France 1995.
8. Proceedings of 1ECSC (First European Conference on Structural Control), Barcelona, Spain 1996.
9. Soong T.T., Dargush G.F. "Passive Energy Dissipation Systems in Structural Engineering" John Wiley & Sons 1996.
10. Proceedings of the International Post-SMiRT Conference Seminar "Seismic Isolation, Passive Energy Dissipation and Active Control of Seismic Vibrations of Structures", Taormina, Italy 1997.
11. Proceedings of EuroMech 373 Colloquium "Modelling and Control of Adaptive Mechanical Structures", Magdeburg, Germany 1998.
12. Proceedings of 2WCSC (Second World Conference on Structural Control), Kyoto, Japan 1998.

DECENTRALIZED SLIDING MODE CONTROL OF A TWO-CABLE-STAYED BRIDGE

N. LUO, M. DE LA SEN
Dept. of Electricity and Electronics, Faculty of Science
University of Basque Country, 48940 Leioa, Bizkaia, SPAIN

J. RODELLAR
Dept. of Applied Mathematics III, School of Civil Engineering
Technical University of Catalunya, 08034 Barcelona, SPAIN

M.E. MAGAÑA
Department of Electrical and Computer Engineering
Oregon State University, Corvallis, Oregon 97331, USA

1. Introduction

One of the main objectives in the design of cable–stayed bridge structures is to mitigate the hazards caused by external forces such as traffic, heavy winds and earthquakes and thus to lengthen its life (see [1] and [2]). Both linear and nonlinear mathematical models were developed to describe the dynamic behaviour of the cable–stayed bridges in [2] and [3]. In the recent years, several active tendon control schemes have been proposed to reduce the transversal vibration of the bridge deck, among which the centralized linear active control in [4], [5] and [6] and decentralized nonlinear active control in [3] and [7] have received a great of attention. In this paper, a new robust stabilizing decentralized controller is presented to attenuate the transversal vibration of a two–cable–stayed bridge structure induced by seismic excitation. Control design is made based on the principle of sliding mode, in which the information on the exact value of the system parameters, structural disturbances and the seismic excitation does not need to be known *a priori*. Conditions for the generation of sliding motion are given. Stability and robustness of the closed–loop system are analized.

J. Holnicki-Szulc and J. Rodellar (eds.), Smart Structures, 183–192.

2. Problem Formulation

Consider a two–cable–stayed bridge structure being built on a hard rock foundation and supported at its left and right. The tower, where the stay cables are attached to, does not provide direct support to the deck and is assumed to be rigid. The ith stay cable is anchored at the location d_i $(i = 1, 2)$ from the fixed left end of the bridge deck and sensors are put at the same points of anchor to measure the relative displacements and velocities of the vertical motion of the bridge deck, respectively. The control forces are generated by the actuators to increase or decrease the effective length of the stay cables in the presence of seismic excitation. The finite element modeling technique using concentrated masses is adopted to derive the dynamic mathematical model of the cable–stayed bridge structure by Magaña and Rodellar [3] such that the decentralized control schemes [8] can be used. It is assumed that: (a) The deck segment is approximated by a uniform beam of constant mass; (b) The mass of the stay cable that affects degree of freedom i is small compared to the ith concentrated mass of such a degree of freedom; (c) The stay cables are used as active tendons and operate in the linear elastic regime. Their sags are minimal and therefore neglected. (d) The structural damping is very small (in general less than 3%) and can be neglected.

Denote $x_i^d(t)$, $x_i^v(t)$ as the relative transversal displacement and velocity of the bridge deck, respectively, measured at locations d_i where the ith stay cable is anchored, and $\boldsymbol{x}_i(t) = [x_i^d(t), x_i^v(t)]^T$ as the state vector $(i = 1, 2)$. Then, the dynamic behaviour of the bridge segment, accounting for the effect of the stay cables, is described by the following mathematical model being decomposed into two susbsystems S_i $(i = 1, 2)$:

$$S_i : \quad \dot{\boldsymbol{x}}_i(t) = A_i \boldsymbol{x}_i(t) + \boldsymbol{b}_i(t) u_i(t) + \boldsymbol{d}_i [f_{ni}(t) + f_{di}(t) + f_{ei}(t)] \qquad (1)$$

$$A_i = \begin{pmatrix} 0 & 1 \\ -k_{ii}/m_c & 0 \end{pmatrix} \ ; \ \boldsymbol{b}_i(t) = \begin{pmatrix} 0 \\ -c_{ai} c_{ei} sin\theta_i(t)/m_c c_{li} \end{pmatrix} \ ; \ \boldsymbol{d}_i = \begin{pmatrix} 0 \\ 1/m_c \end{pmatrix} \qquad (2)$$

$$f_{ni}(t) = c_{ai} c_{ei} [\sin \theta_i(t) - \sin \theta_{pi}] \qquad (3)$$

$$f_{d1}(t) = -(k_{12}, 0)^T \boldsymbol{x}_2(t) + h_1[\boldsymbol{x}(t)] \ ; \ f_{d2}(t) = -(k_{21}, 0)^T \boldsymbol{x}_1(t) + h_2[\boldsymbol{x}(t)] \quad (4)$$

$$f_{ei}(t) = m_c a_{qk}(t) \qquad (5)$$

where $\boldsymbol{x}(t) =: [\boldsymbol{x}_1(t), \boldsymbol{x}_2(t)]^T$, m_c is the concentrated mass and k_{ij} $(i, j = 1, 2)$ are stiffness parameters of the controlled concentrated mass that include the effect of the stay cables. Positive scalars c_{ai}, c_{ei} and c_{li} are the cross–sectional area, $Young$ modulus of elasticity and unstressed length, respectively, of the ith stay cable.

$u_i(t)$ is a control force applied to the ith stay cable to increase or decrease the its effective length. The scalar function $f_{ni}(t)$ describes the nonlinearities due to the cable–beam geometry. $\theta_i(t)$ is the angle between the ith stay cable and the local horizontal at the point where the stay cable is anchored to the bridge deck. θ_{pi} is the same angle, but at the prestressed condition. In eqn.(4), the unknwon scalar function $h_i[\boldsymbol{x}(t)]$ represents the structural disturbances or the influence of residual modes appeared at the point of anchor and the scalar functions $(k_{12}, 0)^T \boldsymbol{x}_2(t)$ and $(k_{21}, 0)^T \boldsymbol{x}_1(t)$ describe the interconnection effects among the local subsystems. $f_{ei}(t)$ represents a distributed effective seismic force seen by the bridge deck due to the supports motion. $a_{qk}(t)$ is the vertical component of an unknwon earthquake force.

Assumption 1. The system parameters are upperly and lowerly bounded such that $m_c \in [m_c^-, m_c^+]$, $c_{ai} \in [c_{ai}^-, c_{ai}^+]$, $c_{ei} \in [c_{ei}^-, c_{ei}^+]$, $c_{li} \in [c_{li}^-, c_{li}^+]$ and $k_{ij} \in [k_{ij}^-, k_{ij}^+]$ with $c_{ai}^-, c_{ai}^+, c_{ei}^-, c_{ei}^+, c_{li}^-, c_{li}^+, m_c^-, m_c^+, k_{ij}^-, k_{ij}^+$ $(i, j = 1, 2)$ being known positive constants, respectively.

Assumption 2. The vertical component of an unknwon earthquake acceleration to the bridge deck $a_{qk}(t)$ is uniformly bounded such that $a_{qk}(t) \in [a_{qk}^-, a_{qk}^+]$ for all $t \geq 0$ with a_{qk}^- and a_{qk}^+ being known non–zero constants.

Assumption 3. The angle between the ith stay cable and the local horizontal at the point of anchor is uniformly bounded such that $\theta_i(t) \in [\theta_i^-, \theta_i^+]$ and $\theta_{pi} \in [\theta_i^-, \theta_i^+]$ $(i = 1, 2)$ for all $t \geq 0$ with θ_i^- and θ_i^+ being known positive real numbers that lie in the range $(0, \pi/2)$ radians.

Assumption 4. $f_{d1}(t)$ and $f_{d2}(t)$, defined in eqn.(4), are unknown scalar functions that represent the unknown structural disturbances appeared at the point of anchor of the ith stay cable and the interconnection effects among the local subsystems such that the following relationships hold:

$$|f_{di}(t)| \leq \alpha_i + \beta_{i1} \|\boldsymbol{x}_1(t)\| + \beta_{i2} \|\boldsymbol{x}_2(t)\| \tag{6}$$

where α_i, β_{i1} and β_{i2} $(i = 1, 2)$ are known positive constants.

Remark 1. Assumptions 3 implies that the functions $sin\theta_i(t)$ $(i = 1, 2)$ remain uniformly bounded in a subinterval $(0, 1)$ for all $t \geq 0$ and increase monotically with the angle $\theta(t)$. Thus, it is known from (1)–(2) that feedback control $u_i(t)$ $(i = 1, 2)$ exist for all $t \geq 0$.

3. Control Design

In the recent years, the technique of sliding mode control has been used successfully in the active protection of civil engineering structures, such as base isolated buildings, in the presence of seismic excitation, by Luo *et al.* in [9] and [10]. The aim of control design is to attenuate the transversal deflections of the cable–stayed bridge deck caused by the vertical component of an unknown earthquake. Many active control schemes have been proposed recently, in which the centralized control in [4], [5] and [6] and decentralized nonlinear control in [3], [7] and [8] have received important attention. In this paper, the control design is done by using a decentralized sliding mode controller.

3.1. SLIDING MODE CONTROL

Now, a robust stabilizing decentralized control scheme based on the sliding mode principle in [11]and [12] is presented to drive the state vector $\boldsymbol{x}(t)$ to zero asymptotically and thus the cable–stayed bridge structure goes back to its equilibrium position in the presence of seismic excitation. In the control design, the following sliding function $\sigma_i(t) \in \mathbb{R}$ is defined for the dynamic subsystems S_i $(i = 1, 2)$

$$\sigma_i(t) = \boldsymbol{\epsilon}_i^T \boldsymbol{x}_i(t) = \epsilon_{di} x_i^d(t) + \epsilon_{vi} x_i^v(t) \tag{7}$$

with $\boldsymbol{\epsilon}_i =: [\epsilon_{di}, \epsilon_{vi}]^T$; $\epsilon_{di} > 0$; $\epsilon_{vi} > 0$ being a parameter vector to be chosen by the designer to guarantee the closed–loop asymptotic stability of the dynamic subsystems S_i $(i = 1, 2)$. Define

$$\eta_i(t) =: -\boldsymbol{\epsilon}_i^T \boldsymbol{b}_i(t) \tag{8}$$

Under the Assumptions 1 and 3 and from the definition of $\boldsymbol{b}_i(t)$ in (2), it is easy to show that the following relationships hold for all $t \geq 0$

$$\eta_i(t) \geq \eta_i^* > 0 \quad ; \quad \eta_i^* = \frac{\epsilon_{iv} c_{ai}^- c_{ei}^- sin\theta_i^-}{m_c^+ c_{li}^+} \tag{9}$$

then

$$\eta_i(t) \left(\eta_i^*\right)^{-1} \geq 1 \tag{10}$$

When $\sigma_i(t) = 0$ $(i = 1, 2)$ for $t \geq t_s$, it is said that a sliding motion [12] is generated in the dynamic subsystems S_i $(i = 1, 2)$. The condition for the generation of a sliding motion is given as follows:

$$\sum_{i=1}^{2} c_i \dot{\sigma}_i(t) sgn[\sigma_i(t)] < 0 \tag{11}$$

where c_i $(i = 1, 2)$ are some positive constants and $sgn(\cdot)$ is a sign function being defined as follows:

$$sgn(\cdot) = \begin{cases} 1 & for \quad (\cdot) > 0 \\ 0 & for \quad (\cdot) = 0 \\ -1 & for \quad (\cdot) < 0 \end{cases} \tag{12}$$

The following decentralized sliding mode controller is proposed to accomplish the condition (11)

$$u_i(t) = (\eta_i^*)^{-1} \{\psi_{i\sigma}\sigma_i(t) + [\psi_{io} + (\psi_{ix} + \psi_{i\beta})||\boldsymbol{x}_i(t)||] \, sgn[\sigma_i(t)]\} \tag{13}$$

where $\psi_{i\sigma}$, ψ_{io}, ψ_{ix}, $\psi_{i\beta}$ $(i = 1, 2)$ are positive constant parameters to be chosen by the designer.

Theorem 1. Under Assumptions 1–4, sliding motion is generated in the dynamic subsystems S_i $(i = 1, 2)$ by using the decentralized sliding mode control law (13) and the state vector $\boldsymbol{x}(t)$ is bounded for all $t \geq 0$ and converges to zero asymptotically as $t \to \infty$ if the following relationships hold

$$\psi_{i\sigma} > 0 \tag{14}$$

$$\psi_{io} \geq \frac{\epsilon_{iv}}{m_c^-} \left[\alpha_i + c_{ai}^+ c_{ei}^+ (sin\theta_i^+ - sin\theta_{pi}) + a_{qk}^+ \right] \tag{15}$$

$$\psi_{ix} \geq \frac{\epsilon_{iv}\beta_{ii}}{m_c^-} + \left[\left(\frac{\epsilon_{iv} k_{ii}^+}{m_c^-} \right)^2 + \epsilon_{id}^2 \right]^{\frac{1}{2}} \tag{16}$$

$$\psi_{1\beta} > \frac{\epsilon_{1v}\beta_{12}}{m_c^-} \quad ; \quad \psi_{2\beta} > \frac{\epsilon_{2v}\beta_{21}}{m_c^-} \tag{17}$$

Theorem 1 can be proved by using the Lyapunov second method and the properties of M–matrices in [13].

3.2. STABILITY AND ROBUSTNESS IN SLIDING MODE

When sliding motion is generated in the dynamic subsystems S_i $(i = 1, 2)$ at the time instant t_s, the following relationship holds:

$$\sigma_i(t) = \boldsymbol{\epsilon}_i^T \boldsymbol{x}_i(t) = \epsilon_{di} x_i^d(t) + \epsilon_{vi} x_i^v(t) = 0 \; ; \; t \geq t_s \tag{18}$$

The equation of motion of the dynamic subsystems S_i $(i = 1, 2)$ in sliding mode can be obtained by using the so–called "technique of equivalent control" in [11] such that $\sigma_i(t) = 0$; $\dot{\sigma}_i(t) = 0$ $(i = 1, 2)$ for $t \geq t_s$. From eqns. (7) and (1), an equivalent control force $u_i^{eq}(t)$ is found as follows

$$u_i^{eq}(t) = -[\boldsymbol{\epsilon}_i^T \boldsymbol{b}_i(t)]^{-1} \boldsymbol{\epsilon}_i^T \boldsymbol{A}_i \boldsymbol{x}_i(t) + [\boldsymbol{\epsilon}_i^T \boldsymbol{b}_i(t)]^{-1} \boldsymbol{\epsilon}_i^T \boldsymbol{d}_i [f_{ni}(t) + f_{di}(t) + f_{ei}(t)] \quad ; \quad t \geq t_s \tag{19}$$

By substituting $u_i(t) = u_i^{eq}(t)$ from (19) into (1), the following closed–loop equation of motion of the dynamic subsystems S_i $(i = 1, 2)$ in sliding mode is obtained:

$$\epsilon_{vi} \dot{\boldsymbol{x}}_i(t) + \epsilon_{di} \boldsymbol{x}_i(t) = 0 \quad ; \quad t \geq t_s \tag{20}$$

Since $\epsilon_{id} > 0$ and $\epsilon_{iv} > 0$, one gets:

$$\boldsymbol{x}_i(t) = \boldsymbol{x}_i(t_s) e^{-\frac{\epsilon_{di}}{\epsilon_{vi}}(t-t_s)} \quad ; \quad t \geq t_s \tag{21}$$

Remark 2. It is seen from (21) that the dynamic subsystems S_i $(i = 1, 2)$ in sliding mode is exponentially stable. Then, the relative displacement $e_i^d(t)$ and the relative velocity $e_i^v(t)$ $(i = 1, 2)$ of the bridge deck are uniformly bounded for all $t \geq t_s$ and converge to zero exponentially as $t \to \infty$. Therefore, the cable–stayed bridge segment goes back to its zero equilibrium position when sliding motion is generated in the dynamic subsystem S_i. Another very important property of the proposed decentralized sliding mode control scheme is the robustness exhibited by the dynamic subsystems S_i $(i = 1, 2)$ in sliding mode. Note from (21) that the functions $f_{ni}(t)$, $f_{ei}(t)$ and $f_{di}(t)$, the matrix $\boldsymbol{A}_i$ and the vectors $\boldsymbol{b}_i$, $\boldsymbol{d}_i$ do not appear in the closed–loop equation of motion of the dynamic subsystems S_i in sliding mode. Thus, the dynamical hehaviour of the controlled cable–stayed bridge structure is robust in the presence of the nonlinearities, structural disturbances, seismic excitation and parametrical uncertainties when sliding motion is generated.

3.3. CONTINUOUS SMC SCHEME

In the implementation of decentralized sliding mode control law (13), non–ideal effects may cause chattering and high control activity which can excite sometimes high–frequency unmodelled dynamics in the system. In order to alleviate the chattering phenomenon, one can make the following change in (13)

$$sgn[\sigma_i(\cdot)] \iff \frac{\sigma_i(\cdot)}{|\sigma_i(\cdot)| + \delta} \tag{22}$$

where δ is a small positive constant to be chosen by the designer. Thus, the following continuous sliding mode control law is obtained:

$$u_i(t) = (\eta_i^*)^{-1} \left\{ \psi_{i\sigma} \sigma_i(t) + [\psi_{io} + (\psi_{ix} + \psi_{i\beta}) \|\boldsymbol{x}_i(t)\|] \frac{\sigma_i(\cdot)}{|\sigma_i(\cdot)| + \delta} \right\} \tag{23}$$

4. Simulation

Consider a scale model of the bridge with the following properties:

Beam (material: steel)

$$
\begin{aligned}
L &= 3\ m \\
A &= (0.1\ m \times 0.02\ m) = 2 \times 10^{-3}\ m^2 \\
E &= 2.06 \times 10^{11}\ N/m^2 \\
I &= 6.7 \times 10^{-8}\ m^4 \\
m_c &= 7.86\ Kg \\
k_{11} &= 1.5596 \times 10^6\ N/m \\
k_{12} &= k_{21} = -1.1735 \times 10^6\ N/m \\
k_{22} &= 1.5957 \times 10^6\ N/m \\
\rho &= 7860\ Kg/m^3
\end{aligned}
$$

Cables (material: stainless steel wires, SUS304JIS)

$$
\begin{aligned}
c_{l1} &= 3.58\ m; \quad c_{l2} = 3.327\ m \\
c_{a1} &= c_{a2} = 2.1 \times 10^{-7}\ m^2 \\
c_{e1} &= c_{e2} = 1.706 \times 10^{11} N/m^2 \\
\theta_{p1} &= 56.04^{\circ}; \quad \theta_{p2} = 63.2^{\circ}
\end{aligned}
$$

In the simulation, the seismic excitation has been the scaled Taft earthquake shown in Figure 2. The upper and lower bounds for the system parameters corresponding to the Assumptions 1–4 are defined as the values with $\pm 10\%$ variation respect to their nominal ones. Also, $\theta_1^- = \theta_2^- = 40^{\circ}$, $\theta_1^+ = \theta_2^+ = 80^{\circ}$, $\alpha_1 = \alpha_2 = 1000$, $\beta_{11} = 2.3 \times 10^5$, $\beta_{12} = \beta_{21} = 1.8 \times 10^5$, $\beta_{22} = 2.6 \times 10^5$. The continuous active decentralized controller (23) has been used with $\epsilon_{di} = 3$, $\epsilon_{vi} = 1$, $\eta_1^* = 548$, $\eta_2^* = 590$, $\psi_{1\sigma} = 0.182$, $\psi_{2\sigma} = 0.169$, $\hat{\psi}_{1o} = \hat{\psi}_{2o} = 2$, $\psi_{1e} = \psi_{2e} = 500$, $\psi_{1p} = \psi_{2p} = 375$, $\delta = 0.1$. The time histories of the displacements of the bridge deck in presence of the scaled earthquake excitation are shown in Figures 3 and 4.

Acknowledgement

This research is supported by the the projects "NATO collaborative research grant CRG 971534" and "DGES PB96-0257" of the Spanish Government.

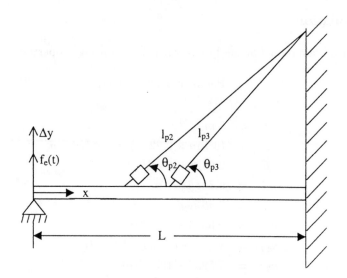

Figure 1. Cable stayed bridge structure.

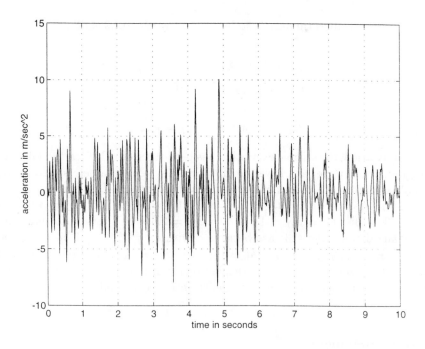

Figure 2. Scaled accelerogram of Taft earthquake vertical component.

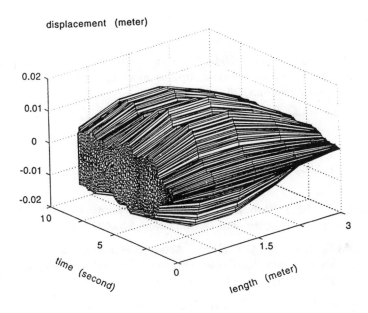

Figure 3. Time response of the bridge without control.

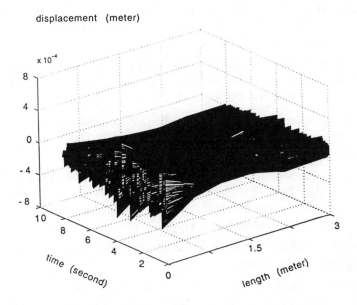

Figure 4. Time response of the bridge with decentralized control.

References

1. Fujino, Y., Susumpou, T. (1994), An experimental study on active control of in–plane cable vibration by axial support motion, *Earthquake Engineering and Structural Dynamics*, **23**, 1283–1297.

2. Achkire, Y., Preumont, A. (1996), Active tendon control of cable–stayed bridges, *Earthquake Engineering and Structural Dynamics*, **25**, 585–597.

3. Magaña, M.E., Rodellar, J. (1997), Nonlinear decentralized active tendon control of cable–stayed bridges, to appear in the *Journal of Structural Control*.

4. Conrad, F. (1990), Stabilization of beams by pointwise feedback control, *SIAM Journal on Control and Optimization*, **28**, 423–437.

5. Habib, M.S., Radcliffe, C.J. (1991), Active parametric damping of distributed parameter beam transverse vibration, *Journal of Dynamic Systems, Measurements and Control*, **113**, 295–299.

6. Oshumi, A., Sawada, T. (1993), Active control of flexible structures subject to distributed and seismic disturbances, *Journal of Dynamic Systems, Measurements and Control*, **115**, 649–657.

7. Magaña, M.E., Volz, P., Miller, T (1997), Robust nonlinear decentralized control of a flexible cable–stayed beam structure, to appear in *ASME Journal of Acoustics and Vibrations*.

8. Sandell, Jr., H.R., Varaiya, P., Athans, M., Safonov, M.G. (1978), Suvey of decentralized control methods for large scale systems, *IEEE Trans. on Automatic Control*, **AC-23**, 108–128.

9. Luo, N., Rodellar, J., De la Sen, M. (1998), "Composite Robust Active Control of Seismically Excited Structures with Actuator Dynamics", *Earthquake Engineering and Structural Dynamics*, Vol. 27, 3, pp. 301–311.

10. Luo, N., De la Sen, M., Rodellar, J. (1998), "Composite Adaptive SMC of Nonlinear Base Isolated Buildings with Actuator Dynamics", *Journal of Applied Mathematics and Computer Science, Special Issue: Adaptive Learning and Control using Sliding Mode*, Vol. 8, 1, 183–197.

11. Utkin, V.I. (1992), *Sliding Modes in Control and Optimization*, Springer–Verlag, Berlin, Heidelberg.

12. Hung, J.Y., Gao, W.-B., Hung, J.C. (1993), "Variable structure control: a survey", *IEEE Trans. on Industrial Electronics*, Vol. 40, 1, pp. 2–22.

13. Araki, M. (1978), "Stability of large–scale nonlinear systems — quadratic-order theory of composite–system method using M–matrices", *IEEE Trans. on Automatic Control*, Vol. AC–23, 2, pp. 129–142.

ACTIVE CONTROL OF CABLE-STAYED BRIDGES

M.E. MAGAÑA
Department of Electrical and Computer Engineering
Oregon State University, Corvallis, Oregon 97331, USA

J. RODELLAR, J. R. CASAS, J. MAS
School of Civil Engineering
Technical University of Catalunya, 08034 Barcelona, SPAIN

1. Introduction

In this work we propose the design of an automatic control system for a cable-stayed bridge to actively counteract external environmental forces generated by earthquakes. The introduction of this intelligent feature in the structure turns an otherwise passive bridge into a smart bridge, capable of reducing the amplitude of the deflections of its deck down to a desired level automatically. The control strategy proposed here uses a subset of the stay cables as active tendons to provide control forces through appropriate actuators. Each individual actuator is controlled by a decentralized controller that uses local linear velocity and local linear relative displacement information only. The effectiveness of the control algorithm herein proposed is tested on a three-dimensional model of the Quincy Bayview bridge [1,2]. This bridge has been chosen because it is fairly representative of the new generation of cable-stayed bridges. The mathematical model is derived using a finite element approach. Although some work has been done in this field by other researchers [3-6], to our knowledge, no results that use three-dimensional full-scale models have been published in the literature.

2. Finite element bridge model

As shown in Figure 1, the deck of the bridge is supported at the two ends and at two intermediate points where two towers are located. The length of the main span is 274 m, and the length of each of the side spans is 134 m, for a total length of 542 m. The width of the deck is 12.2 m (from cable center to cable center) and the tower height is 70.7 m (from water line). The bridge has a total of 56 stay cables. Other bridge data can be found in [1].

The bridge deck is modeled by a flexible and massless spine with infinitely stiff ribs whose end points are used to attach the stay cables [1]. As in [1], the model uses 213 three-dimensional two-node beam elements (six degrees of freedom per

J. Holnicki-Szulc and J. Rodellar (eds.), Smart Structures, 193–202.

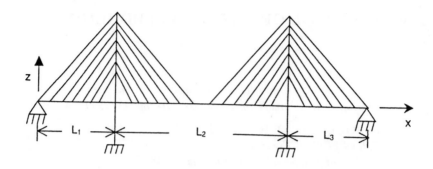

Figure 1. Quincy Bayview bridge model.

node) for the deck and towers, and 56 one-dimensional truss elements. The total number of nodes is 206. The FEM program used for this purpose generates the discretized model equations of motion in the form

$$M\ddot{x} + Kx = f_{eff}, \quad x, f_{eff} \in \mathbb{R}^{1236}. \tag{1}$$

The FEM program computes an inconsistent mass matrix M to expedite simulation time, that is, M is a diagonal matrix.

The first 10 modal frequencies (in Hz) computed from the finite element dynamic model of the bridge are the following: 0.38, 0.45, 0.52, 0.55, 0.60, 0.65, 0.70, 0.74, 0.80, 0.82. These values are very close to those obtained in [1,2] for the real Quincy Bayview bridge.

Because the center spine of the deck is modeled as a massless and flexible beam, the masses and moments of inertia of the 6 degrees of freedom associated with the first 90 nodes are equal to zero. This means that the first 540 dof's have zero mass or moment of inertia. The model itself has another 120 dof's throughout the structure that have either zero mass or moment of inertia due to movement constrainst. So, the total number of dof's with either zero mass or moment of inertia is equal to 660. Since the total number of dof's is equal to 1236, the dimension of the stiffness matrix is 1236×1236 and the dimension of the nonzero portion of the mass matrix is 576×576.

Let P be a 1236×1236 permutation matrix and let $x = P^T z$. Then, from equation (1), we can write

$$\hat{M}\ddot{z} + \hat{K}z = \hat{f}_{eff}, \tag{2}$$

where $\hat{M} = PMP^T$, $\hat{K} = PKP^T$ and $\hat{f}_{eff} = Pf_{eff}$.

By an appropriate choice of matrix P, we can get permutations of the rows and columns of M and K so that

$$\hat{M} = \begin{pmatrix} \hat{M}_{11} & 0 \\ 0 & 0 \end{pmatrix}, \quad \hat{K} = \begin{pmatrix} \hat{K}_{11} & \hat{K}_{12} \\ \hat{K}_{21} & \hat{K}_{22} \end{pmatrix}, \quad \hat{f}_{eff} = \begin{pmatrix} \hat{f}_{eff1} \\ 0 \end{pmatrix}. \tag{3}$$

Let $z = [z_1, z_2]^T$, then $z_2 = -\hat{K}_{22}^{-1}\hat{K}_{21}z_1$ and the reduced-order system is given by

$$\hat{M}_{11}\ddot{z}_1 + (\hat{K}_{11} - \hat{K}_{12}\hat{K}_{22}^{-1}\hat{K}_{21})z_1 = \hat{f}_{eff1}, \tag{4}$$

where $\hat{M}_{11} \in \mathbb{R}^{576 \times 576}$ is a diagonal matrix, $z_1 \in \mathbb{R}^{576}$ and $z_2 \in \mathbb{R}^{660}$.

Notice that $\hat{f}_{eff1}$ contains the components of f_{eff}, the vector of external forces, that affect to the dof's with non-zero mass or moment of inertia. The vector z_1 contains total displacement components of the structure as well as ground support displacement components. Among them, we can separate the dof's of the supports and the dof's of the bridge structure itself. In the FEM model we have identified 10 support nodes that represent the points of contact of the structure and the ground at the ends of the deck and the bottom of the towers. By an appropriate permutation matrix, the matrices $\hat{K}_{11} - \hat{K}_{11}\hat{K}_{22}^{-1}\hat{K}_{21}$ and $\hat{M}_{11}$ can be transformed and the bridge dynamics can be expressed in the form

$$\begin{pmatrix} \tilde{M}_{11r} & 0 \\ 0 & \tilde{M}_{gg} \end{pmatrix} \begin{pmatrix} \ddot{z}_{1t} \\ \ddot{z}_g \end{pmatrix} + \begin{pmatrix} \tilde{K}_{11r} & \tilde{K}_g \\ \tilde{K}_g^T & \tilde{K}_{gg} \end{pmatrix} \begin{pmatrix} z_{1t} \\ z_g \end{pmatrix} = \begin{pmatrix} 0 \\ \hat{f}_g \end{pmatrix}, \tag{5}$$

where $z_{1t} \in \mathbb{R}^{516}$ is the vector of total displacements of the bridge structure, $z_g \in \mathbb{R}^{60}$ is the vector of (support) ground displacements and $\hat{f}_g \in \mathbb{R}^{60}$ is the vector of ground forces.

Let the total displacement vector of the bridge structure be given by

$$z_{1t} = z_{1r} + z_{1s} \tag{6}$$

where $z_{1r} \in \mathbb{R}^{516}$ is the vector of relative displacements of the structure and $z_{1s} \in \mathbb{R}^{516}$ is the vector of quasi-static displacements that can be computed from [8]

$$\begin{pmatrix} \tilde{K}_{11r} & \tilde{K}_g \\ \tilde{K}_g^T & \tilde{K}_{gg} \end{pmatrix} \begin{pmatrix} z_{1s} \\ z_g \end{pmatrix} = \begin{pmatrix} 0 \\ \hat{f}_{gs} \end{pmatrix}, \tag{7}$$

where $\hat{f}_{gs}$ is the vector support forces necessary to statically impose displacements z_g that vary with time. From the last equation,

$$z_{1s} = \gamma z_g, \tag{8}$$

where $\gamma = -\tilde{K}_{11r}^{-1}\tilde{K}_g$ is the influence coefficient matrix. Finally, the equation of relative motion of the cable-stayed bridge structure has the form

$$\tilde{M}_{11r}\ddot{z}_{1r} + \tilde{K}_{11r}z_{1r} = -\tilde{M}_{11r}\gamma a_e, \quad a_e = \ddot{z}_g. \tag{9}$$

3. Control system

The control strategy proposed here uses a subset of the stay cables as active tendons to provide control forces through appropriate actuators. Thus the above finite element model has to be augmented with the effective control forces applied to the dof's of the nodes where the actively controlled stay cables are anchored at. Our objective is to design a decentralized control scheme in which each individual actuator is driven by a controller that uses local linear velocity and local linear relative displacement information only. Thus the overall model is broken into a set of interconnected subsystems. Let Ω_n be the subset of the degrees of freedom directly controlled by the active tendon n. Then for such degrees of freedom the dynamics are approximately described by

$$(\tilde{M}_{11r})_i(\ddot{z}_{1r})_i + (\tilde{K}_{11r})_i(z_{1r})_i + \sum_{\substack{j=1 \\ j \neq i}}^{516} (\tilde{K}_{11r})_{ij}(z_{1r})_j = -(\tilde{M}_{11r})_i \sum_{l=1}^{60} \gamma_{il} a_{el} -$$

$$\frac{A_n E_n}{l_{0n}} g(\theta_n) u_n, \qquad (10)$$

for $i \in \Omega_n$, where A_n, E_n and l_{0n} are, respectively, the cross-sectional area, Young modulus of elasticity, and unstressed length of cable n, θ_n is the angle between cable n and the local horizontal at the point of anchor on the deck, and u_n is the cable elongation produced by the nth actuator as control signal. The function $g(\cdot)$ depends on the controlled degree of freedom, i.e., $g(\theta_n) = \sin \theta_n$ if i denotes a dof with linear displacement in the z direction and $g(\theta_n) = \cos \theta_n$ if i denotes a dof with linear displacement in the x direction (see Figure 1). If i corresponds to a linear displacement in the y direction, then no effective active control can be provided through the stay cables and $g(\theta_n) \approx 0$.

The mathematical model of the bridge dynamics presented earlier neglects structural damping. This is because modern cable-stayed bridges are highly flexible structures whose damping is generally less than 3%. Common sense would suggest that the introduction of artificial active damping could improve the dynamic stability properties of the bridge structure. This could be achieved through the application of control forces that are proportional to velocity measurements. Specifically, in this work the forces are generated by actively changing the effective length of a subset of the stay cables. Although the decentralized control approach proposed here uses local information only, the health of the other actuators and sensors is constantly monitored so that no control forces are applied at one side of the deck when either the actuator or sensor fails on the other side.

Before we proceed with the controller design, we need to assess the uncontrolled bridge performance in the presence of seismic excitation. Let us assume for the moment that the seismic wave reaches the bridge supports at the same time. If only the vertical component of the earthquake is considered, then only 10 out of 60 acceleration inputs will provide excitation to the dynamic model of the bridge. Let such an excitation be provided by the corrected vertical acceleration record of the

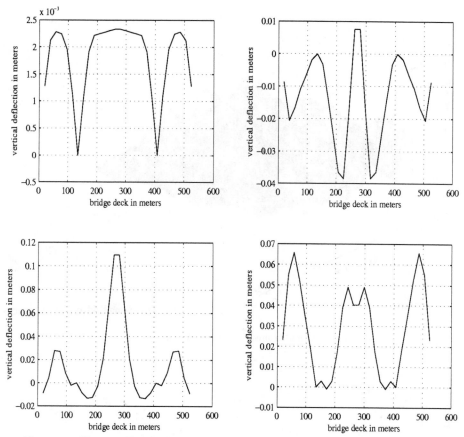

Figure 2. Uncontrolled bridge deck vertical deflections at times $t = 0.2, 2, 10, 20$ seconds.

Taft Earthquake. As discussed in [6], the maximum magnitude of the acceleration is a little over 1 m/s^2, while most of its energy is concentrated in the range of 0 to 13 Hz. This is a good test signal to assess uncontrolled performance, since it can provide a high enough level of excitation to practically every one of the lower vertical modes of the bridge structure. Figure 2 shows the bridge's deck vertical deflections at $t = 0.2, 2, 10$ and 20 seconds.

Figure 3 shows the entire bridge deck deflection over the earthquake duration. The maximum deck vertical deflection is 0.18 meters. Figure 4 shows the deck deflection in the y direction. This is a qualitative measure of the coupling that exists between the vertical and horizontal modes of the bridge structure.

Since in this study only the vertical deflections are significant, only the dof's with displacements in the z direction are controlled, so that function g in equation (10) is $g(\theta_n) = \sin\theta_n$. Let i be the dof with vertical displacement of the node

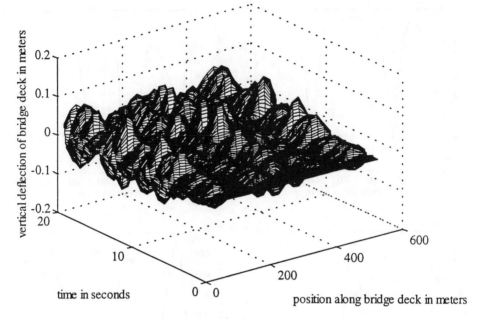

Figure 3. Uncontrolled bridge deck vertical deflections.

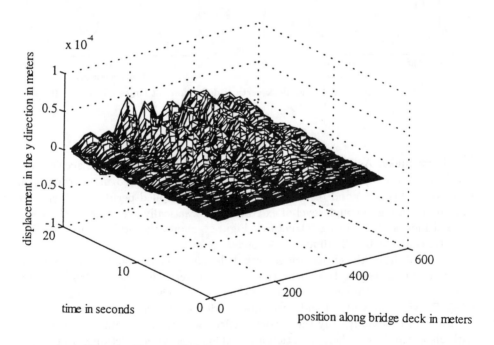

Figure 4. Uncontrolled bridge deck deflections in the y direction.

where cable n is attached to. Then the local n controller has the general form

$$u_n(t) = f_n([z_{1r}]_i, [\dot{z}_{1r}]_i). \tag{11}$$

If the deflections at the points of anchor of active tendons are small compared to the rise and run of them, then $\sin\theta_n(t)$ is approximately constant and the entire controlled cable-stayed bridge structure modeled by equation (10) can now be modeled as a linear and time-invariant dynamic system. For this case, it can be shown that if the local control function $f_n(\cdot)$ is a linear function of the local velocity, the global system will be asymptotically stable.

After observing the vertical deflections of the deck of the uncontrolled bridge caused by the earthquake, it was decided to place 6 couples of active controllers at cables indicated by $n = 5, 10, 13, 16, 19, 24, 33, 38, 41, 44, 47, 52$. Since the dynamic model of the bridge is linear and time-invariant, thus we can use the LQR approach to design the local controllers so that the modes directly controlled by them have a prescribed degree of stability and rate of decay [9].

To design the active control system, a state variable representation of the dynamics of the dof's directly controlled is more suitable. Let $(s_r)_i = [(z_{1r})_i, (\dot{z}_{1r})_i]^T$, then equation (10) for the vertical dof i can be rewritten in the form

$$(\dot{s}_r)_i = \begin{pmatrix} 0 & 1 \\ -\dfrac{(\tilde{K}_{11r})_{ii}}{(\tilde{M}_{11r})_{ii}} & 0 \end{pmatrix} (s_r)_i - \begin{pmatrix} 0 \\ \dfrac{1}{(\tilde{M}_{11r})_{ii}} \displaystyle\sum_{\substack{j=1 \\ j\neq i}}^{516} (\tilde{k}_{11r})_{ij}(z_{1r})_j \end{pmatrix} -$$

$$\begin{pmatrix} 0 \\ \displaystyle\sum_{l=1}^{60} \gamma_{il} a_{el} + \dfrac{A_n E_n}{(\tilde{M}_{11r})_{ii} l_{0n}} \sin\theta_n u_n \end{pmatrix}. \tag{12}$$

Consider the "unperturbed" subsystem i, that is, neglecting coupling and excitation,

$$(\dot{s}_r)_i = A_i(s_r)_i + B_i u_n \tag{13}$$

where

$$A_i = \begin{pmatrix} 0 & 1 \\ -\dfrac{(\tilde{K}_{11r})_{ii}}{(\tilde{M}_{11r})_{ii}} & 0 \end{pmatrix}, \qquad B_i = \begin{pmatrix} 0 \\ -\dfrac{A_n E_n}{(\tilde{M}_{11r})_{ii} l_{0n}} \sin\theta_n \end{pmatrix}. \tag{14}$$

Let the cost functional of the ith controlled vertical dof be given by

$$J_i = \int_0^\infty e^{2\sigma t}[(s_r)_i^T Q_i(s_r)_i + r_i u_n^2]dt, \tag{15}$$

where $Q_i = Q_i^T \geq 0$ and $r_i > 0$.

The local unperturbed second-order subsystems are controllable and observable. Thus, local optimal controllers u_n^* of the form

$$u_n^*(t) = -\beta_1(z_{1r})_i(t) - \beta_2(\dot{z}_{1r})_i(t), \tag{16}$$

where β_1 and β_2 are real and constant gains, can be found using the above procedure. Let the desired rate of decay be $\sigma = 0.7$, that is, it is desired that the two eigenvalues of each of the second-order subsystems lie to the left of the vertical line located at $s = -0.7$, where s is a complex variable. The existence of such an optimal control can be guaranteed for each of the second-order subsystems, because we can always find a matrix D_i such that $D_i D_i^T = Q_i$ and the pair $(A_i + \sigma I, D_i)$ is completely detectable. Alternatively, its existence can also be guaranteed because each subsystem that is directly controlled is completely controllable, that is, the pair (A_i, B_i) is completely controllable [9].

Let us now design the local controllers using the approach outlined above. To accomplish this task, we need to first compute the input distribution matrix B_i of each controlled local subsystem. Table 1 below lists the parameter data associated with the cables used as active tendons [1]. The true modulus of elasticity of the stay cables was used for modeling purposes, that is, the cables were treated as having a completely linear force-deformation relationship described by the true material modulus of elasticity [1]. The actual value used was $2.068 \times 10^{11} N/m^2$. Then the controlled response of the bridge in the presence of the same seismic input (Taft earthquake vertical acceleration component) is shown in Figure 5. Although only local information has been used to control the entire bridge, the deflections have been reduced substantially as compared to the uncontrolled case.

TABLE 1. Active tendon parameter values

n	$A_n\ (m^2) \times 10^{-3}$	$l_{0n}\ (m)$	$\theta_n\ (deg)$
5	3.48	71.3	36.7
10	3.48	69.5	37.9
13	6.77	121	24.9
16	6.77	121	24.9
19	3.48	69.5	37.9
25	3.48	71.3	36.7
33	3.48	71.3	37.7
38	3.48	69.5	37.9
41	6.77	121	24.9
44	6.77	121	24.9
47	3.48	69.5	37.9
53	3.48	71.3	36.7

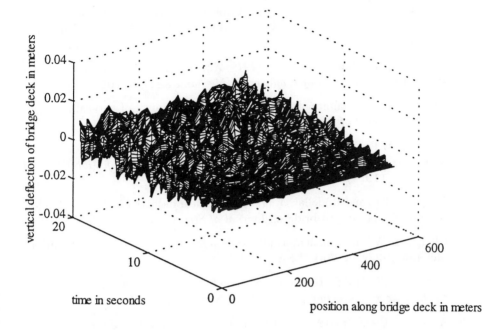

Figure 5. Controlled bridge deck vertical deflections.

4. Conclusions

It has been shown in this paper that active decentralized control algorithms can be effective in reducing the magnitude of the deflections of the deck of a real-life cable-stayed bridge caused by the vertical component of an earthquake. The design methodology is simple and the implementation of the local controllers is very cost effective because only local information is used to generate control signals for each of the actuators.

Acknowledgment

This work has been partially supported by a NATO Collaborative Research Grant (No. CR6971534) and by the US National Science Foundation through grant No. CMS-9301464 with Dr. S. C. Liu as Program Director. Such a support is highly appreciated.

5. References

1. Wilson, J.C. and Gravelle, W. (1991) Modelling of a cable-stayed bridge for dynamic analysis, *Earthquake Engineering and Structural Dynamics*, **20**, 707-721.

202

2. Wilson, J.C. and Liu, T. /1991) Ambient vibration measurements on a cable-stayed Bridge, *Earthquake Engineering and Structural Dynamics*, **20**, 723-747.

3. Yang, J.N. and Giannopolous, F. (1979) Active control and stability of cable-stayed Bridge, *Journal of Engineering Mechanics ASCE*, **105**, 677-694.

4. Yang, J.N. and Giannopolous, F. (1979) Active control of two-cable-stayed bridges, *Journal of Engineering Mechanics ASCE*, **105**, 795-810.

5. Magaña, M.E., Volz, P. and Miller, T. (1997) Robust nonlinear decentralized control of a flexible beam Structure, *ASME Journal of Vibration and Acoustics*, **119(4)**, 523-526.

6. Magaña, M.E. and Rodellar, J. (1998) Nonlinear decentralized active tendon control of cable-stayed bridges, *Journal of Structural Control*, **5(1)**, 45-62.

7. Clough, R.W. and Penzien, J. (1993) *Dynamics of Structures*, 2nd edition, McGraw Hill, New York, USA.

8. Chopra, A.K. (1995) *Dynamics of Structures, Theory and Applications to Earthquake Engineering*, Prentice Hall, New Jersey, USA.

9. Anderson, B.D.O. and Moore, J.B. (1990) *Optimal Control: Linear Quadratic Methods*, Prentice Hall, New Jersey, USA.

STRATEGY OF IMPULSE RELEASE OF STRAIN ENERGY FOR DAMPING OF VIBRATION

Z. MARZEC[1,2], J. HOLNICKI-SZULC[1], F.LOPEZ-ALMANSA[3]
1) Institute of Fundamental Technological Research, PAS,
   Swietokrzyska 21, 00-49 Warsaw, Poland,
2) Institute of Machine Design Fundamentals,
   Warsaw University of Technology
   Narbutta 84, 02-524 Warsaw, Poland
3) Technical University of Catalonia
   Avda. Diagonal, 649; 08028 Barcelona; Spain

1. Introduction

The goal of this paper is to present an impulse release of strain energy strategy for damping of free vibration in adaptive structures (equipped with actively controlled dissipaters). The concept can be applied to skeletal structures as well as to two-layer beams or plates, where stored strain energy available for dissipation through eventual „delamination effect" is relatively high.

For simplicity of presentation, the concept is demonstrated on ideal quasi-static model of dissipative effect in the first section and then, fully dynamic simulation of the adaptive structure behaviour is demonstrated on numerical examples in the second section.

The main concept has been already formulated and discussed on the classical elastic beam model (Ref.1)and a particular technical application has been proposed (Ref.2). Similar approach to the problem of strain energy dissipation with use of controllable rotational dissipaters in joints of skeletal structures has been also proposed recently (Ref.3).

2. Strategy of Impulse Release of Strain Energy

Let us consider the flexible truss structure shown in Fig.1a equipped with actuators (e.g. hydraulic actuators with controllable valves, Fig.1b) located in members 2, 4, 6, 8, 10, 12, 14, 16, 18, 20. The strain modification caused in the structure due to the unit elongation $\varepsilon^\circ=1$ of the active members 2, 4, 6, 8, 10, 12, 14, 16, 18, 20 can be expressed as follows:

J. Holnicki-Szulc and J. Rodellar (eds.), Smart Structures, 203–210.
© 1999 Kluwer Academic Publishers. Printed in the Netherlands.

$$\varepsilon_i^R = \sum_j D_{ij}\varepsilon_j^o \qquad (1)$$

where the influence matrix D_{ij} denotes deformation of the member i caused by the unit „virtual distortion" provoked in the member j. The corresponding stress modification can be expressed as:

$$\sigma_i^R = \sum_j E_i(D_{ij} - \delta_{ij})\varepsilon_j^o = \sum_j D_{ij}^\sigma \cdot \varepsilon_j^o \,, \qquad (2)$$

where E_i and δ_{ij} denote the elasticity coefficient and the Kronecker's symbol, respectively. Superimposing the above initial states of compatible strains (1) and self-equilibrated stresses (2) with externally caused strains ε_i^L and stresses σ_i^L, the „modified" response of actively controlled structure

$$\varepsilon_i = \varepsilon_i^L + \varepsilon_i^R \quad, \quad \sigma_i = \sigma_i^L + \sigma_i^R \qquad (3)$$

can be reached.

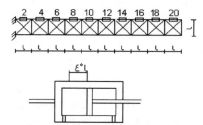

Figure1. Truss beam equipped with dissipaters   *Figure2.* Characteristic of dissipater No. 2

A structure can be called an „adaptive structure" if there is no need for external energy sources to generate movements of actuators. It means that actuators are purely dissipative (semi-active interfaces allowing controlled dissipation of stored strain energy) what can be formally postulated by the following condition:

$$\sigma_i \Delta\varepsilon_i^o \geq 0 \qquad (4)$$

where $\Delta\varepsilon_i^o$ denotes increment of virtual distortions due to yielding of actuators.

The corresponding strain energy dissipation in a chosen actuator (dissipater) can be expressed as follows:

$$U = \int_0^{\varepsilon'} \left(\sigma^L + D^\sigma\varepsilon^o\right)d\varepsilon^o = \sigma^L\varepsilon' + \frac{1}{2}D^\sigma(\varepsilon')^2 \qquad (5)$$

where ε' is the maximum value to be taken by ε^o.

To maximise U, let us calculate the gradient

$$\frac{dU}{d\varepsilon'} = \sigma^L + D^\sigma \varepsilon'$$

(6)

tending to zero for : $\varepsilon' = -\sigma_L / D^\sigma$.

(7)

Therefore, it can be demonstrated, substituting (7) to (5), that the maximal value for energy dissipation

$$U = U\big|_{\varepsilon'=-\sigma_L/D^\sigma} = \frac{1}{2}\sigma^L \varepsilon' = -\frac{\left(\sigma^L\right)^2}{2D^\sigma} = \frac{\left(\sigma^L\right)^2}{2\left|D^\sigma\right|}$$

(8)

corresponds to zero stresses in active elements (substituting (7) to (2) and $(3)^2$). Consequently, for free vibration problem, the activity based on the impulse release of structural interface connections (due to virtual distortion (7) generation causing vanishing of stresses in active members) at the instant of extremal loading of the structure can be proposed.

Having more dissipaters available and applying the same impulse release strategy requiring instantaneous vanishing of local stresses in active elements, the corresponding distortions increments $\Delta\varepsilon_i^o$ generated authomatically in dissipaters can be determined from the following set of k linear equations (describing stresses in active members), where k denotes the structural redundancy:

$$\sigma_i^L + \sum_j D_{ij}^\sigma \varepsilon_j^o + \sum_j D_{ij}^\sigma \Delta\varepsilon_j^o = 0$$

(9)

where σ_i^L and ε_j^o are known, actual values. Note that rank($\mathbf{D}^\sigma$)=k and therefore, the maximal efficiency of vibration control for the truss structure shown in Fig.1a is installation of 10 dissipaters at the positions determined according to the maximal measures (8), what depends on the chosen mode of vibration to be controlled. Constraining our control activity to the first mode of vibration, the best location of actuators in our testing structure is demonstrated in Fig.1a. The corresponding impulse release control strategy is the following:

a) Impulse full opening of valves in k dissipaters (and full locking after a short time interval)
   if $u \cdot \ddot{u} < 0$ and $\dot{u} = 0$ (extremal deflection of the beam) where u denotes vertical movement of the node A.

b) Unloading of the introduced distortions ε_i° whenever $\sigma_i \varepsilon_i^\circ < 0$ (release of the stored prestress if $u \cdot \ddot{u} > 0$ and $\dot{u} = 0$ (unloading stopped by the stored prestress)

3. Numerical Examples

Let us demonstrate the effect of impulse release of strain energy strategy for damping of free vibration on two examples of adaptive structures.

a) The free vibration of truss-beam structure shown in Fig.3 and exicited by the impulse force application to the tip point A has been demonstrated in Fig.4 (line uA shows oscilation of the tip point A). The structure response to the impulse release strategy of damping is shown in Fig.4 (line uAd), while the distribution of total kinetic (line Uk), strain (line Us) and dissipated (line Ud) energy in time is shown in Fig.5. Very high damping effect can be observed. The vibration is almost completely damped after two activations of dissipaters. The strain-stress histeretic loop for the chosen dissipater (No. 2) is shown in Fig.6.

b) The double layer beam Fig.7. with the third layer composed of switchable elements able to release instantly the accumulated shear stresses can realise a controlled „delamination effect". The free vibration response to the force impulse applied to the tip point A has been demonstrated in Fig. 8 (line uA) while the structure response after application of the impulse release strategy of damping is shown in Fig. 8 (line uAd). The corresponding distribution of energies has been demonstrated in Fig. 9., while the strain-stress histeretic loop for interface element has been shown in Fig. 10.

The percentage of energy disspation effect (compared with the initially introduced strain energy) during first four half-cycles (four activations of dissipaters) for the two examples discussed above is shown in Table 1.

TABLE 1. Energy Dissipation Effect

Activations	Model a	Model b
1 halfcycle	52.2	16.0
2 halfcycle	69.5	57.5
3 halfcycle	70.0	72.0
4 halfcycle	72.0	75.2

Acknowledgment

The research presented in this paper was supported by the NATO Collaborative Research Grant No.OUTREACH CRG 950316 and by the grant No. KBN.8T11F00196C/2778. from the Institute of Fundamental Technological Research, funded by the National Research Committee. This paper presents a part of the Ph.D. thesis of the first author, supervised by the second author. Numerical calculations has been performed on the computer CRAY CS64000 at the Warsaw Technical University Computer Centre (ABAQUS package).

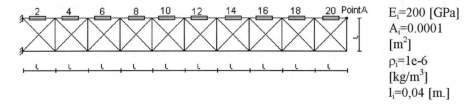

E_i=200 [GPa]
A_i=0.0001 [m^2]
ρ_i=1e-6 [kg/m^3]
l_i=0,04 [m.]

Figure 3. Adaptive truss model

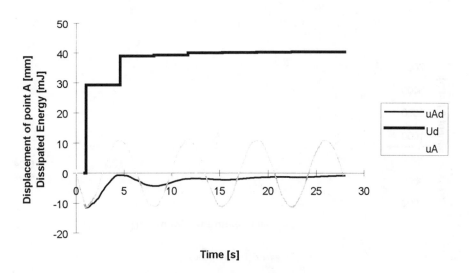

Figure 4. Tip point A oscillations

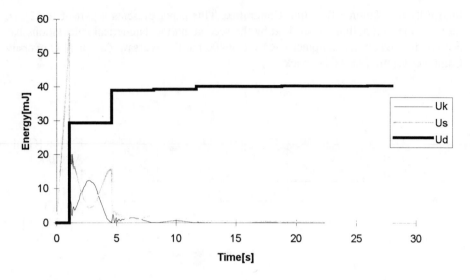

Figure 5. Distribution of energy components

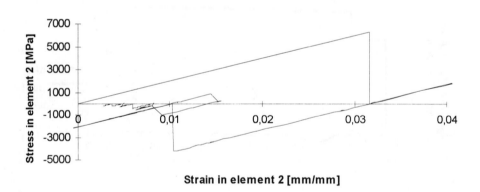

Strain in element 2 [mm/mm]

Figure 6. Histeresis loops

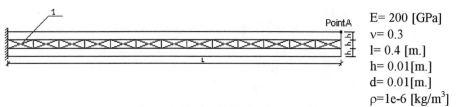

E= 200 [GPa]
ν= 0.3
l= 0.4 [m.]
h= 0.01[m.]
d= 0.01[m.]
ρ=1e-6 [kg/m³]

Figure 7. Adaptive double layer beam model

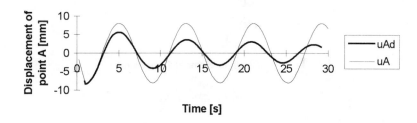

Figure 8. Tip point A oscillations

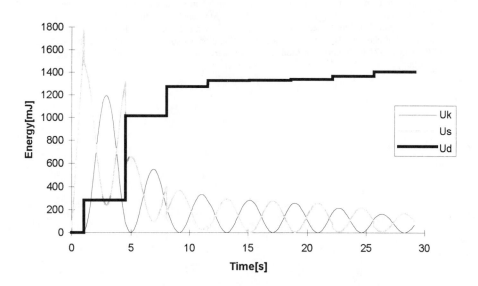

Figure 9. Distribution of energy components

210

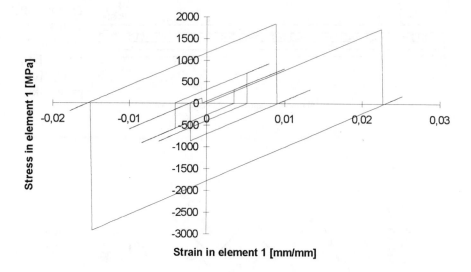

Figure 10. Histeresis loops

References

1. J.Holnicki-Szulc and Z.Marzec, Adaptive Structures with Semi-Active Interface, Proc. of the EUROMECH 373 Colloquium on Modelling and Control of Adaptive Mechanical Structures, March 1998, Magdeburg
2. J.Holnicki-Szulc, Panel for Impact Energy Dissipation, Patent Licence applied for, Warsaw, 1997
3. L.Gaul, R.Nitsche and D.Sachau, Semi-Active Vibration Control of Flexible Structures, Proc. of the EUROMECH 373 Colloquium on Modelling and Control of Adaptive Mechanical Structures, March 1998, Magdeburg

ANALYSIS OF A HYBRID SYSTEM FOR NOISE REDUCTION

SORIN MICU

Departamento de Matemática Aplicada, Facultad de Ciencias Matemáticas, Univ. Complutense, 28040, Madrid, Spain

Abstract. We consider a simple model arising in the control of noise. We assume that the two-dimensional cavity $\Omega = (0, 1) \times (0, 1)$ is occupied by an elastic, inviscid, compressible fluid. The potential Φ of the velocity field satisfies the linear wave equation. The boundary of Ω is divided into two parts Γ_0 and Γ_1. The first one, Γ_0, is flexible and occupied by a Bernoulli-Euler beam. On Γ_0 the continuity of the normal velocities of the fluid and the beam is imposed. The subset Γ_1 of the boundary is assumed to be rigid and therefore, the normal velocity of the fluid vanishes. We analyse the possibility of changing the dynamics of the system by acting only on the flexible part of the boundary. The problems of stabilization, control and existence of periodic solutions are considered (the damping term, the control and the periodic non-homogeneous term respectively acting on Γ_0).

1. Introduction

A great number of works, both in the mathematical and technical literature, deal with the problem of active control of noise generated in acoustic cavities by means of the vibrations of their flexible walls. For instance, such studies were motivated by the development of a new class of turboprop engines which are very fuel efficient but also very loud. In this case the low frequency high magnitude acoustic fields produced by these engines cause vibrations in the fuselage which in turn generate unwanted interior noise.

Many control techniques were proposed and studied for this kind of problems. One of the most interesting ones consists of a control implemented via piezo-ceramic patches embedded in the boundary of the cavity which are excited in a manner so as to produce pure bending moments. Because of the coupling between the interior acoustic field and the struc-

J. Holnicki-Szulc and J. Rodellar (eds.), Smart Structures, 211–220.

tural vibrations, this provides a means of controlling the interior acoustic pressure levels through the control of dynamics of the boundary structure.

In this context we consider a simplified model inspired in and related to that of H. T. BANKS, W. FANG, R. J. SILCOX and R. C. SMITH (see [3]). It consists of a two-dimensional cavity filled with a compressible fluid. A part of the boundary (the structure) is supposed to be flexible. We try to explain the nature of the fluid-structure coupling and to give an answer to the following question: how does the dynamics of such a system change when we act on the structural part?

The aim of this paper is to describe briefly the main mathematical results related with this question and obtained in our previous works. We consider three different types of actions on the structure. In section three we analyse the effect produced by a damping term concentrated on the active part of the boundary. In the next section we discuss the possibility of controlling the level of noise by acting on the flexible part of the boundary. In section five we address the problem of existence of periodic solutions when an exterior force acts on the structure. All the results are based on spectral analysis and indicate a weak fluid-structure coupling at high frequencies.

2. The mathematical model

As we mentioned in the Introduction, we consider a simplified model inspired in and related to that of BANKS et al. [3]. We assume that $\Omega = (0,1) \times (0,1)$ is filled with an elastic, inviscid, compressible fluid whose velocity field $\vec{v}$ is given by the potential $\Phi = \Phi(x, y, t)$, $(\vec{v} = \nabla \Phi)$. By linearization we assume that the potential Φ satisfies the linear wave equation in $\Omega \times (0, \infty)$. The boundary $\Gamma = \partial \Omega$ of Ω is divided in two parts: $\Gamma_0 = \{(x, 0) : x \in (0, 1)\}$ and $\Gamma_1 = \Gamma \setminus \Gamma_0$. The subset Γ_1 is assumed to be rigid and we impose zero normal velocity of the fluid on it. The subset Γ_0 is supposed to be flexible and occupied by a flexible beam that vibrates under the pressure of the fluid in the plane where Ω lies. The displacement of Γ_0, described by the scalar function $W = W(x, t)$, obeys the one-dimensional dissipative beam equation. On the other hand, on Γ_0 we impose the continuity of the normal velocities of the fluid and the beam. The beam is assumed to satisfy Neumann boundary conditions on its extremes.

Under natural initial conditions for Φ and W, the linear motion of this system is described by means of the following coupled equations (by ν we denote the unit outward normal to Ω):

$$\begin{cases} \Phi_{tt} - \Delta\Phi = 0 & \text{in} \quad \Omega \times (0, \infty), \\ \frac{\partial\Phi}{\partial\nu} = 0 & \text{on} \quad \Gamma_1 \times (0, \infty), \\ \frac{\partial\Phi}{\partial\nu} = -W_t & \text{on} \quad \Gamma_0 \times (0, \infty), \\ W_{tt} + W_{xxxx} + W_t + \Phi_t = 0 & \text{on} \quad \Gamma_0 \times (0, \infty), \\ W_x(0, t) = W_x(1, t) = 0 & \text{for} \quad t > 0, \\ W_{xxx}(0, t) = W_{xxx}(1, t) = 0 & \text{for} \quad t > 0, \\ \Phi(0) = \Phi^0, \Phi_t(0) = \Phi^1 & \text{in} \quad \Omega, \\ W(0) = W^0, W_t(0) = W^1 & \text{on} \quad \Gamma_0. \end{cases} \tag{1}$$

We define the energy associated with this system by

$$E(t) = \frac{1}{2}\int_\Omega (|\nabla\Phi|^2 + (\Phi_t)^2) + \frac{1}{2}\int_{\Gamma_0} ((W_{xx})^2 + (W_t)^2). \tag{2}$$

The system has a dissipative nature. Indeed, multiplying in (1) the first equation by Φ_t, the fourth equation by W_t and integrating by parts, we get, formally, that:

$$dE(t)/dt = -\int_{\Gamma_0} (W_t)^2 \le 0. \tag{3}$$

We define the space of finite energy corresponding to (1) by

$$\mathcal{X} = H^1(\Omega) \times L^2(\Omega) \times H_N^2(\Gamma_0) \times L^2(\Gamma_0),$$

where $H_N^2(\Gamma_0) = \{v \in H^2(0, 1) : v_x(0) = v_x(1) = 0\}$.

$\mathcal{X}$ with the natural inner product is a Hilbert space.

We define in $\mathcal{X}$ the unbounded operator $(\mathcal{D}(\mathcal{A}), \mathcal{A})$ in the following way:

$$\mathcal{A}(\Phi, \Psi, W, V) = (-\Psi, -\Delta\Phi, -V, W_{xxxx} + V + \Psi),$$

$$\mathcal{D}(\mathcal{A}) = \{U = (\Phi, \Psi, W, V) \in \mathcal{X} : \mathcal{A}(U) \in \mathcal{X}\}.$$

Remark 1 *By using classical regularity results it is easy to show that* $\mathcal{D}(\mathcal{A}) \subseteq H^2(\Omega) \times H^1(\Omega) \times H^4(\Gamma_0) \times H^2(\Gamma_0)$. *Therefore* $\mathcal{D}(\mathcal{A})$ *is compact in* $\mathcal{X}$.

We can consider now the following abstract Cauchy formulation of (1):

$$\begin{cases} U_t + \mathcal{A}U = 0, \\ U(0) = U^0, \\ U(t) = (\Phi, \Phi_t, W, W_t)(t) \in \mathcal{D}(\mathcal{A}). \end{cases} \tag{4}$$

The following classical result of existence, uniqueness and stability for (4) can be obtained directly from the definition of the operator $\mathcal{A}$.

THEOREM 1 : *i) $\mathcal{A}$ is a maximal monotone operator in $\mathcal{X}$ generating a strongly continuous semigroup of contractions, $\{S(t)\}_{t\geq 0}$, in $\mathcal{X}$.*

ii) Strong solutions: If $U^0 = (\Phi^0, \Phi^1, W^0, W^1) \in \mathcal{D}(\mathcal{A})$ then (4) has a unique strong solution

$$S(t)U^0 = U \in \mathcal{C}([0, \infty), \mathcal{D}(\mathcal{A})) \cap \mathcal{C}^1([0, \infty), \mathcal{X}).$$

iii) Weak solutions: If $U^0 = (\Phi^0, \Phi^1, W^0, W^1) \in \mathcal{X}$ then (4) has a unique solution

$$S(t)U^0 = U \in \mathcal{C}([0, \infty), \mathcal{X}).$$

For any weak solution, the associated energy (2) satisfies (3).

The terminology we use is the same as in [5].

The aim of the following section is to study the effect of the damping term (concentrated in the beam equation) on the asymptotic dynamics of the whole system. We shall prove that the dissipation can force the strong stabilization but it cannot ensure a uniform decay rate.

3. Stabilization results

Concerning the asymptotic behavior of solutions, the following results hold.

THEOREM 2 : *For each initial data $U^0 = (\Phi^0, \Phi^1, W^0, W^1)$ in $\mathcal{X}$ the corresponding weak solution of (4) tends asymptotically towards the equilibrium point $(c_1, 0, c_2, 0)$, where $c_1 = \int_\Omega \Phi^1 + \int_{\Gamma_0} W^1 + \int_{\Gamma_0} \Phi^0$ and $c_2 = \int_{\Gamma_0} W^0 + \int_\Omega \Phi^1$.*

The proof is more or less classic. The main tools are an extension of the well known Invariance Principle of La Salle and Holmgren's Uniqueness Theorem (see [5] and [7]). The details are given in [10].

Remark 2 *We obtain that the pressure of the fluid, the velocities of the fluid and the beam go to zero whereas the position of the beam tends to some constant functions that is uniquely determined by the initial data.*

THEOREM 3 : *The rate of decay of the energy is not exponential in $\mathcal{X}$, i.e. there is no $C > 0$ and $\omega > 0$ such that $E(t) \leq CE(0)e^{-\omega t}$ for all weak solutions.*

The proof consists in constructing solutions with decay rates slower than any preassigned exponential function (see [12] for a detailed analysis of the spectrum of the corresponding operator).

Remark 3 *Results like this are typical for linear hybrid systems in which the dissipation is very weak: it can force the strong stabilization but it cannot assure the uniform decay.*

We remark that this result is not surprising in view of the structure of the damping region. Indeed, as Bardos, Lebeau and Rauch prove in [4], in the context of the control and stabilization of the wave equation in bounded domains, if one characteristic ray escapes to the dissipative region we can not expect a uniform decay to hold (see also Ralston [13]). In our case each segment $\{(x,a),\ x \in (0,1)\}, 0 < a < 1$ is such a ray and therefore the decay rate may not be uniform.

Nevertheless, in our problem, the lack of uniform decay is fundamentally due to the hybrid structure of the system. Indeed, the nature of the coupling between the acoustic and elastic components of the system (i.e. the boundary conditions on Γ_0)allows to build solutions with arbitrarily slow decay rate and with the energy distributed in all of the domain and not only along some particular ray of geometrical optics as in [13] (see [12]).

Let us remark that, in the classical wave equation with dissipation on the boundary,

$$\begin{cases} u_{tt} - \Delta u = 0 & \text{in } \Omega \times (0,\infty), \\ \frac{\partial u}{\partial \nu} + u_t = 0 & \text{on } \Gamma_0 \times (0,\infty), \\ u = 0 & \text{on } \Gamma_1 \times (0,\infty), \end{cases} \qquad (5)$$

the exponential decay holds if the frequency of vibration in the $x-$direction is fixed. This is not the case for system (1) where we can construct solutions with frequency of vibration in the $x-$direction fixed and with arbitrarily decay rate. Remark that the decay rate of (5) vanishes as the frequency of vibration in the $x-$direction tends to infinity.

Remark 4 We also remark that, in the case of linear semigroups, the exponential decay is equivalent to the uniform decay. Therefore, if a linear semigroup $\{S(t)\}_{t \geq 0}$ does not have exponential decay then there are initial data U^0 such that $S(t)U^0$ decays arbitrarily slowly to zero. More precisely, if $\psi : [0,\infty) \longrightarrow \infty$ is a continuous decreasing function such that $\psi(t) \to 0$ as $t \to \infty$ then there exist an initial data $U^0 \in \mathcal{X}$ and a sequence $(t_k)_{k \geq 0}$ tending to infinity such that $\|S(t_k)U^0\| > \psi(t_k)$ (see [8]).

4. Controllability results

As we mentioned before, one of the most interesting applications for this type of systems is the possibility of controlling the level of noise by acting on the flexible part of the boundary of the domain, Γ_0. The control is given by a scalar function $\beta = \beta(t,x)$ in the space $H^{-2}(0,T;L^2(\Gamma_0))$. Of course this is an arbitrary choice and many others make sense. However, this is the most natural one when solving the control problem by means of the Hilbert Uniqueness Method (HUM) introduced by J. L. Lions in [7], as we will do. The controlled system reads as follows

$$
\begin{cases}
\Phi_{tt} - \Delta\Phi = 0 & \text{in} \quad \Omega \times (0, \infty), \\
\frac{\partial\Phi}{\partial\nu} = 0 & \text{on} \quad \Gamma_1 \times (0, \infty), \\
\frac{\partial\Phi}{\partial y} = -W_t & \text{on} \quad \Gamma_0 \times (0, \infty), \\
W_{tt} + W_{xxxx} + \Phi_t = \beta & \text{on} \quad \Gamma_0 \times (0, \infty), \\
W_x(0, t) = W_x(1, t) = 0 & \text{for} \quad t > 0, \\
W_{xxx}(0, t) = W_{xxx}(1, t) = 0 & \text{for} \quad t > 0, \\
\Phi(0) = \Phi^0, \Phi_t(0) = \Phi^1 & \text{in} \quad \Omega, \\
W(0) = W^0, W_t(0) = W^1 & \text{on} \quad \Gamma_0.
\end{cases}
\tag{6}
$$

The problem of controllability can be formulated as follows: Given $T > 2$, find the space of initial data $(\Phi^0, \Phi^1, W^0, W^1)$ that can be driven to an equilibrium in time T by means of a suitable control $\beta \in H^{-2}(0, T; L^2(\Gamma_0))$.

In [3] the control acts on the system through a finite number of piezo-ceramic patches located on Γ_0. This restricts very much the set of admissible controls, that are essentially second derivatives of Heaviside functions, and much weaker controllability results have to be expected. In [3] the controllability problem is not addressed. Instead, they consider a quadratic optimal control problem. More recently in [2], a Riccati equation for the optimal control is derived. The problem of the controllability of one-dimensional beams with piezoelectric actuators has been successfully addressed in [14]. However, to our knowledge, there are no rigorous results on the controllability of fluid-structure systems under such controls.

The propagation of singularities for the wave equation on any segment parallel to Γ_0 proves that the space of controlled functions will be small. It will not contain all functions of finite energy.

To prove the controllability problem we use separation of variables and solve the corresponding one-dimensional problems. Indeed, let us decompose the control β, the solutions Φ, W and the initial data in the following way:

$$
\begin{cases}
\beta = \sum_{n=0}^{\infty} \beta_n(t) \cos(n\pi x), \\
(\Phi, W) = \sum_{n=0}^{\infty} (\psi_n(y, t), v_n(t)) \cos(n\pi x), \\
(\Phi^0, \Phi^1, W^0, W^1) = \sum_{n=0}^{\infty} (\psi_n^0(y), \psi_n^1(y), v_n^0, v_n^1) \cos(n\pi x).
\end{cases}
\tag{7}
$$

With this decomposition, system (6) can be split into the following sequence of one-dimensional controlled systems for $n = 0, 1, \ldots$:

$$
\begin{cases}
\psi_{n,tt} - \psi_{n,yy} + n^2\pi^2\psi_n = 0 & \text{for} \quad y \in (0,1), t > 0, \\
\psi_{n,y}(1,t) = 0 & \text{for} \quad t > 0, \\
\psi_{n,y}(0,t) = -v_t(t) & \text{for} \quad t > 0, \\
v_{n,tt}(t) + n^4\pi^4 v_n(t) + \psi_{n,t}(0,t) = \beta_n(t) & \text{for} \quad t > 0, \\
\psi_n(0) = \psi_n^0, \psi_{n,t}(0) = \psi_n^1 & \text{in} \quad (0,1), \\
v_n(0) = v_n^0, v_{n,t}(0) = v_n^1.
\end{cases}
\tag{8}
$$

First we will study the controllability of system (8) by using classical methods that combine HUM, multiplier techniques (see [7]) and Ingham type inequalities (see [6]). The most important result is the following observability inequality.

THEOREM 4 : *For each $T > 2$ and $n > 0$ there exists a constant $C(n,T)$ such that any solution of the observation problem*

$$
\begin{cases}
\Psi_{tt} - \Psi_{yy} + n^2\pi^2\Psi = 0 & \text{for } y \in (0,1), t > 0, \\
\frac{\partial\Psi}{\partial y}(1) = 0 & \text{for } t > 0, \\
\frac{\partial\Psi}{\partial y}(0) = V_t & \text{for } t > 0, \\
V_{tt} + n^4\pi^4 V - \Psi_t = 0 & \text{for } t > 0, \\
\Psi(0) = \Psi^0, \Psi_t(0) = \Psi^1 & \text{for } y \in (0,1), \\
V(0) = V^0, V_t(0) = V^1,
\end{cases}
\tag{9}
$$

with initial conditions in $\mathcal{Y}$, satisfies

$$
\|(\Psi^0, \Psi^1, V^0, V^1)\|_{\mathcal{Y}}^2 \leq C(n,T) \int_{-T}^{T} |V_{tt}(0,t)|^2 \, dt,
\tag{10}
$$

where $\mathcal{Y} = H^1(0,1) \times L^2(0,1) \times \mathbb{C} \times \mathbb{C}$ and $\mathcal{Y}'$ is its dual space.

From (10), by using HUM, it follows that system (8) is exactly controllable in $\mathcal{Y}$ with controls $\beta_n \in L^2(0,T)$. Next we combine the resulting controls getting in this way a control for the initial problem. Nevertheless some very restrictive conditions have to be imposed in order to ensure that the resulting control is in $H^{-2}(0,T;L^2(\Gamma_0))$. The precise result is the following.

THEOREM 5 : *All initial data from the space*

$$
H = \left\{ \sum_n (\Psi^0, \Psi^1, V^0, V^1) \cos(n\pi y) \,\middle|\, (\Psi^0, \Psi^1, V^0, V^1) \in \mathcal{Y} \text{ such that} \right.
$$

$$
\left. \sum_n C(n,T)(\|(\Psi^0, \Psi^1, V^0, V^1)\|_{\mathcal{Y}'} + |\Psi^0(0)|^2) < \infty \right\}
$$

can be driven to an equilibrium point in time T by using a control $\beta = \beta(y,t) = \sum_{n=0}^{\infty} \beta_n(t) \cos(n\pi x) \in H^{-2}(0,T;L^2(\Gamma_0))$.

Remark that the space H depends on the constants $C(n,T)$: when $C(n,T)$ "increases", H becomes smaller. Therefore it is very important to estimate the constants $C(n,T)$ and see how do they depend on T and n. The following result is proved in [1]:

THEOREM 6 : *For T and n big enough and for any positive real number q, there is a constant C_q such that*

$$C(n,T) \leq Ce^{\alpha(T)|n|} \text{ with } \alpha(T) \leq \frac{C_q}{T^{1-q}}. \tag{11}$$

The proof the Theorem 6 relies on a detailed spectral analysis of the corresponding operator and uses the moments theory to obtain the inequality (10) with an explicit constant $C(n,T)$.

Remark 5 *We obtain that any initial condition whose Fourier coefficients in y decrease like $e^{-|n|\alpha}$ can be controlled if T is greater than $T(\alpha) = \sqrt[1-q]{\frac{C_q}{\alpha}}$. This condition on the Fourier coefficients means that the initial condition is analytic with respect to y and that it can be continued as a holomorphic function over the complex strip $|\Im m\, y| < \alpha$.*

This means that any initial condition of finite energy that is analytic with respect to y can be controlled in a finite time (which is not uniform).

5. Existence of periodic solutions

Let us now consider the system

$$\begin{cases} \Phi_{tt} - \Delta\Phi = 0 & \text{in} & \Omega \times (0,\infty), \\ \frac{\partial\Phi}{\partial\nu} = 0 & \text{on} & \Gamma_1 \times (0,\infty), \\ \frac{\partial\Phi}{\partial y} = -W_t & \text{on} & \Gamma_0 \times (0,\infty), \\ W_{tt} + W_{xxxx} + W_t + \Phi_t = f & \text{on} & \Gamma_0 \times (0,\infty), \\ W_x(0,t) = W_x(1,t) = 0 & \text{for} & t > 0, \\ W_{xxx}(0,t) = W_{xxx}(1,t) = 0 & \text{for} & t > 0. \end{cases} \tag{12}$$

We study the existence of periodic solutions of system (12) under the assumption that the non homogeneous term f is periodic in time with period T, $f(t+T) = f(t)$, for all $t > 0$. We remark that this is a natural hypothesis when f models an exterior source of noise which is a frequent situation in the problems of noise reduction. Observe that

$$\frac{dE}{dt}(t) = -\int_{\Gamma_0} (W_t)^2 + \int_{\Gamma_0} f\,W_t. \tag{13}$$

The term W_t in the beam equation produces the dissipativity of the system. Since, as we saw in the third section, the dissipation is weak at high frequencies, one can expect that some additional conditions on the Fourier modes of the function f have to be imposed in order to ensure the existence of finite energy periodic solutions. Our aim is to characterize, in terms of Fourier series, the space of functions f for which (12) admits a periodic solution. This space consists of functions in $H^1(0, T; L^2(0, 1))$ whose Fourier coefficients, roughly, decay exponentially as the frequency of vibration in the $x-$direction increases. From this point of view, the result is similar to the one we have obtained in the previous section for the control problem.

As in the control problem, we use the Fourier decomposition method. Thus we develop the non-homogeneous term f in Fourier series:

$$f(t, x) = \sum_{n=0}^{\infty} f_n(t) \cos(n\pi x). \tag{14}$$

By using a perturbation argument (see [9]) we prove that each of the corresponding one-dimensional systems

$$\begin{cases} \psi_{n,tt} - \psi_{n,yy} + n^2\pi^2\psi_n = 0 & y \in (0, 1), \ t > 0, \\ \psi_{n,y}(1, t) = 0 & t > 0, \\ \psi_{n,y}(0, t) = -v_{n,t}(t) & t > 0, \\ v_{n,tt} + n^2\pi^2 v_n + v_{n,t} + \psi_{n,t}(0) = f_n(t) & t > 0, \end{cases} \tag{15}$$

has a finite energy periodic solution (ψ_n, v_n) if $f_n \in H^1(0, T)$. Notice that we need then one more derivative on f what is needed to ensure the well-posedness of the initial boundary-value problem. Combining the one-dimensional results with the Fourier decomposition (14), a periodic solution, $\sum_n (\psi_n, v_n) \cos(n\pi x)$, for (12) is obtained. Nevertheless, in order to ensure the convergence of the series which defines the periodic solution, we need to impose several conditions on the regularity of the non-homogeneous term f. The main result concerning the existence of periodic solutions of (12) is the following (see [9]).

THEOREM 7 *If $f \in H^1((0, T); L^2(0, 1))$ is a periodic function of period T such that $\int_0^T f_0(s)ds = 0$ and which satisfies:*

$$\sum_{n=1}^{\infty} n^5 \int_0^T (f_{n,t})^2 \, dt \, exp \left(\frac{3\pi}{2}n\right) < \infty, \tag{16}$$

$$\sum_{n=1}^{\infty} n^9 \int_0^T (f_n)^2 \, dt \, exp \left(\frac{3\pi}{2}n\right) < \infty, \tag{17}$$

then (12) has a periodic solution of finite energy.

Remark 6 *We have obtained that equation (12) has a periodic solution if the Fourier coefficients in the $x-$variable of the function f decay exponentially. This is due once more to the fact that there are solutions of (12) for which the energy concentrated on the beam decays exponentially as the frequency of vibration in the $x-$direction increases. In this sense this result is strongly related to the results on the controllability of the system.*

6. Conclusions

The results we have presented show that the coupling between the fluid and the structure is week at high frequencies. This phenomenon is due to the localization of the damping term in a relatively small part of the boundary, and also to the effect of the hybrid structure of the system (or, more precisely, to the boundary coupling conditions of the system).

References

1. Allibert, B. and Micu, S. (1998) Controllability of analytic functions for a wave equation coupled with a beam, to appear in *Revista Matem. Iberoamericana.*
2. Avalos, G. and Lasiecka, I. (1997) A differential Riccati equation for the active control of a problem in structural acoustics, to appear in *JOTA.*
3. Banks, H. T., Fang, W., Silcox, R. J. and Smith, R. C. (1993) Approximation Methods for Control of Acustic/Structure Models with piezo-ceramic Actuators, *Journal of Intelligent Material Systems and Structures,* **4**, pp. 98-116.
4. Bardos, C., Lebeau, G. and Rauch J. (1992) Sharp sufficient conditions for the observation, control and stabilization of waves from the boundary, *SIAM J. Control Optim.,* **30**, pp. 1024-1065.
5. Cazenave, T. and Haraux, A. (1990) *Introduction aux problèmes d'évolution semi-linéaires,* Mathématiques et Applications, 1, Ellipses, Paris.
6. Ingham, A. E. (1936) Some trigonometrical inequalities with applications to the theory of series, *Math. Z.,* **41**, pp. 367-369.
7. Lions, J. L. (1988) *Contrôlabilité exacte, perturbations et stabilization de systèmes distribués. Tome 1. Contrôlabilité exacte,* Masson RMA 8, Paris.
8. Littman, W. and Marcus, L., Some Recent Results on Control and Stabilization of Flexible Structures, *Univ. Minn., Mathematics Report 87-139.*
9. Micu, S (1998), Periodic solutions for a bidimensional hybrid system arising in the control of noise, to appear in *Adv. Diff. Eq.*
10. Micu, S. and Zuazua, E. (1994), Propriétés qualitatives d'un modèle hybride bi-dimensionnel intervenant dans le cotrôle du bruit, *C. R. Acad. Sci. Paris,* **319**, pp. 1263-1268.
11. Micu, S. and Zuazua, E. (1997) Boundary controllability of a linear hybrid system arising in the control of noise, *SIAM J. Control Optim.,* **35**, pp. 1614-1637.
12. Micu, S. and Zuazua, E. (1998) Asymptotics for the spectrum of a fluid/structure hybrid system arising in the control of noise, to appear in *SIAM J. Math. Anal.*
13. Ralston, J., (1969) Solutions of the wave equation with localized energy, *Comm. Pure Appl. Math.,* **22**, pp. 807-823.
14. Tucsnak, M. (1996) Regularity and Exact Controllability for a Beam with Piezo-electric Actuators, *SIAM J. Cont. Optim.,* **34**, pp. 922-930.

EFFECTIVENESS OF SMART STRUCTURE CONCEPTS TO IMPROVE ROTORCRAFT BEHAVIOUR

JANUSZ P. NARKIEWICZ
Warsaw University of Technology,
Politechnika, ITLiMS, ul. Nowowiejska 24, 00-665 Warsaw, Poland

1. Introduction

Since the late 1980's, a concept of smart structures is widely explored in both fixed- and rotary-wing aircraft technology. Intelligent devices are investigated for application in different parts of aircraft (such as fuselage, wings, empennage, engines, avionics) for various tasks: vibration reduction, performance improvement, noise reduction, health and usage monitoring, etc.

The specific parts of rotorcraft are main and tail rotors, and improvement of their behaviour is a key factor for enhancement the overall rotorcraft performance.

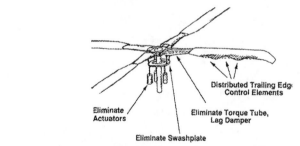

Figure 1. Concept of a smart rotor [1]

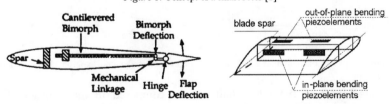

Figure 2. Tab at blade trailing edge   *Figure 3.* Concept of blade spar deflection

The long term objective of smart structure application to improve the helicopter rotor can be illustrated in a somehow futuristic manner as in Fig.1. [1]. The complicated mechanism of swash plate and hub hinges is replaced by various adaptive structure type elements to reduce total number of parts and allowing active control.

J. Holnicki-Szulc and J. Rodellar (eds.), Smart Structures, 221–228.
© *1999 Kluwer Academic Publishers. Printed in the Netherlands.*

222

From the variety of practical concepts considered and investigated in many research centres and universities, two seem to be the most promising: actively controlled devices mounted on the blade, such as a trailing edge tab, Fig.2 [2] and shape adaptive blades, Fig.3 [3].

For a tab mounted at the blade trailing edge the main difficulty in practical realisation is to design a driving mechanism, which would work in rotating and periodically varying environment at the blade as well as to choose a proper control strategy. To control the blade shape, all problems concerning active composites and aeroelastic tailoring should be considered simultaneously.

In this paper effectiveness of two control strategies in time domain for trailing edge tab is considered. The study is based on computer simulation using the general blade model developed for investigation of different concepts of rotorcraft blades.

2. General blade model

The computer model of an individual rotor blade [4] is aimed for investigation of various hub and blade designs. The most general case of the model with all hinges and blade deflections which can be included into analysis is presented in Fig.4.

A helicopter rotor in steady flight is considered. The angular velocity Ω of the rotor shaft is constant.

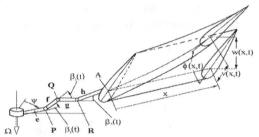

Figure 4. General model of the blade

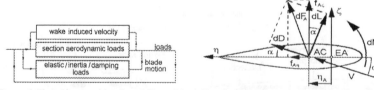

Figure 5. Flow decomposition for a 2-D model   *Figure 6.* Aerodynamic loads on a blade section

A blade attachment to the rotor shaft can be composed in many different ways, including from none to three hinges in arbitrary sequence, and up to four stiff segments described as translation vectors $\mathbf{e}$, $\mathbf{f}$, $\mathbf{g}$, $\mathbf{h}$ linking the blade via hinges with the rotor shaft. Length and orientation of the segments relative to the shaft allow for different hinge placement within the hub. At the end of each segment a flap, lag or pitch hinge can be placed which is reflected by the matrices of rotations denoted as $\mathbf{P}$, $\mathbf{Q}$ and $\mathbf{R}$.

Blade pitch control angle $\theta(t)$ is added to the pitch hinge rotation q in the form:

$$\theta(t) = \theta_0 + \theta_1 \cos(\Omega t) + \theta_2 \sin(\Omega t). \tag{1}$$

The rigid or deformable blade is attached to the hub at the point A at the end of the last hub segment. The blade is treated as a slender body (beam) having arbitrary planform along the span, and geometrically twisted about the straight axis of the last hub segment in the rigid case. If a deformable blade it is considered to twist about its straight elastic longitudinal axis and bend in two perpendicular directions. Blade elastic loads are derived assuming that there is neither cross-section distortion nor section warping. The blade deflections are considered to be small and its curvature moderate.

The blade displacement vector $\mathbf{p} = [p_i] = [p_j, \beta_i]$ $j = 1,..,N_v + N_w + N_f$, $i = 0/1/2/3$ is comprised of elastic degrees of freedom resulting from discretization of blade deflections by free vibration rotating modes, namely: in-plane bending $v(x,t)$, out-of-plane bending $w(x,t)$ and torsion $f(x,t:)$

$$v(x,t) = \sum_{i=1}^{N_v} \eta_i(x)p_i(t), \quad w(x,t) = \sum_{i=N_v+1}^{N_v+N_w} \eta_i(x)p_i(t), \quad \phi(x,t) = \sum_{i=N_v+N_w+1}^{N_v+N_w+N_f} \eta_i(x)p_i(t), \tag{2}$$

and rigid degrees of freedom corresponding to the rotations $\beta_i(t)$ at the hinges.

The equations of motion are obtained from Hamilton's principle. The blade deformations are discretized at an early stage of equation derivation, allowing to express the equations in matrix form and to transfer most of the algebraic manipulations to the computer program.

Two kinds of equations of motion are obtained:

- for elastic degrees of freedom $(j=1,...,N_g$, $N_g = N_v+N_w+N_f)$

$$\int_R \int_A \left[\frac{d}{dt}\left(\frac{\partial T}{\partial \dot{q}_j}\right) - \frac{\partial T}{\partial q_j} + \frac{\partial U}{\partial q_j} \right] \eta_j(x) dA dR +$$

$$+ \int_R \int_A \left\{ \left[\frac{d}{dt}\left(\frac{\partial T}{\partial \dot{q}_j'}\right) - \frac{\partial T}{\partial q_j'} + \frac{\partial U}{\partial q_j'} \right] \eta_j'(x) + \frac{\partial U}{\partial q_j''} \eta''(x)_j \right\} dA dR = \int_R \int_A Q_j^s \eta_j(x) dA dR, \tag{3}$$

where the integration extends over blade cross-sections A and over the blade length R,

- for rigid degrees of freedom $(j=N_g +1,...,N_g +N_p)$:

$$\frac{d}{dt}\left(\frac{\partial T}{\partial \dot{q}_j}\right) - \frac{\partial T}{\partial q_j} + \frac{\partial U}{\partial q_j} = Q_{Pj}. \tag{4}$$

The blade forms an unsteady, periodically excited aeroelastic system, where the main loads are aerodynamic, elastic and inertial. With the general forms of equations motion

(1) and (2), the expressions which model different kinds of loads can be considered separately, allowing easy modification and extension of the model.

Elastic loads come from variation of the potential energy and depend on the elastic model of the blade. Inertia loads result from variation of the kinetic energy. Nonconservative loads are calculated from the work done by aerodynamic and damping forces.

The crucial part in the rotor blade is modelling of aerodynamic loads. In this study the aerodynamic loads are calculated by the strip theory using a two-dimensional model. The flow (Fig.5) is decomposed into an internal part, where the aerodynamic loads on the aerofoil section are calculated, and an external part manifesting itself as the induced velocity generated by rotor wake. The aerodynamic loads acting in aerodynamic centre of blade cross-section profile are shown in Fig.6.

The flow velocity vector V results from the velocity of helicopter flight, blade motion relative to a helicopter fuselage including shaft rotation, angles at the hinges, blade deformation and induced velocity v_i. The component of air flow velocity relative to the helicopter can be allowed to vary with time, which enables the inclusion of gust into the analysis. The section angle of attack a is calculated using the components of the resultant velocity vector .

The aerofoil aerodynamic coefficients for drag $C_D(\alpha)$, lift $C_L(\alpha)$ and moment $C_M(\alpha)$ are obtained for a section at instantaneous angles of attack α by a table look-up procedure, which allows for nonlinear aerofoil aerodynamic characteristics.

The induced velocity v_i is calculated from the Glauert-type formula. There is only one component of the induced velocity vector, perpendicular to the plane of rotation.

Collecting together the expressions for loads calculation, the equations of motion are put in the form of the system of ordinary differential equations:

$$\mathbf{B(q)\ddot{q}} = -2\mathbf{C(q)\dot{q}} - \mathbf{g(\dot{q},q)} - \mathbf{f(q)} + \mathbf{Q_A}(t,\dot{q},q) + \mathbf{Q_D}(\dot{q},q) - \mathbf{Q_E(q)}, \qquad (5)$$

where matrices $\mathbf{B(q)}, \mathbf{C(q)}, \mathbf{g(\dot{q},q)}$ and $\mathbf{f(q)}$ result from inertia loads, $\mathbf{Q_A}(t,\dot{q},q)$ describe aerodynamic loads, $\mathbf{Q_D(\dot{q},q)}$ are damping and $\mathbf{Q_E(q)}$ are elastic loads.

The objective of the research undertaken in this part of the study is to investigate the efficiency of the controlled tab placed at the blade trailing edge. The general blade model described above was adjusted to the case considered by including unsteady aerodynamic loads for airfoils with tab (stall regime included) at the part of the blade.

3. Rotor performance improvement

The control algorithms optimising the rotor performance is investigated in this part of the study.

The goal of the control was to minimise the blade performance index defined as

$$\eta = \frac{C_{Mz} - C_{Mz0}}{C_T}, \qquad (6)$$

where C_{Mz} is rotor torque moment coefficient and C_T is the rotor thrust coefficient.

During optimisation process the thrust should be kept constant.

The algorithm constructed utilises two control variables: blade collective pitch angle θ_0 and tab deflection angle δ. The algorithm consists of two parts: *thrust stabilisation* by tab deflection and *minimisation of the blade torque moment* by using the collective blade pitch. The details are given in [5]. The sample results of calculations are given in Fig.7, where ε is recursion of the thrust value from the required value.

Calculations were done for hover and forward flight, different flap angles and different blade pitch. The tab extended from 0.69 to 0.91 of the rotor radius. The results shown in Fig.7 were obtained for constant collective pitch $\theta_0=20°$ for different advance ratios. The performance index improvement obtained varied from 0% to 5% in high speed flight.

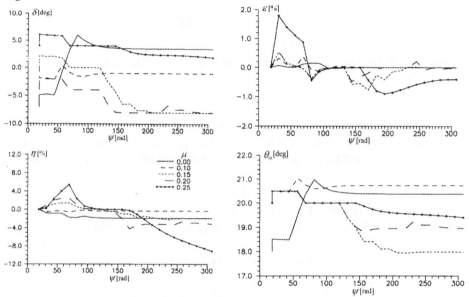

Figure 7. Results of rotor performance optimisation using two control variables.

4. Application of the learning algorithm

The objective of the second part of the study was to investigate the possibility of the application of a time domain „learning algorithm" for controlling a tab mounted at the trailing edge of a blade to obtain the assumed blade motion. The control algorithm applied during this study is a modification of that developed in [6].

The control method concerns a discrete, linear system, periodic with respect to time with scalar control $u(k)$:

$$\mathbf{x}(k+1) = \mathbf{C}(k)\mathbf{x}(k) + \mathbf{D}(k)u(k) + \mathbf{d}(k) , \tag{7}$$

where the subscript $k=1,2,..,M$ denotes time intervals and subscript i describes the period number. Matrices $\mathbf{C}(k)$ and $\mathbf{D}(k)$ are periodic with respect to time, i.e. for all k $\mathbf{C}_{i+1}(k) = \mathbf{C}_i(k)$ $\mathbf{D}_{i+1}(k) = \mathbf{D}_i(k)$, and $\mathbf{d}(k)$ is a periodic disturbance bounded all k and i.

The sequence of control applied $u_i(k)$, $i=1,2,...$ should provide that, starting from some point of time, the trajectory $\mathbf{x}_i(k)$ of the system will satisfy the condition:

$$\left\| \mathbf{x}(k) - \mathbf{x}_d(k) \right\| \le \varepsilon_0 \tag{8}$$

where the state vector $\mathbf{x}_d(k)$ is a realisable, periodic trajectory and ε_0 is the assumed tolerance bound.

It was showed in [6], that the control defined as:

$$u_{i+1}(k) = u_i(k) + \lambda\,[\hat{\mathbf{D}}_i^+(k) : -\hat{\mathbf{D}}_i^+(k)\hat{\mathbf{C}}_i(k)] \times [\mathbf{e}_i(k+1)^T : \mathbf{e}_i^T(k)]^T,$$
$$\mathbf{e}_i(k) = \mathbf{x}_d(k) - \mathbf{x}_i(k), \tag{9}$$

fulfils the learning condition if, for initial error $\mathbf{e}_i(0) = 0$, the estimate of matrix $\mathbf{D}(k)$ satisfies the condition

$$\left| 1 - \lambda\,\hat{\mathbf{D}}_i^+(k)\mathbf{D}(k) \right| < 1. \tag{10}$$

The mathematical model (5) of a helicopter rotor blade was transferred into a nonlinear system of ordinary differential equations periodic with respect to time, with scalar control $u(t)$ corresponding to the angle of deflection of the trailing edge tab:

$$\dot{\mathbf{x}} = \mathbf{f}(t, \mathbf{x}, u). \tag{11}$$

For the assumed nominal tab control $u_d(t)$, the desired periodic solution for this equation is $\mathbf{x}_d(t)$. To obtain control law according to (2), the systems linearized about $\mathbf{x}_d(t)$:

$$\dot{\mathbf{x}} = \mathbf{A}(t)\mathbf{x} + \mathbf{B}(t)u(t) + \mathbf{R}(\mathbf{x}, \mathbf{x}_d, u(t), u_d(t), t),$$

$$\mathbf{A} = [A_{ij}] = \left[\frac{\partial f_i}{\partial x_j}\right]_{\mathbf{x}_d, u_d}, \quad \mathbf{B} = [B_i] = \left[\frac{\partial f_i}{\partial u}\right]_{\mathbf{x}_d, u_d}, \tag{12}$$

and $\mathbf{R}(\mathbf{x}, \mathbf{x}_d, u(t), u_d(t), t)$ contains the higher order terms, discretized with respect to time by means of approximating the time derivative by the forward finite difference

$$\dot{\mathbf{x}} = \frac{\mathbf{x}(t + \Delta t) - \mathbf{x}(t)}{\Delta t}. \tag{13}$$

Substitution of (11) into (9) transfers the linearized equation to the discrete time domain:

$$\mathbf{x}(t + \Delta t) = [\mathbf{I} + \mathbf{A}(t)\Delta t]\mathbf{x} + \mathbf{B}(t)\Delta t u(t) + \mathbf{R}(\mathbf{x}, \mathbf{x}_d, u(t), u_d(t), t)\Delta t \tag{14}$$

and by the substitutions

$$t = k\Delta t, \ \mathbf{C}(k) = [\mathbf{I} + \mathbf{A}(k\Delta t)\Delta t], \ \mathbf{D}(k) = \mathbf{B}(k\Delta t)\Delta t,$$

$$\mathbf{d}(k) = \mathbf{R}(\mathbf{x}(k\Delta t), \mathbf{x}_d(k\Delta t), u(k\Delta t), u_d(k\Delta t), k\Delta t, \Delta t),$$

$$\text{(15)}$$

equation (12) can be reformulated to the form of (7) and control (9) may be applied. Application of the algorithm to the nonlinear case consists of:

I. Dividing the time period into M steps by prescribing the points of time $t_k = k\Delta t$, k=1,2,..,M and calculating at these points:

1. The desired solution $\mathbf{x}_d(k)$,
2. Matrices $\mathbf{A}(k)$, $\mathbf{B}(k)$ of the linearized, continuous system (12),
3. Matrices $\mathbf{C}(k)$, $\mathbf{D}(k)$ according to (14),
4. The gain matrix $\mathbf{G}(k)$ according to the formula $\mathbf{G}(k) = [\hat{\mathbf{D}}_i^+(k) : -\hat{\mathbf{D}}_i^+(k)\hat{\mathbf{C}}_i(k)]^T$.

II. Assuming the value of control gain λ, as the theory gives no indication for selecting its value.

III. Starting from $\mathbf{e}_0(k)=\mathbf{0}$, the system is controlled in each period of time according to (9).

The algorithm was applied to obtain the required motion of a helicopter rotor blade with an attempt to suppress the prescribed harmonics of steady motion.. The effectiveness of the control strategy depends both on the plant properties and the control algorithm. As one of the assumptions of the method comprises controllability of the system, the tab mounted at the blade should produce aerodynamic loads sufficient to influence the blade motion. A tab would most likely influence the blade twisting moment, so this was the reason for selecting this blade degree of freedom for investigation, and the hingeless blade stiff in bending and elastic in torsion was selected to test the control algorithm. As the first result of numerical simulation it was found that due to high fundamental torsional frequency of the blade, a tab of chord 0.1c, which can influence the blade twisting deflection should elongate from 23.3% to 95% of the blade span.

The control constant l should be adjusted by trial and errors and in the case considered, the smallest value of λ which was found to be effective was 0.05.

For the chosen tab chord and control parameter, the sample results of blade control are given in the figures for helicopter advance ratios of 0.15 and 0.35. These show the motion of the nonlinear system after 10 rotations (which is regarded as the blade steady motion), the required motion for the case considered and the controlled motion after 10 rotations of the algorithm being applied.

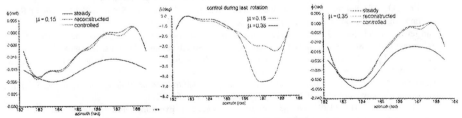

Figure 8. Required motion composed of selected harmonics

228

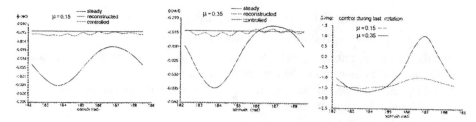

Figure 9. Required motion of constant values.

Two cases required motion of the blade are considered. In the first case shown in Fig.8, the required motion is reconstructed from the Fourier coefficients of steady motion up to the seventh order but without the third and fourth harmonics. In the second case, Fig.9 , the constant component of steady motion was forced by the control algorithm. In both cases the control algorithm proved to be effective, driving the blade twist to the vicinity of the required motion. The required tab deflections are within the acceptable limits, although the time dependence varies with the type of the motion required.

5. Conclusions

These are the first attempts of application of the control algorithms in *the time domain* to control the helicopter rotor blade using „smart devices". The first algorithm used for rotor optimisation applied two control variables, and the second – the new learning control technique. In computer simulation both methods were efficient fulfilling the required tasks.

Acknowledgement
This research was done under Polish State Committee for Scientific Research (KBN) grant No. 9T12C 00911, „Additional control of helicopter rotor blades"

References

1. Ormiston, R.A. (1991) *Can Smart Materials Make Helicopter Better?*, Fourth Workshop on Dynamics and Aeroelastic Stability Modelling of Rotorcraft Systems, The University of Maryland.
2. Ben Zeev, O. and Chopra, I. (1996) Development of an Improved Helicopter Rotor Model Employing Smart Trailing Edge for Vibration Suppression, *Smart Materials and Structures* 5,
3. Narkiewicz, J. and Done, G.T.S. (1994) *An Overview of Smart Structure Concepts for the Control of Helicopter Rotor*, Second European Conference on Smart Structures and Materials, Glasgow.
4. Narkiewicz, J., (1994) *Rotorcraft Aeromechanical and Aeroelastic Stability*, Scientific Reports of Warsaw University of Technology, Mechanics, No. 158, (PL).
5. Narkiewicz, J. (1998) *Application of Smart Structure Concept to Rotorcraft Additional Control*, Scientific Reports of Warsaw University of Technology, Mechanics, No. 169 (PL).
6. Park, H.J. and Cho, H.S. (1992) On the realisation of an accurate hydraulic servo through an iterative learning control, *Mechatronics* 2, 75-88.

CONTROL SYSTEM FOR HIGHWAY LOAD EFFECTS

ANDRZEJ S. NOWAK AND JUNSIK EOM
Department of Civil and Environmental Engineering
University of Michigan, Ann Arbor, MI 48109-2125, USA

1. Introduction

The objective of this paper is to investigate the feasibility of the development of an efficient system to control the highway load effects and bridge deterioration. Such an integrated system may involve the monitoring and control of various parameters including:

- vehicle weight, axle loads, axle configuration, speed, multiple presence
- traffic volume (average daily truck traffic, ADTT)
- vehicle position within the width of the bridge
- load effects (moment, shear force, stress, strain and deflection) in bridge components
- identification of critically overloaded vehicles
- minimum load carrying capacity of the bridge.

Therefore, an important part of the project is to develop a procedure for evaluation of live load spectra on bridges. Truck weights, including gross vehicle weight (GVW), axle loads and spacing, are measured to determine the statistical parameters of the actual live load. Stress is measured in various components of girder bridges to determine component-specific load spectra. Minimum load carrying capacity is verified by proof load tests. The research involved analytical development of the models for live load and field verification through weigh-in-motion (WIM) measurements, truck counts, truck surveys, proof load tests and diagnostic tests.

2. Weigh-in-Motion Measurements

The primary bridge live load is truck traffic. In the past, truck load data were collected by surveys. The most common survey method consisted of weighing trucks with static scales, present at weigh stations along fixed locations on major highways. The usefulness of this data is limited, however, because many drivers of overloaded trucks intentionally avoid the scales. This results in a load bias toward lighter trucks.

A weight-in-motion test attempts to gather unbiased truck traffic data, which includes axle weight, axle spacing, vehicle speed, multiple truck presence on the

J. Holnicki-Szulc and J. Rodellar (eds.), Smart Structures, 229–236.
© *1999 Kluwer Academic Publishers. Printed in the Netherlands.*

bridge, and average daily truck traffic (ADTT). Sensors measure strains in girders, and these data are then used to calculate the truck parameters at the given traffic speed. Beneath the deck, the WIM system is invisible to the truck drivers, and so overloaded trucks do not avoid the bridge. Unbiased results can thus be obtained. The system is portable and easily installed to obtain site-specific truck data.

The WIM results can be presented in a traditional histogram (frequency or cumulative). However, this approach does not allow for an efficient analysis of the extreme values (upper or lower tails) of the considered distribution. Therefore, results of gross vehicle weight (GVW) WIM measurements can also be shown as cumulative distribution functions (CDFs) on the normal probability paper as shown in Fig. 1. The measurements are shown for selected 7 bridges. The horizontal axis is the considered truck parameter (e.g. gross vehicle weight, axle weight, lane moment or shear force). The vertical axis represents the probability of a particular point being exceeded, p. The probability of being exceeded (vertical scale) is then replaced with the inverse standard normal distribution function.

The distribution of truck type by number of axles will typically bear a direct relationship to the GVW distribution; the larger the population of multiple axle vehicles (greater than five axles) the greater the GVW load spectra. Past research has indicated that 92% to 98% of the trucks are four and five axle vehicles. Three and four axle vehicles are often configured similarly to five axle vehicles, and when included with five axle vehicles, this group accounts for 55% to 95% of the truck population. in Michigan, between 0% and 7.4% of the trucks have eleven axles. Most states in the US allow a maximum GVW of 355 kN, where up to five axles per vehicle are permitted. The State of Michigan legal limit allows for an eleven axle truck of up to 730 kN, depending on axle configuration.

Potentially more important for bridge fatigue and pavement design are the axle weights and axle spacings of the trucks passing over the bridge. Fig. 2 presents the distributions of the axle weights of the measured vehicles.

Once truck data has been collected by a WIM test, the results can be used to generate the expected moments and shears for any bridge span. Each truck in the data base is analytically driven across the desired span (using influence lines) to determine the maximum static bending moment and shear per lane. The cumulative distribution functions of these load effects for the same span are then determined. As an example, the resulting CDF's for a specific span (27m) are shown in Fig. 3 for lane moment and Fig. 4 for lane shear. As a point of reference, the calculated load effects are divided by the values resulting from using the AASHTO LRFD (1994) design loads (the standard 320kN design truck and a 9.3kN/m uniform lane load). The results indicate that there is a wide variation of truck load distribution among the example bridges investigated. Maximum values of lane-moment-to-LRFD-moment ratio vary from 0.6 at M153/M39 to 2.0 at I94/M10. The variation of lane shears in Fig. 4 is similar to that of lane moments. For I94/M10, the extreme value exceeds 2.0. For the other bridges, the maximum shears can be seen to vary from 0.65 at M153/M39 to 1.5 at I94/I75.

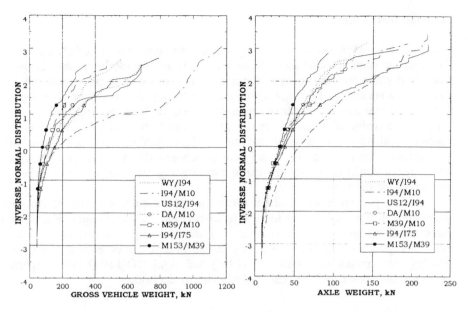

Figure 1. Examples of CDF of Gross Vehicle Weight

Figure 2. Examples of CDF of Axle Weigh

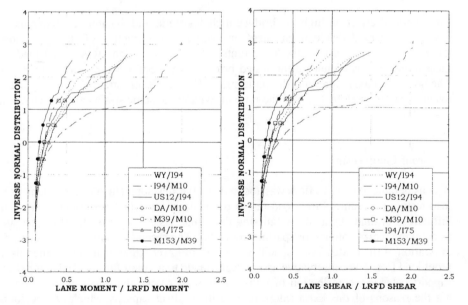

Figure 3. CDFs of Lane Moment for the Span Length of 27 m

Figure 4. CDFs of Lane Shear for the Span Length of 27m

3. Load Distribution

A considerable effort was also directed to establish component-specific load spectra. The measurements indicate that fatigue loads vary from girder to girder. Therefore, truck load effect control should be focused on the most loaded components. A field procedure was developed for measurement of load effect in components. The objective is to identify the critical members which determine the load carrying capacity of the bridge. Knowledge of a realistic girder distribution factor (GDF) is essential for a rational evaluation of existing structures. Test results confirmed some prior observations that live load is strongly component-specific. Code-specified values of GDF are conservative for longer spans and larger girder spacings, but rather too permissive for short spans.

The portion of live load per girder can be measured in terms of the girder distribution factor GDF. Field tests were carried out on five bridges. Examples of measured GDFs for the selected structures are shown in Fig. 5 for two trucks side-by-side. The vehicles used in tests are fully loaded 11 axle trucks, each weighing about 650 kN. For comparison, AASHTO (1996) and AASHTO LRFD (1994) specified GDFs are also shown for a one lane and two lane girder bridges.

4. Dynamic Loads

The live load effect on highway bridges includes static and dynamic components. It has been observed that dynamic load, measured as the fraction of static live load, decreases for increasing truck weights. This observation was confirmed by measurements carried out on the tested bridges. The results are presented in Fig. 6. The horizontal axis represents the static strain. For each static strain the corresponding dynamic strain is plotted so that the ratio of dynamic to static strains are shown on the vertical axis.

5. Proof Load Tests

A considerable research effort was focused on the development of an efficient proof load testing procedure. Proof load test is particularly important for bridges which are difficult to evaluate by analysis (missing drawings, visible signs of deterioration such as cracking, corrosion and/or spalling concrete). The second reason is, that theoretical evaluation of the bridge load capacity can result in lower value, than determined using proof load test. The conservative assumptions made to account for materials properties, support behavior, contribution of non-structural members and effect of deterioration are the reasons of this extra safety reserve in the load capacity which can be found performing proof load test. In certain cases, this extra safety reserve in the load capacity can be utilized to prove that the bridge is adequate, thus avoiding replacement or rehabilitation. Proof load tests were carried out on several bridges. It was difficult to provide a load which is considerably higher than legal Michigan truck load. Therefore, military tanks were used from the National Guard. An M-60 tank weighs about 60 Tons and the length of tracks is about 4.2m. In general, analytically predicted deflections were larger than the actual deflections. The tested bridges clearly showed that there is a considerable safety reserve in the load carrying capacity. A typical linear relationship between strain vs. applied moment, and strain vs. deflection, is shown in Fig. 7.

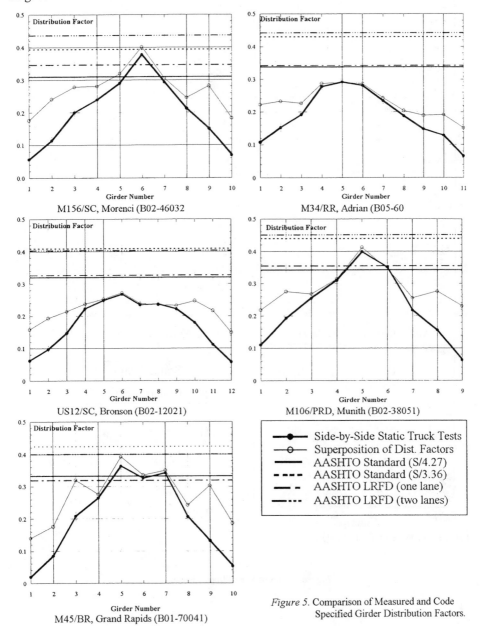

M156/SC, Morenci (B02-46032

M34/RR, Adrian (B05-60

US12/SC, Bronson (B02-12021)

M106/PRD, Munith (B02-38051)

M45/BR, Grand Rapids (B01-70041)

Side-by-Side Static Truck Tests
Superposition of Dist. Factors
AASHTO Standard (S/4.27)
AASHTO Standard (S/3.36)
AASHTO LRFD (one lane)
AASHTO LRFD (two lanes)

Figure 5. Comparison of Measured and Code Specified Girder Distribution Factors.

234

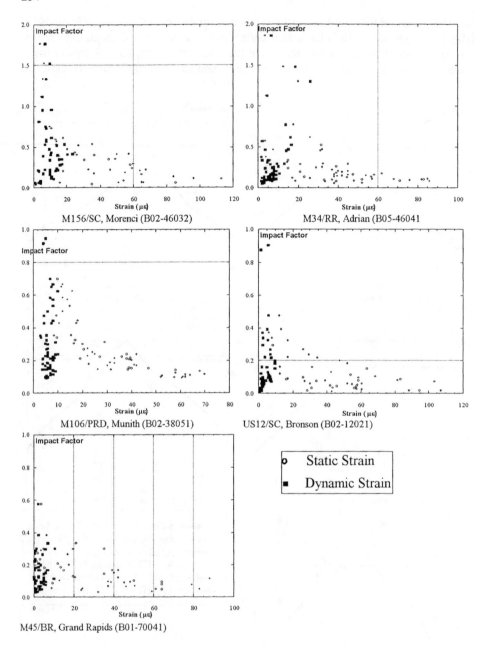

Figure 6. Strains Versus Impact Factors.

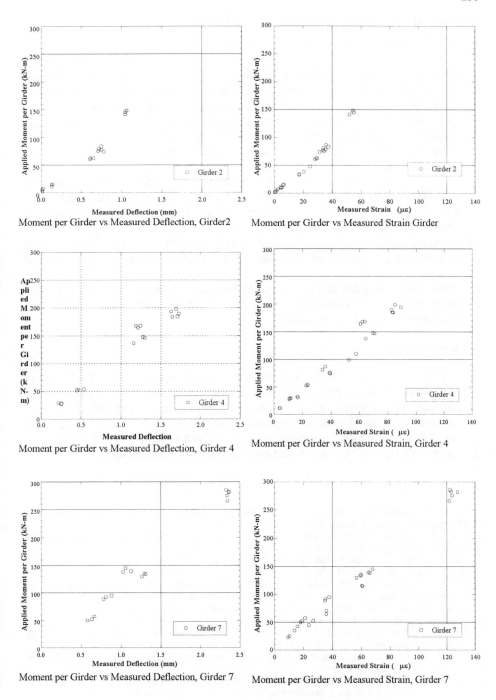

Figure 7 Moment per Girder vs. Measured Deflection and Strains for Selected Girders.

6. Conclusions

The most important conclusions are:

(a) The observed response is linear, which is confirmed by superposition of truck effects. The comparison of strain values for a single truck indicates that for two trucks side-by-side, the results are equal to superposition of single truck results.

(b) The absolute value of measured strains is lower than expected. For a single truck, the maximum observed strain is less than 80 microstrains. This corresponds to 16 Mpa. For two trucks side-by-side, the maximum strain is 110 microstrains, which corresponds to 23 Mpa.

(c) Girder distribution factors observed in the tests are close to the values specified by AASHTO (1996). The maximum measured values for two trucks side-by-side were close to the specified values in three bridges. However, the absolute values of stresses were rather low (less than 23 MPa).

(d) Dynamic load is lower than specified value by AASHTO (1996). For two trucks side-by-side it is about 0.10. Dynamic load decreases with increasing static load effect.

(e) Proof load tests confirmed that the tested bridges are adequate to carry the normal truck traffic. The measured deflections and strains were relatively low, and considerably lower than expected.

Acknowledgments

The research presented in this paper has been partially sponsored by the Michigan Department of Transportation, the Great Lakes Center for Truck Transportation Research at the University of Michigan, National Science Foundation, National Cooperative Highway Research Program SHRP/IDEA, and NATO Cooperative Research Program which is gratefully acknowledged. The presented results and conclusions are those of the authors and not necessarily those of the sponsors.

References

1. AASHTO, "Standard Specifications for Highway Bridges", American Association of State Highway and Transportation Officials, Washington, DC, 1996.
2. Kim, S-J. and Nowak, A.S., "Load Distribution and Impact Factors for I-Girder Bridges", *ASCE Journal of Bridge Engineering*, Vol. 2, No. 3, August 1997, pp. 97-104.
3. Kim, S-J., Sokolik, A.F., and Nowak, A.S., 1997, "Measurement of Truck Load on Bridges in the Detroit Area", *Transportation Research Record*, No. 1541, pp. 58-63.
4. Laman, J.A., and Nowak, A.S., "Site-Specific Truck Loads on Bridges and Roads", Proceedings of the Institution of Civ. Engrs, *Transport Journal*, No. 123, May 1997, pp. 119-133.
5. Laman, J.A., and Nowak, A.S., 1996, "Fatigue Load Models for Girder Bridges", *ASCE Journal of Structural Engineering*, Vol. 122, No. 7, pp. 726-733.
6. Nowak, A.S., Kim, S. and Saraf, V.K., "Evaluation of Existing Bridges using Field Tests" (in Polish), *Inzynieria i Budownictwo*, No. 10, 1997, pp. 499-502.
7. Saraf, V., Sokolik, A.F., and Nowak, A.S., 1997, "Proof Load Testing of Highway Bridges", *Transportation Research Record*, No.1541, pp. 51-57.
8. Saraf, V. and Nowak A.S., "Proof Load Testing of Deteriorated Steel Girder Bridges", *ASCE Journal of Bridge Engineering*, Vol. 3, No. 2, May 1998, pp. 82-89.
9. Saraf, V. and Nowak A.S., "Field Evaluation of a Steel Girder Bridge", *Transportation Research Record*, No. 1594, 1997, pp. 140-146.

OPTIMAL DELAYED FEEDBACK VIBRATION ABSORBER FOR FLEXIBLE BEAMS

N. OLGAC AND N. JALILI

Department of Mechanical Engineering, The University of Connecticut, Storrs, Connecticut 06269-3139 U. S. A.

Abstract

A new tunable vibration absorber, Delayed Resonator (DR), is considered on a flexible beam. The point of absorber attachment (PAA) on the beam is shown to be quieted (i.e., an artificial node is introduced at this point). This is achieved if the excitation forcing is simple harmonic. The DR absorber is tuned to this frequency and this tuning can track a relatively fast varying excitation frequency. There is, however, a short-fall of DR absorber: as they eliminate the vibration completely at the tuning frequency they introduce some vulnerable side frequencies. This scenario suggests a trade-off: less perfection in suppressing the tuning frequency, with reduced vulnerability at the side frequencies. In order to resolve this trade-off an optimization procedure is suggested which forms the main contribution of this text. The dynamics of the beam, exciter, and absorber setting is studied in modal coordinates. This constrained optimization problem is solved utilizing direct update method, one of the constraints being the **stability assurance** of the proposed active system. Example simulations are presented to show the benefits of this trade-off.

1. Introduction

The DR vibration absorber is a delayed feedback control strategy which offers some attractive features in eliminating tonal vibrations from the objects to which it is attached [1, 2, and 3]. Time delay is implemented on a partial state feedback, which can be position, velocity, or acceleration. This feedback structure lends itself very favorably to "pole placement" of the absorber, otherwise known as "sensitization" of the absorber. The objective is to create a resonator out of the absorber section which becomes "absorbent" at its most sensitive frequency. This resonance frequency of the so-called "Delayed Resonator" (DR) is under the control of the user and can be tuned in real-time. Therefore, DR can suppress oscillations of varying frequencies. Some important issues such as the tuning speed and, the system stability are discussed in another publication [4]. In this study, the primary focus is on the optimum design and operation of the DR absorber.

J. Holnicki-Szulc and J. Rodellar (eds.), Smart Structures, 237–246.

The trade-off which requires an optimization arises as follows: a perfectly tuned DR can absorb the vibratory energy at its resonant frequency, in other words it acts like a "notch filter" at that frequency. That is, the system exhibits zero frequency response at ω (the excitation frequency). In the same time two side frequencies appear at which peak responses are observed, one on each side of the tuning frequency. These peak responses are not desirable and should be minimized. The fear is the possibility of contamination of single harmonic excitation with some side frequency components. One way to alleviate this is by adjusting the control. The gain and the delay of the feedback can be selected such that the peak of the frequency response is minimized while the suppression quality at the tuning frequency is (slightly) sacrificed. This optimization problem forms the main contribution of this paper.

The dynamic model of the beam and the discrete attachments (i.e., the exciter and absorber) is presented in Ritz-Galerkin modal format. For this modal analysis realistic boundary conditions (BC's) for a fixed-fixed beam are considered. Some small deflections and slopes are assumed at the end points instead of zero. Consequently some unconventional BC's are obtained. This aspect is also treated experimentally on a realistic structure. An optimal control on the absorber is suggested using these BC's. The benefit/loss outlook is demonstrated numerically.

The composition of the paper is as follows. Section 2 contains an overview of the DR absorber using acceleration feedback. Section 3 is on the modeling of the beam with exciter and absorber. Section 4 explains the identification of unconventional BC's. Section 5 addresses the optimization problem and Section 6 presents analytical findings and numerical results which show the comparison of the performance for passive, DR, and optimal absorber. Concluding remarks are in Section 7.

2. Delayed Resonator (DR), An Overview

The core idea of the DR absorption technique was recently introduced in Olgac and Holm-Hansen [1]. DR is an actively tuned vibration absorber which uses proportional feedback with a time delay, as shown in Figure 1b (acceleration feedback is considered here). This feedback sensitizes the dissipative structure of Figure 1a and converts it into a conservative (or marginally stable) one (Figure 1b) with a designated resonance frequency ω_c. The corresponding system dynamics is

$$m_a\ddot{x}_a(t)+c_a\dot{x}_a(t)+k_ax_a(t)-g\ddot{x}_a(t-\tau)=0 \tag{1}$$

The characteristic equation of this system is transcendental and it is given as

$$m_as^2+c_as+k_a-gs^2e^{-\tau s}=0 \tag{2}$$

It is shown that selecting the control parameters as

$$g_c=\frac{1}{\omega_c^2}\sqrt{\left(c_a\omega_c\right)^2+\left(m_a\omega_c^2-k_a\right)^2} \tag{3}$$

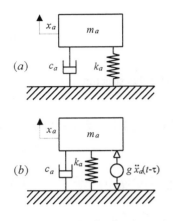

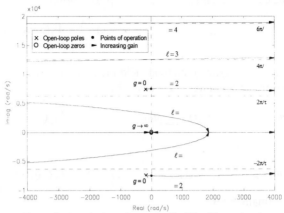

Figure 1. (a) Passive absorber, and (b) DR with acceleration feedback.

Figure 2. A typical root locus plot for DR with acceleration feedback (τ is fixed).

$$\tau_c = \frac{1}{\omega_c}\left\{\tan^{-1}\left[\frac{c_a\omega_c}{m_a\omega_c^2 - k_a}\right] + 2(\ell-1)\pi\right\} , \quad \ell = 1,2,\ldots \tag{4}$$

two of the characteristic roots of (2) are placed at $\mp\omega_c j$, where $j = \sqrt{-1}$. There are numerous stability issues to be resolved in pursuit to this proposition. We will suppress them here and refer the reader to earlier publications [2, 5].

The variable parameter ℓ in equation (4) refers to the branch of root loci that happens to cross the imaginary axis at ω_c (Figure 2). Each branch carries one root. Notice that, τ_c is a multi valued function of ω_c while g_c is single valued (equations 3 and 4). That means in order to sensitize the absorber at ω_c a particular gain, g, must be used, the delay τ, however ,can be selected among many. The preference in this selection is generally on the lower branch numbers (i.e., small ℓ) which yield smaller delay, τ, and less congestion in the branch distributions (see the location of the asymptotes in Figure 2).

A point to note is that the control logic does not require any information from the primary structure, thus it is decoupled from it. It is clear, however, that the time delay is a destabilizing factor for the combined system. This concern is kept in-view throughout the work presented here.

3. DR Vibration Absorber on Flexible Beams

Consider a general beam as the primary system with DR, subjected to a harmonic force excitation, as shown in Figure 3. The point excitation is at b, and the absorber is placed at a. Uniform cross section is considered for the beam and Euler-Bernoulli assumptions are made. The beam properties are all assumed to be constant and uniform. We use $y(x,t)$ to characterize the elastic deformation from the undeformed natural axis of the beam.

The conventional Ritz-Galerkin approximation

$$y(x,t)=\sum_{i=1}^{n} \Phi_i(x)q_{bi}(t) \tag{5}$$

and Euler-Lagrange method yield the following dynamics for this structure [4].
Absorber dynamics is governed by

$$m_a\,\ddot{q}_a(t)+c_a\Big\{\,\dot{q}_a(t)-\sum_{i=1}^{n}\Phi_i(a)\dot{q}_{bi}(t)\,\Big\}+k_a\Big\{\,q_a(t)-\sum_{i=1}^{n}\Phi_i(a)q_{bi}(t)\,\Big\}-g\,\ddot{q}_a(t-\tau)=0 \tag{6}$$

and **exciter**

$$m_e\,\ddot{q}_e(t)+c_e\Big\{\,\dot{q}_e(t)-\sum_{i=1}^{n}\Phi_i(b)\dot{q}_{bi}(t)\,\Big\}+k_e\Big\{\,q_e(t)-\sum_{i=1}^{n}\Phi_i(b)q_{bi}(t)\,\Big\}=-f(t) \tag{7}$$

and finally **beam**

$$N_i\ddot{q}_{bi}(t)+S_iq_{bi}(t)+c_a\Big\{\sum_{i=1}^{n}\Phi_i(a)\dot{q}_{bi}(t)-\dot{q}_a(t)\Big\}\Phi_i(a)+c_e\Big\{\sum_{i=1}^{n}\Phi_i(b)\dot{q}_{bi}(t)-$$

$$\dot{q}_e(t)\Big\}\Phi_i(b)+k_a\Big\{\sum_{i=1}^{n}\Phi_i(a)q_{bi}(t)-q_a(t)\Big\}\Phi_i(a)+k_e\Big\{\sum_{i=1}^{n}\Phi_i(b)q_{bi}(t)- \tag{8}$$

$$q_e(t)\Big\}\Phi_i(b)+g\,\Phi_i(a)\ddot{q}_a(t-\tau)=f(t)\Phi_i(b) \qquad\qquad i=1,2,...,n$$

where $\Phi_i(x)$ and q_{bi} are the respective mode shape and time dependent generalized coordinate for the i-th mode of the beam.

It is clear that when the absorber subsection is tuned to resonate at ω_j, its point of attachment, a, will have no residual response at this frequency, i.e., steady state $y(a,t)=0$.

The stability of this controlled structure, however, is a crucial issue to resolve. Relevant detailed treatment can be found in [4, 5]. It is shown that, the characteristics equation of the combined system is :

$$\det\{A=\begin{pmatrix} m_a s^2+c_a s+k_a-gs^2 e^{-\tau s} & 0 & -\Phi_1(a)(c_a s+k_a) & \cdots & -\Phi_n(a)(c_a s+k_a) \\ 0 & m_e s^2+c_e s+k_e & -\Phi_1(b)(c_e s+k_e) & \cdots & -\Phi_n(b)(c_e s+k_e) \\ m_a\Phi_1(a)s^2 & m_e\Phi_1(b)s^2 & N_1 s^2+cs+S_1(1+j\delta) & \cdots & 0 \\ \vdots & \vdots & \vdots & \ddots & \vdots \\ m_a\Phi_n(a)s^2 & m_e\Phi_n(b)s^2 & 0 & \cdots & N_n s^2+cs+S_n(1+j\delta) \end{pmatrix}\}=0 \tag{9}$$

The stability analysis is built on the premise that, the characteristic roots of this equation all have negative real parts.

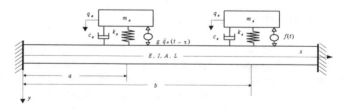

Figure 3. Beam-absorber-exciter system configuration.

4. Unconventional Boundary Conditions

The beam dynamics alone is considered as

$$\rho\frac{\partial^2 y}{\partial t^2} + EI\frac{\partial^4 y}{\partial x^4} = 0 \tag{10}$$

Substituting the Galerkin approximation of equation (5), into (10), yields an expression for the eigenfunction $\Phi_i(x)$ as

$$\Phi_i(x) = \alpha_1\sin\kappa_i x + \alpha_2\cos\kappa_i x + \alpha_3\sinh\kappa_i x + \alpha_4\cosh\kappa_i x \quad i=1,2,...,n \tag{11}$$

where κ_i is the modified spatial frequency which is defined as:

$$\kappa_i^2 = \omega_i\sqrt{\frac{\rho}{EI}} \qquad i=1,2,...,n \tag{12}$$

For non-ideal clamping conditions, Figure 4, k_{TR} and $k_{\theta R}$, k_{TL} and $k_{\theta L}$ are taken as the transverse stiffness, and the torsional stiffness constants, at the left and right end of the beam, respectively. When these spring constants are selected as combinations of infinity or zero, a variety of ideal BC's result, for instance, fixed-fixed, fixed-free, free-free, and etc. For symbolic simplification, we define the following non-dimensional spring constants

$$\beta_1 = \frac{k_{\theta L}L}{EI}, \quad \beta_2 = \frac{k_{TL}L^3}{EI}, \quad \beta_3 = \frac{k_{\theta R}L}{EI}, \quad \beta_4 = \frac{k_{TR}L^3}{EI} \tag{13}$$

The associated unconventional BC's for the governing differential equation of the system (10), are stated in [6] as

$$\text{at } x = 0 \begin{cases} \Phi''(x) - \dfrac{\beta_1}{L}\Phi'(x) = 0 \\ \Phi'''(x) + \dfrac{\beta_2}{L^3}\Phi(x) = 0 \end{cases} \text{and at } x = L \begin{cases} \Phi''(x) + \dfrac{\beta_3}{L}\Phi'(x) = 0 \\ \Phi'''(x) - \dfrac{\beta_4}{L^3}\Phi(x) = 0 \end{cases} \tag{14}$$

Inserting the eigenfunctions (11) into the BC's (14), we obtain a set of four homogenous equations in terms of C_1, C_2, C_3, and C_4, given by

$$(\mathbf{u} + \mathbf{bv})\{\mathbf{C}\} = 0 \tag{15}$$
$$(\mathbf{u} + \mathbf{bv})\{\mathbf{C}\} = 0$$

where

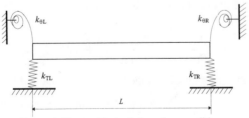

Figure 4. Beam with elastic boundary conditions.

$$\mathbf{u} = \begin{pmatrix} 0 & 1 & 0 & 0 \\ 0 & 0 & 0 & -1 \\ -C+Ch & -C-Ch & -S+Sh & -S-Sh \\ S+Sh & S-Sh & -C+Ch & -C-Ch \end{pmatrix}, \quad \mathbf{v} = \begin{pmatrix} 0 & 0 & \dfrac{1}{\kappa L} & 0 \\[2mm] \dfrac{1}{(\kappa L)^3} & 0 & 0 & 0 \\[2mm] \dfrac{(S-Sh)}{\kappa L} & -\dfrac{(C+Ch)}{\kappa L} & -\dfrac{(C+Ch)}{\kappa L} & -\dfrac{(C-Ch)}{\kappa L} \\[2mm] -\dfrac{(C+Ch)}{(\kappa L)^3} & -\dfrac{(C-Ch)}{(\kappa L)^3} & -\dfrac{(S+Sh)}{(\kappa L)^3} & \dfrac{(S+Sh)}{(\kappa L)^3} \end{pmatrix}$$

$$\mathbf{b} = \mathrm{diag}\,[\beta_1, \beta_2, \beta_3, \beta_4], \quad \mathbf{C} = \{C_1, C_2, C_3, C_4\}^T$$

κL is the non-dimensional natural frequency and $C = \cos(\kappa L)$, $S = \sin(\kappa L)$, $Ch = \cosh(\kappa L)$, $Sh = \sinh(\kappa L)$. The determinant of the coefficient matrix in equation (15) is set to zero to obtain the characteristic equation, from which κL can be determined. Consequently, the eigenfunctions necessary for the Galerkin representation of equation (5) appear. The BC's are needed (i.e., β_i, $i = 1,...,4$) on case specific basis. An experimental method for this is presented below:

The first four natural frequencies of the beam, ω_i, are experimentally determined. The determinant of the coefficient matrix (15) is set to zero for each one of these four ω_i's. Ensuing four nonlinear simultaneous equations are solved for β_i, $i = 1,...,4$.

These nonlinear simultaneous equations are

$$\det(\mathbf{u} + \{\mathbf{b}\}\mathbf{v}) = 0 \quad i = 1,...,4 \tag{16}$$

The desired solution should be all positive, $\beta_i > 0$, $i = 1,...,4$. This problem, although possible to solve by some powerful packages (such as MAPLE, [7]) needs a starting point which should be very close to the solution. Therefore, we adopt a simplifying assumption here by setting

$$\beta_1 = \beta_3, \quad \beta_2 = \beta_4$$

which suggests symmetricity of the BC's and reduces the unknowns to two. This assumption is utilized in the simulations of Section 6.

5. Optimization on DR Parameters

The objective is to minimize the peaks of the frequency response at the PAA on the beam. The displacement $q_{bi}(t)$, $i=1,2,...,n$ can be evaluated from equations (6)-(8) using Laplace transformation and Cramer's rule. A general non-dimensional transfer function is defined between the excitation force and the response at PAA as:

$$TR(s) = \frac{Y(a,s)}{F(s)/k_e} = \frac{1}{\det(\mathbf{A})} \sum_{i=1}^{n} (-1)^{i+1}\, \Phi_i(a)\, M_{2,i+2} \tag{17}$$

where $F(s)/k_e$ represents a quasi-displacement of the beam at the exciter location, and M_{ij}, the minor of a_{ij} ($\mathbf{A} = \{a_{ij}\}$), is the determinant of $\mathbf{A}$ without row and column containing a_{ij} (i.e. $(n+1) \times (n+1)$ matrix). For the scope of this paper, it suffices to know that the expression $TR(s)$ is fully defined dimensionless transfer function between F and Y.

With a slight diversion we refer, to the results which are obtained in Section 6, qualitatively only. The frequency response of (17) is given in Figure 7, for variety of cases. First, with passive absorber ($g=0$) there is a small peak shown at point **A**. Second, when the DR is tuned to a certain frequency the total suppression occurs (Point **B**). Third is the optimum DFVA (Delayed Feedback Vibration Absorber). Notice that the name does not have DR any more.

For the optimization the following objective function is selected

$$\underset{g,\tau}{Min} \left\{ \underset{\omega_{min} \le \omega \le \omega_{max}}{Max} [\| TF(g,\tau,\omega j) \|] \right\} \qquad (18)$$

It is necessary that, g and τ are selected such that the combined system (i.e., beam, exciter, and DR) is stable. The detailed discussions on the stability zone of (g,τ) are presented in Olgac and Jalili [4]. An example topography is given in Figure 6 which defines a **constraint** as,

$$(g,\tau) \in (g,\tau) \big|_{\text{stable zone}} \qquad (19)$$

This numerical constraint is well defined for a given set of (m_a, k_a, c_a) parameters.

The optimization scheme used to minimize (18) subject to constraint (19) is a **direct update method**: BFGS (Broyden-Fletcher-Goldfarb-Shanno) method [8, 9]. The following section presents a case study on this implementation.

6. Numerical Results

For simulation purposes, the primary structure, shown in Figure 5, is taken from Olgac *et al.* [5]as the test bed. It is a 3/8"x1"x12" steel beam (2) clamped at both ends to a granite bed (1). A piezoelectric actuator with reaction mass (3 and 4) are used to generate the periodic disturbance on the beam. A similar actuator-mass setup constitutes the DR absorber (5 and 6). They are located symmetrically at one quarter of the length along the beam from the center. The feedback signal used to implement the DR is obtained from the accelerometer (7) mounted on the reaction mass of the absorber (6). The other accelerometer (8) attached to the beam is only to monitor the performance of DR absorber in suppressing the beam oscillations.

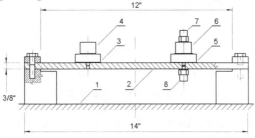

Figure 5. Experimental structure, steel E=210 Gpa.

244

The numerical values for this beam-absorber-exciter setup are given below, including those mass-stiffness-damping properties contributed by the piezoelectric actuator:

Absorber: $m_a = 0.183$ kg, $k_a = 10130$ kN/m, $c_a = 62.25$ N.s/m, $a = L/4$,

Beam: $E = 210$ Gpa, $\rho = 1.8895$ kg/m;

Exciter: $m_e = 0.173$ kg, $k_e = 6426$ kN/m, $c_e = 3.2$ N.s/m $b = 3L/4$

Utilizing the scheme described in Section 4 the BC's are determined for this set-up as :

$$\beta_1 = \beta_3 = 22.3565, \quad \beta_2 = \beta_4 = 4530.1978 \tag{20}$$

These values are reached based on the first two natural frequencies of the beam measured as

$$\omega_1 = 466.4 \text{ Hz}, \quad \omega_2 = 1269.2 \text{ Hz} \tag{21}$$

Only 3 modes are considered for this example. This is selected because higher order modes are numerically found non-contributory.

Figure 6 depicts the stable operating region of (g, τ) plane for this set-up. The light line is the DR operating curve. For instance, the tuned DR for absorption at $\omega = 1270$ Hz is found at

$$g_1 = 0.0252 \text{ Kg}, \quad \tau_1 = 0.8269 \text{ msec} \tag{22}$$

When the optimization steps are taken a global optimum Delayed Feedback Vibration Absorber (OP-DFVA) is found at

$$g^* = 0.0598 \text{ Kg}, \quad \tau^* = 1.4000 \text{ msec} \tag{23}$$

for which the search bounds of $\omega_{min} = 200$Hz and $\omega_{max} = 2200$Hz were used.

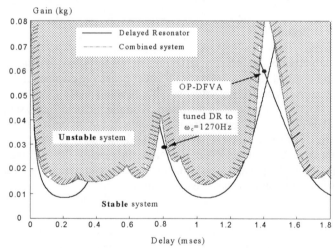

Figure 6. Stability Chart for the combined system and operating points.

The corresponding frequency response curves are given in Figure 7. It is clear that, at 1270 Hz the DR absorber performs the best, but to the expense of a peak response elsewhere (about 637Hz at the level of 23). The OP-DFVA on the other hand yields a peak at about 652Hz at the level of 12: It is therefore fair to judge that, for this example the OP-DFVA reduces the max frequency response by approximately 50%. This is substantial for most practical applications. The stability envelope is depicted in Figure 6, where the tuned DR and OP-DFVA are marked. It is clear that both operating points fall in the feasible regions.

Figure 8 displays the tuned DR absorption at 1270 Hz. The passive absorber is functioning in the first 5 msec, immediately after, the DR tuning is triggered. An apparent suppression takes effect within approximately 200 msec, and no residual acceleration is observed. The OP-DFVA, on the other hand, would yield some residual oscillations. The oscillation amplitudes for both tuned DR (point **B**) and OP-DFVA (point **B'**) are indicated in Figure 7 as 0 and 0.55 (dimensionless), respectively. The latter is the consequence of the compromise brought by the trade-off.

7. Conclusions

A modal analysis representation of beam-exciter-absorber dynamics is taken into account while implementing the Delayed Feedback Vibration Absorber (DFVA). Feedback gain and time delay, the only two control parameters, are systematically varied to search for a global optimum such that the peak frequency response is minimized. System stability is always used as a constraint. The results prove successful as an example case shows.

Acknowledgment

This research is sponsored in part by Consortium on Variable Frequency Vibration Elimination, and UCONN Research Foundation (Grant No. : 441211).

References

1. Olgac, N. and Holm-Hansen, B. (1994) A novel active vibration absorption technique: Delayed Resonator, *J. of Sound and Vibration* **176**, 93-104.
2. Olgac, N. and Holm-Hansen, B. (1995) Tunable active vibration absorber: The Delayed Resonator, *ASME, J. of Dynamic Systems, Measurement, and Control* **117**, 513-519.
3. Olgac, N., Elmali, H., and Vijayan, S. (1996) Introduction to Dual Frequency Fixed Delayed Resonator (DFFDR), *J. of Sound and Vibration* **189**, 355-365.
4. Olgac, N. And Jalili, N. (1998) Modal analysis of flexible beams with Delayed Resonator vibration absorber: Theory and experiment, *in print, J. of Sound and Vibration*.
5. Olgac, N., Elmali, H., Hosek, M., and Renzulli, M. (1997) Active vibration control of distributed systems using Delayed Resonator with acceleration feedback, *ASME, J. of Dynamic Systems, Measurement, and Control* **119**, 380-389.
6. Lee, S. Y. and Lin, S. M. (1996) Dynamic analysis of nonuniform beams with time-dependent elastic boundary conditions, *ASME, J. of Applied Mechanics* **63**, 474-478.
7. MAPLE V, release 5, 1998, Waterloo Maple Inc., Ontario, Canada.

8. Arora, J. S. (1989) *Introduction to optimization design*, McGraw-Hill Book Company.
9. Gill, P. E., Murray, W. and Wright, M. H. (1981) *Practical optimization*, Academic Press, New York.

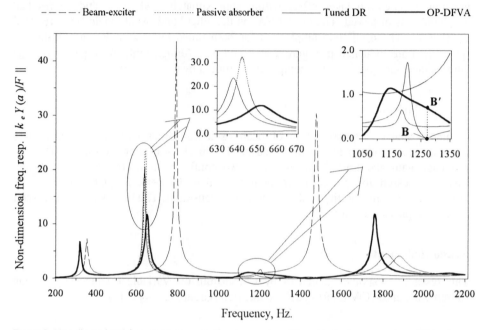

Figure 7. Non-dimensional frequency response of various systems (the wide frequency versions of this plot at the peak and tuning frequency are shown separately).

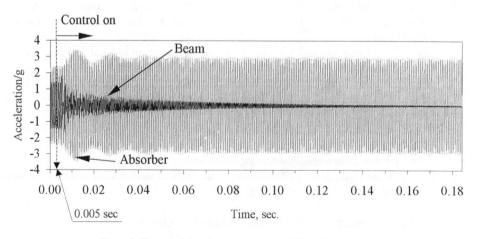

Figure 8. Beam and absorber response to 1270 Hz disturbance.

ACTIVE PARAMETRIC MODIFICATION OF LINEAR VIBRATING SYSTEMS

A. OSSOWSKI

*Institute of Fundamental Technological Research PAS,
ul. Swietokrzyska 21, 00-049 Warsaw, POLAND*

1. Introduction

Dynamical systems with variable parameters are usually considered in the context of parameteric resonances and instabilities. However, if the system parameters are changed suitably, then an improvement of dynamical properties of a given system can be observed. If this is the case, then we say that an active parametric modification is applied to the system.

In mechanical design we very often face the problems of stability and vibrations. We usually want the vibrations of a given system or structure excited by external interactions to be relatively small [1,2]. In other words, we want the excited states $\mathbf{x}(t)$ of the system to be stable and convergent to the equilibrium $\mathbf{x} = \mathbf{0}$ with a maximal rate. One way to achieve this goal is to ensure optimality of system parameters $\mathbf{p}$.

In the classical approach to parametric optimization we usually apply *passive parametric modification* (PPM), i.e. we look for optimal constant parameters of the system that ensure the best system properties [3]. However, there are no theoretical reasons for which the system parameters should be constant. It is possible that variable parameters can ensure better dynamical properties of a given system. That is why, in mechatronics, we consider *active parametric modification* (APM) i.e. we look for optimal variable system parameters $\mathbf{p} = \mathbf{p}(t)$. This is possible, if certain parameters of the real system can be changed. In practical realizations it is convenient to apply a suitable state feedback $\mathbf{p} = \mathbf{p}(\mathbf{x})$. The feedback realization of the APM enables us to ensure system optimality for any excited motion and at any time instant.

J. Holnicki-Szulc and J. Rodellar (eds.), Smart Structures, 247–254.
© *1999 Kluwer Academic Publishers. Printed in the Netherlands.*

We consider in this paper the class of linear dynamical systems that can be described by the equations of the form

$$\dot{\mathbf{x}} = \mathbf{A}(\mathbf{p})\mathbf{x}, \tag{1}$$

where $\mathbf{x} \in R^n$ is the state vector, $\mathbf{A} = \mathbf{A}(\mathbf{p})$ is an $n \times n$ matrix dependent on the vector of accessible system parameters $\mathbf{p}$ belonging to a compact region $P \subset R^k$. In this case, the general APM feedback system can be illustrated by the following block diagram (Fig. 1):

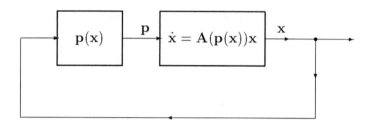

Figure 1. The block scheme of the general APM feedback system.

At each time instant t the above feedback system ensures optimal values of parameters $\mathbf{p}(\mathbf{x})$ according to the actual state $\mathbf{x}(t)$ measured at the output.

The main problem in the active parametric modification is to find an optimal feedback function $\mathbf{p}(\mathbf{x})$ for a given system (1) (see e.g. [4]), and then to estimate dynamical properties of the modified system. It is also necessary to propose a practical realization of the optimal feedback.

It is obvious that the dynamics of the APM closed-loop system can be described by the vector equation $\dot{\mathbf{x}} = \mathbf{A}(\mathbf{p}(\mathbf{x}))\mathbf{x}$ which is nonlinear in spite of linearity of the original system (1). Thus, the problem of the active parametric modification is nonlinear in its nature.

The general problem of active parametric modification of linear discrete dynamical systems with respect to exponential stability is formulated in this paper. A Lyapunov approach to the problem is described and applied to multidimensional vibrating systems. Effective methods of the active modification of system parameters (e.g. damping and stiffness coefficients) that ensure optimal vibration damping are studied. A neural network realizing the required state-feedback system of modification is proposed. The possibility of application of the obtained results to intelligent mechanical systems, such like actively controlled suspension systems or seismic structures, is pointed out.

2. Lyapunov Approach to Parametric Modification

Particularly interesting from the practical point of view is the problem of parametric modification of the system (1) with respect to stability properties. It can be formulated, if there exists such a compact P that the matrix $\mathbf{A}(\mathbf{p})$ is stable for every $\mathbf{p} \in P$.

To solve the above problem we apply the Lyapunov approach to stability analysis based on the concept of optimal Lyapunov functions and the index of stability (see e.g. [5-8]). It is known that if the stability index

$$\gamma(\mathbf{S}, \mathbf{p}) = -\sup_{\mathbf{x} \neq 0} \frac{\mathbf{x}^T \mathbf{S} \mathbf{A}(\mathbf{p}) \mathbf{x}}{\mathbf{x}^T \mathbf{S} \mathbf{x}} \tag{2}$$

is positive for a positive definite $n{\times}n$ matrix $\mathbf{S}$, then the system (1) is exponentially stable with respect to the norm $\| \mathbf{x} \|_{\mathbf{S}} = \sqrt{\mathbf{x}^T \mathbf{S} \mathbf{x}}$ with the exponent equal to $\gamma(\mathbf{S}, \mathbf{p})$. Then the quadratic form $\mathbf{x}^T \mathbf{S} \mathbf{x}$ is a Lyapunov function of the system and the following estimate is satisfied:

$$\| \mathbf{x}(t) \|_{\mathbf{S}} \leq \| \mathbf{x}(t_0) \|_{\mathbf{S}} \, exp[-\gamma(\mathbf{S}, \mathbf{p})(t - t_0)] \tag{3}$$

for the excited motion $\mathbf{x}(t) = \mathbf{x}(t, t_0, x_0)$ starting from any initial conditions $(t_0, \mathbf{x}_0)$. To improve passively the stability properties of (1) with respect to a given norm $\| \mathbf{x} \|_{\mathbf{S}}$, we can perform the parametric optimization, for example

$$\max_{\mathbf{p} \in P} \gamma(\mathbf{S}, \mathbf{p}) = \hat{\gamma}. \tag{4}$$

As the result we obtain an optimal constant vector $\hat{\mathbf{p}} = \hat{\mathbf{p}}(\mathbf{S})$ and the corresponding optimal stability index $\hat{\gamma}(\mathbf{S}, \hat{\mathbf{p}}(\mathbf{S}))$ that can be optimized with respect to $\mathbf{S}$.

It follows from (2) that the expression under the supremum in (2) for a given $\mathbf{x}$ can be interpreted as the exponential rate of convergence of a system trajectory at the point $\mathbf{x}$. Therefore, to improve actively the stability properties of a given system (1), it is logical to look for optimal parameters $\mathbf{p}$ that minimize the exponential rate of convergence of the system trajectories at each point $\mathbf{x}$ of the state space. It means that the structure of optimal feedback can be derived (in the first step of optimization) from the following maximization problem:

$$\max_{\mathbf{p} \in P} \left[-\frac{\mathbf{x}^T \mathbf{S} \mathbf{A}(\mathbf{p}) \mathbf{x}}{\mathbf{x}^T \mathbf{S} \mathbf{x}} \right], \tag{5}$$

solved at each point $\mathbf{x} \neq \mathbf{0}$ of the state space. Since the feedback function $\mathbf{p}(\mathbf{x}, \mathbf{S})$ obtained in this way is usually non-linear, so is the dynamics of the corresponding closed-loop system $\dot{\mathbf{x}} = \mathbf{A}(\mathbf{p}(\mathbf{x}, \mathbf{S}))\mathbf{x}$.

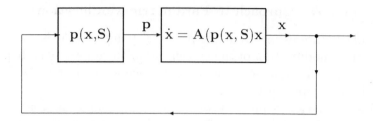

Figure 2. The block scheme of the optimal APM feedback system.

The corresponding APM feedback system can be illustrated by the block diagram shown in **Fig.2**. Exponential convergence of the trajectories of the above system can be estimated globally by the following index:

$$\gamma(\mathbf{S}) = -\sup_{\mathbf{x} \neq 0} \frac{\mathbf{x}^T \mathbf{S} \mathbf{A}(\mathbf{p}(\mathbf{x}, \mathbf{S}))\mathbf{x}}{\mathbf{x}^T \mathbf{S} \mathbf{x}}. \tag{6}$$

It is clear that $\gamma(\mathbf{S}) \geq \gamma(\mathbf{S}, \mathbf{p})$ for any constant vector of system prameters $\mathbf{p} \in P$ and any positive definite matrix $\mathbf{S}$. The active parametric modification is effective if there exist such $\mathbf{S} > 0$ that $\gamma(\mathbf{S}) > \gamma(\mathbf{S}, \mathbf{p})$.

The optimal feedback $\mathbf{p}(\mathbf{x}, \mathbf{S})$ is not completely determined at this stage yet since it is dependent on the matrix $\mathbf{S}$ of the Lyapunov function; the function $\mathbf{p}(\mathbf{x}, \mathbf{S})$ determines in fact a class of suboptimal APM systems. Therefore, having in mind stability properties of the system, we can now perform further optimization (Lyapunov function optimiazation) with respect to $\mathbf{S}$. Finally, we obtain an optimal Lyapunov function $\mathbf{x}^T \hat{\mathbf{S}} \mathbf{x}$ and the corresponding optimal feedback function $\hat{p}(\mathbf{x}) = \mathbf{p}(\mathbf{x}, \hat{\mathbf{S}})$ of the active parametric modification.

3. Active Parametric Modification of Linear Vibrating Systems

In this section, we apply the described approach to the problem of the active parametric modification of multidimensional linear vibrating systems. To do that, let us consider a general linear vibrating system [9]

$$\ddot{\mathbf{x}} + \mathbf{C}(\mathbf{p})\dot{\mathbf{x}} + \mathbf{D}(\mathbf{q})\mathbf{x} = 0, \mathbf{x} \in R^n, \tag{7}$$

where the matrices $\mathbf{C}(\mathbf{p})$, $\mathbf{D}(\mathbf{q})$ are linearly dependent on certain system parameters $\mathbf{p} = [p_1, p_2, ..., p_k]^T \in R^k$, $\mathbf{q} = [q_1, q_2, ..., q_l]^T \in R^l$, respectively i.e. $\mathbf{C}(\mathbf{p}) = \mathbf{C}_0 + p_1\mathbf{C}_1 + ... + p_k\mathbf{C}_k$, $\mathbf{D}(\mathbf{q}) = \mathbf{D}_0 + q_1\mathbf{D}_1 + ... + q_l\mathbf{D}_l$. We assume that there are certain nominal values of parameters $\mathbf{p}_0$, $\mathbf{q}_0$ such that $\mathbf{p} = \mathbf{p}_0 + \mathbf{u}$, $\mathbf{q} = \mathbf{q}_0 + \mathbf{v}$ and $\mid u_i \mid \leq \alpha_i > 0, \mid v_j \mid \leq \beta_j > 0$ for $i = 1, ..., k, j = 1, ..., l$.

The system (7) can be described as the so-called bilinear control system of the form (see [10])

$$\dot{\mathbf{Z}} = \mathbf{A}_0\mathbf{Z} + u_1\mathbf{A}_1\mathbf{Z} + \cdots + u_k\mathbf{A}_k\mathbf{Z} + v_1\mathbf{B}_1\mathbf{Z} + \cdots + v_l\mathbf{B}_l\mathbf{Z}, \qquad (8)$$

where $\mathbf{Z} = [\mathbf{x}, \mathbf{y} = \dot{\mathbf{x}}]^T \in R^{2n}$ and

$$\mathbf{A}_0 = \begin{bmatrix} \mathbf{0} & \mathbf{I} \\ -\mathbf{C}(\mathbf{p}_0) & -\mathbf{D}(\mathbf{q}_0) \end{bmatrix}, \mathbf{A}_i = \begin{bmatrix} 0 & 0 \\ 0 & -\mathbf{C}_i \end{bmatrix}, \mathbf{B}_j = \begin{bmatrix} 0 & 0 \\ -\mathbf{D}_j & 0 \end{bmatrix}$$

for $i = 1, \ldots, k, j = 1, \ldots, l$. It is easy to see that the optimal APM feedback for (8) derived from criterion (5) has the following form

$$u_i = -\alpha_i sign[\mathbf{Z}^T\mathbf{S}\mathbf{A}_i\mathbf{Z}], v_j = -\beta_j sign[\mathbf{Z}^T\mathbf{S}\mathbf{B}_j\mathbf{Z}] \qquad (9)$$

where $i = 1, \ldots, k, j = 1, \ldots, l$. Thus, the optimal parametric modification of (8) is achieved by bang-bang controls with quadratic switching functions. A given system parameter can be modified actively, if the corresponding switching function is not of a constant sign.

It can be easily proved that damping and stiffness parameters of linear systems of interacting oscillators always satisfy this condition. Indeed, applying a quadratic Lyapunov function

$$V_\mathbf{S}(\mathbf{Z}) = \mathbf{Z}^T\mathbf{S}\mathbf{Z} = \mathbf{x}^T\mathbf{S}_1\mathbf{x} + 2\mathbf{y}^T\mathbf{S}_3\mathbf{x} + \mathbf{y}^T\mathbf{S}_2\mathbf{y} \qquad (10)$$

to (9) we can conclude that the switching functions for $u_i, v_i, i = 1, ..., k$ can be described by the following products of linear functions: $\mathbf{Z}^T\mathbf{S}\mathbf{A}_i\mathbf{Z} = (\mathbf{g}_i^T\mathbf{y})(\mathbf{h}_i^T\mathbf{Z})$, $\mathbf{Z}^T\mathbf{S}\mathbf{D}_i\mathbf{Z} = (\mathbf{f}_i^T\mathbf{x})(\mathbf{h}_i^T\mathbf{Z})$, respectively, where $\mathbf{f}_i, \mathbf{g}_i \in R^n$, $\mathbf{h}_i \in R^{2n}, i = 1, ..., k$ are certain vectors such that $\mathbf{h}_i^T\mathbf{Z} = (\mathbf{S}_3\mathbf{x} + \mathbf{S}_2\mathbf{y})_i$. Thus, the optimal switchings of the damping and stiffness parameters of linear vibrating systems take place when a system trajectory crosses certain hyperplanes in the state space.

The bang-bang controls (9) require ideal switchings that cannot be realized in practice, especially for small input signals. Therefore, it is important to answer the question about the behaviour of APM systems in small neighbourhoods of the origin of the state space. If nonideal switches with signoidal continuous characteristics are applied, then, in a sufficiently small neighbourhood of the origin, the modifying controls u_i , $u_{j,i} = 1, ..., k, j = 1, ...l$, take the form of the quadratic functions $u_i \simeq a_i \cdot \mathbf{Z}^T\mathbf{S}\mathbf{A}_i\mathbf{Z}$, $v_j \simeq b_j \cdot \mathbf{Z}^T\mathbf{S}\mathbf{B}_j\mathbf{Z}$, where $a_i, b_j, i = 1, ..., k, j = 1, ..., l$, are the amplifications of the corresponding switches for small signals. Putting the above controls into (8) it is easy to see that the system dynamics can be described by the equation $\dot{\mathbf{Z}} = \mathbf{A}_0\mathbf{Z} + o(3)$. Thus, the real APM system near the origin behaves as the system with the matrix $\mathbf{A}_0 = \mathbf{A}(\mathbf{p}_0)$ i.e. the system stability with the nominal constant values of the parameters is guaranteed.

4. Application of neural networks

It is important in practical realizations of systems of the active parametric modification to ensure a sufficiently small feedback delay and high system reliability. Neural network control systems seem to be an attractive alternative in comparison with e.g. computer systems [11,12]. This is possible for example in the case of the active parametric modification of systems of k interacting oscillators. Since the optimal switching functions of stiffness and damping parameters are products of linear state functions, the optimal control switches for a given pair of controls $u_i, v_i, i = 1, ..., k$ can be realized by a simple three-layer neural network shown in **Fig.3**,

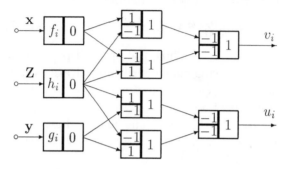

Figure 3. Neural network realization of the optimal APM feedback.

where the symbol $\mathbf{x} \circ\!\!\!-\!\!\!\boxed{\mathbf{w}\,|\,s}\!\!\!-\!\!\!\rightarrow y$ denotes the static cell with the input vector $\mathbf{x}$, the vector of weights $\mathbf{w}$, the threshold s and the single output $y = sign[\mathbf{w}^T\mathbf{x} - s]$.

Hence, the active modification of stiffness and damping parameters in the system containing k pairs of strings and attenuators requires a three-layer neural network composed of $9k$ cells. The input layer of the network plays the role of an A-D converter while the hidden and the output layers work as a logical system [11]. The thresholds and weights of the logical system are completely determined. Only $4nk$ input weights of the network are to be determined. Thus, the process of the nework learning reduces in this case to the optimization of the input weights which are determined by vectors $\mathbf{f}_i, \mathbf{g}_i, \mathbf{h}_i, i = 1, ...k$. Since the input weights depend on the matrix $\mathbf{S}$, this problem is equivalent to the Lyapunov function optimization. Moreover, since the Lyapunov functions with $2n \times 2n$ matrices $\mathbf{S}$ and $\lambda\mathbf{S}$ are equivalent (i.e. give the same stability estimates) for any $\lambda > 0$, at most $(2n - 1)(n + 1)$ parameters can be optimized.

It is easy to prove that the above network for small input signals ensures the system stability (in small neighbourhoods of the origin) with the maximal constant values of system parameters.

5. Example

To illustrate the usefulness of the presented approach to active parametric modification we consider a simple system of two interacting oscillators, what is shown in **Fig.4**. Five system parameters can be actively modified, namely k_1, k_2, k_3, l_1, l_2. We consider the particular case of active modification of the damping parameters $p_i = \frac{l_i}{m_i}, i = 1, 2$. Let us assume the following nominal values of parameters: $m_1 = m_2 = m, k_1 = k_2 + 2k_3 = k, l_1 = l_2 = l, q_1 = q_2 = 2q_3 = \frac{k}{m} = 1, p_1 = p_2 = \frac{l}{m} = 0.5$.

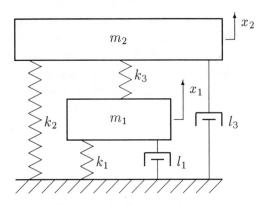

Figure 4. Example of a vibrating system of two degres of freedom.

Then the system dynamics can be described by the equation of the form (8) where

$$\mathbf{A}_0 = \begin{bmatrix} 0 & 1 & 0 & 0 \\ -\frac{3}{2} & -1 & \frac{1}{2} & 0 \\ 0 & 0 & 0 & 1 \\ \frac{1}{2} & 0 & -\frac{3}{2} & -1 \end{bmatrix}, \mathbf{A}_i = \begin{bmatrix} 0 & 0 & 0 & 0 \\ 0 & -2\delta_{1i} & 0 & 0 \\ 0 & 0 & 0 & 0 \\ 0 & 0 & 0 & -2\delta_{2i} \end{bmatrix}, i = 1, 2.$$

Applying formulas (9) and the optimal Lyapunov function $V_{\mathbf{S}}(\mathbf{Z}) = \mathbf{Z}^T \mathbf{S} \mathbf{Z}$ with the matrix

$$\mathbf{S} = \begin{bmatrix} 3 & 1 & -1 & 0 \\ 1 & 2 & 0 & 0 \\ -1 & 0 & 3 & 1 \\ 0 & 0 & 1 & 2 \end{bmatrix},$$

we finally obtain the optimal modifying feedback controls of the form

$$u_i = \alpha_i \cdot sign[y_i(x_i + 2y_i), i = 1, 2.$$

The effectiveness of vibration damping in the above feedback system is dependent on the initial conditions and can be characterized e.g. by the time instant at which the norm $\| \mathbf{x}(t) \|_{\mathbf{s}}$ of a given transient process $\mathbf{x}(t)$ decreases by 50%. Simple numerical analysis, performed for control amplitudes $\alpha_1 = \alpha_2 = 0.4$, shows that the attenuation time T_{MOD} achieved in the modified system is by about (4-15)% shorter in comparision with the corresponding system without modification. The obtained results can be improved, if all the five system parameters will be actively modified.

6. Conclusions

The described method of the active parametric modification enables us to improve significantly the stability properties of dynamical systems. Since the active control of systems parameters requires relatively small amounts of energy and does not use external forces, it can be an attractive alternative or suplementary solution in practical systems of the active control of vibrations with actuators such as actively controlled suspension systems or seismic structures. Neural networks as feedback controllers can be recommended for the active parametric modification of multidimensional linear vibrating systems.

References

1. Warburton, A.G.R.(1992) Reduction of Vibrations, *John Wiley and Sons Ltd.*
2. Pantelides, C.P.(1990) Optimum design of actively controlled structures,*Earthquake Eng. Struct. Dyn.* **19**, 583-596.
3. Kaniathra, J.N., Speckhart, F.H.(September 1975) A technique for determining damping values and damper locations in multi-degree-of-freedom systems,*Design Engineering Technical Conference, Washington,D.C*,pp.17-19.
4. Ossowski, A.(1997) Active parametric modification of a linear oscillator, *5th Polish-German Workshop on Dynamical Problems in Mechanical Systems, Zakopane, August 31st-September 6th.*
5. Muszy/nska, A., Radziszewski, B. (1981) Exponential stability as a criterion of parametric modification in vibration control, *Nonlinear Vibration Problems*,**20**, PWN, pp. 175-191.
6. Ossowski, A. (1989) On the exponential stability of non-stationary dynamical systems,*Nonlinear Vibration Problems*, **21**, PWN, pp. 109-121.
7. Ossowski, A. (1994) Nonlinear stabilization of linear systems, *Archives of Control Sciences*, Vol.**3**,(XXXIX), No.**1-2**, pp. 69-84.
8. Kalman, R., Beltram, J. (1960) Control system analysis and design via the "second method of Lyapunov", I-Continuous time systems, *Trans. ASME. J. of Basic Engineering*, pp.371-393.
9. Lubiner, E., Elishakoff, I. (1986) Random vibrations of systems with finitely many degrees of freedom and several coalescent frequencies, *Int. J. of Eng. Science.* Vol.**24**, No.**4**.
10. Mohler, R.R. (1991) Non-linear systems, Vol.II, Application to bilinear control, *Prentice Hall*, Englewood Cliffs, New Jersey.
11. Ossowski, A. (1992) Application of neural networks to stabilization of dynamical systems, *IFTR reports* **32**.
12. Cheung, J.Y., Mulholland, J. (1989) Using neural network as a feedback controller, *Proc. of 32nd Midwest Symposium on Circuits and Systems*, Washington.

LASER CONTROL OF FRAME MICROSTRUCTURES

A. OSSOWSKI AND J. WIDLASZEWSKI
Institute of Fundamental Technological Research PAS
Swietokrzyska 21, 00-049 Warsaw, POLAND

Keywords: *structure control, frame microstructures, laser forming, laser positioning, laser adjustment.*

1. Introduction

Well known is the fact that a pulsed or moving laser beam applied to a metallic body can cause its permanent deformation. There is a considerable amount of papers concerning practical applications of the above phenomenon to the so-called *laser forming* (e.g.[1]–[14]). Although laser forming, as a technological process, has gained succesful industrial implementations (e.g. in electronic industry [15]–[20]), it is still difficult to give its precise theoretical description due to complexity of thermo-elasto-plastic phenomena involved.

Laser radiation can also be applied to deform multicomponent mechanical systems like frames or trusses (e.g. [20]). The deformation of a given frame system can be achieved by laser heating of selected structure beams that will be bent and//or shortened, according to the laser forming mechanisms occuring [9]. The process of laser deformation of a structure can be easily described if only certain geometric characteristics of the structure are of interest. It is especially justified in the case of frames or trusses, the geometric state of which can be well described by space coordinates of a finite number of distinguished structure points.

The application of laser pulses to induce controlled deformations of a structure from a given initial state to a required final state is called *laser control of the structure*. The process of laser control can be succesfully applied to microsystems. Therefore, we consider in this paper the general problem of laser control of metal frame microstructures and then apply the presented approach to certain class of micromechanical systems that can be used, for example, in micropositioning or precise adjustment.

J. Holnicki-Szulc and J. Rodellar (eds.), Smart Structures, 255–264.

2. Laser control of frame microstructures

Let us consider a frame structure composed of rigid bodies $B_1, ..., B_a$ and flexible branches 1,2,...,b, as it is shown in Fig. 1. We assume that the structure can be represented by a finite nonoriented and simply connected graph with branches and nodes corresponding to the structure beams and rigid bodies, respectively. We also assume that the whole structure is attached to a fixed rigid base B.

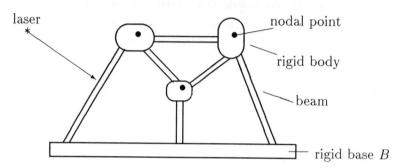

Figure 1.   General frame structure

Laser control of a frame structure consists in the application of laser pulses of suitable parameters to selected beams (branches) $1, 2, ..., k \leq b$ of the structure in an appropriate order. To describe the control process it is useful to introduce a vector of geometric state of the structure $\mathbf{X} = [x_1, ..., x_p]^T \in R^p$ as a collection of space coordinates of nodes in a co-ordinate system associated with the base B. We consider in this paper a class of practical control processes such that the parameters of the laser beam are fixed and only one branch is heated at each process step. Then, any control process consisting of n pulses can be described by a sequence of irradiated branches $(i_1, ..., i_n)$ and the corresponding state evolution $\mathbf{X}_0 \to \mathbf{X}_1 \to ... \to \mathbf{X}_n$. In the case of a complex process, when certain branch sequences are repeated $r_1, r_2, ...$ times, the process will be denoted by $(i_1, i_2, ..., i_{n_1})^{r_1} (j_1, j_2, ..., j_{n_2})^{r_2} ...$, where $n_1 + n_2 + ... = n$.

In many practical situations only certain geometric characteristics of the structure are essential (e.g. space coordinates of a selected point of the structure). Therefore, we assume a state vector function $\mathbf{Y} = \mathbf{G}(\mathbf{X}) \in R^q$ as the output function of the structure. If we also define the input as the control vector $\mathbf{u} = [u_1, u_2, ..., u_k]^T$ such that $u_i = 1$, if the i-th branch is irradiated and $u_i = 0$, if not, then the frame structure can be considered as a discrete time control system (see Fig. 2).

The main objective of the laser control is to achieve the required final output $\mathbf{Y}_n$ of the structure starting from a given initial state $\mathbf{X}_0$. This task

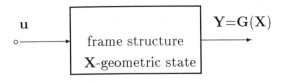

Figure 2. Frame structure as a control system

is feasible if there exist a final state $\mathbf{X}_n$, (achievable from $\mathbf{X}_0$ by using n laser pulses of the assumed parameters) such that $\mathbf{Y}_n = \mathbf{G}(\mathbf{X}_n)$.

Any process of laser control with the fixed beam parameters can be well characterized by the maximal number of laser pulses n and the number of activated structure branches k. Moreover, it is reasonable to introduce a limit on the maximal number l of pulses applied to any branch $1, 2, ..., k$. The three natural numbers k, l, n determine a class of control processes which will be denoted by $\{k, l, n\}$.

It is clear that applying n pulses to k branches of a given structure, each being activated by no more than l pulses, we can achieve a number $N(k, l, n)$ of final states $\mathbf{X}_n$ dispersed in a neighbourhood of the initial state $\mathbf{X}_0$. The combinatoric factor $N(k, l, n)$, as the number of all $\{k, l, n\}$ processes, grows rapidly with k, l, n. Therefore, we have a great number of possibilities of laser control even for relatively simple systems (i.e. for small k, l, n).

In order to formulate the objectives of laser control of a frame microstructure, it is useful to introduce the following definition:

Definition: Achievability set of a frame structure is the collection $A(\mathbf{X}_0, k, l, n)$ of all final states $\mathbf{X}_n$ of the structure achievable from the initial state $\mathbf{X}_0$ by using $\{k, l, n\}$ laser control processes.

On this basis we can determine the set of achievable outputs of the structure $\mathbf{G}(A) = \mathbf{G}(A(\mathbf{X}_0, k, l, n)) = \{\mathbf{Y}_n = \mathbf{G}(\mathbf{X}_n) : \mathbf{X}_n \in A(\mathbf{X}_0, k, l, n)\}$ as the set of all outputs achievable from the initial output $\mathbf{Y}_0 = \mathbf{G}(\mathbf{X}_0)$ by using $\{k, l, n\}$ processes.

If the required final output $\mathbf{Y}_n \in \mathbf{G}(A)$ can be achieved by different sequences of laser pulses, then we can choose a sequence optimal e.g. with respect to the control energy (or the total number of pulses n). Thus the problem of optimal laser control can be considered.

At least three problems essential for the laser control of microstructures and its applications can be recognized:

1. To determine the achievability set $A(\mathbf{X}_0, k, l, n)$ for a given structure.
2. To design a structure with the assumed achievability set.

3. To find optimal control to achieve the required final output $\mathbf{Y}_n$ of the structure starting from the initial state $\mathbf{X}_0$.

Geometric properties of the achievability sets A, $\mathbf{G}(A)$ of a given structure depend on material properties and dimensions of the structure beams as well as on the structure topology. Furthermore, the process of laser control of a structure proves to be nonlinear and history dependent. Therefore, the above problems are easily solvable in simple cases only. To overcome the above difficulties it is useful to apply the concept of multimodule structures described in the next section.

3. Micropositioning of multimodule structures

In micropositioning it is essential that the range of achievability should be large enough and//or the achievable outputs should be dispersed as uniformly as possible in an assumed neighbourhood of $\mathbf{Y}_0 = \mathbf{G}(\mathbf{X}_0)$. However, the distribution of achievable states of the structure is usually nonuniform due to the existing system nonlinearities. For this reason, it is convenient to evaluate the structures with respect to achievable resolution and//or accuracy of the micropositioning.

To design a system satisfying such achievability requirements it is usefull to apply the so-called *multimodule structures* composed of elementary structures (modules) $M_1, M_2, \ldots, M_m$ that can be easily analysed and controlled indepedently. Then, the achievability sets A, $\mathbf{G}(A)$ of an m-module structure can be deduced from the achievability sets of the corresponding structure modules. It is easy to see that the number of final states of such systems is large enough (even for small m) to achieve high resolution.

To illustrate the usefulness of multimodule structures we consider the particular case of a serial multimodule structure composed of modules $M_1, M_2, \ldots, M_m$ installed one over another by using rigid bases (tables) $B_0, B_1, \ldots, B_m$, as shown in Fig. 3b.

The module independence of such system enables us to apply independent control processes $(i_1, i_2, \ldots, i_{n_1}) \in \{k_1, l_1, n_1\}$, $(j_1, j_2, \ldots, j_{n_2}) \in \{k_2, l_2, n_2\}$, ... at each level of the structure. Suppose that the output function of the system is determined by space coordinates of a selected point C on the top table B_m (i.e. $\mathbf{Y} = [x_C, y_C]^T$). Then, any output achieved by using control processes at levels $M_1, M_2, \ldots, M_s, s < m$, is split by a process applied to the next module M_{s+1}. Hence, the total number of achievable outputs, as the product $N(k_1, l_1, n_1) \cdot N(k_2, l_2, n_2) \cdots N(k_m, l_m, n_m)$, grows rapidly with the number of modules m of the system. In the particular case, when all modules $M_1, M_2, \ldots, M_m$ are topologically equivalent, the set of achievable outputs $\mathbf{Y} = [x_C, y_C]^T$ of the m-module system exhibits the property of m-stage *selfsimilarity*, i.e. it is a *fractal*.

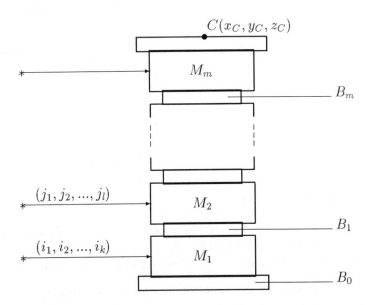

Figure 3. The block diagram of a serial m-module structure

It is essential for micropositioning applications to ensure the desired achievability properties of a given multimodule system. The obtained distribution of the output points $\mathbf{Y} = [x_C, y_C]^T$ is dependent on the topology and geometric dimensions of each module as well as on the total number m of system modules. Thus, in order to design a multimodule system, optimal in the sense of the assumed achievability properties, we should determine:

- the minimal sufficient number of modules m,
- the optimal topology of each system module,
- optimal dimensions of the modules.

The introduced concepts and notations allow to describe the process of laser control of various frame microstructures. In the following section the presented approach will be illustrated by results of the laboratory experiments performed with an elementary structure and its 3-module expansion.

4. Experimental studies

Experimental studies show that the process of laser forming is nonlinear in general (e.g. [6], [12]). Even simple bending of a cantilever beam exhibits nonlinear dependence of the bending angle on the number of laser irradiations as shown in Fig. 4. When we act upon a single beam of a frame structure, the corresponding deformation characteristic tends practically to saturation due to structure constraints (see Fig. 5).

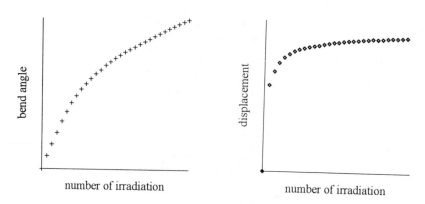

Figure 4.   Typical nonlinear displacement - irradiation number *dependence in*
laser bending of a cantilever beam.

Figure 5.   Typical nonlinear displacement - irradiation number *dependence in*
deformation of a frame structure.

Therefore, the process of deformation of the frame structure is effective
if sequences of at most several pulses are applied to a single beam of the
structure (this justifies the introduced limit l).

We provide in this section some results of laboratory experiments performed with the structures shown in Fig. 6. The constructions of this type
are relatively easily controllable and prove to be very useful, especially in
micropositioning applications. Specimens used in investigations were made
of stainless steel. Dimension of the specimen measured along the diagonal of
the base was 6 mm. Experiments were carried out with the use of Nd:YAG
laser of the power 150 W. The structure deformation was measured by
non-contact method using the Laser Scan Micrometer.

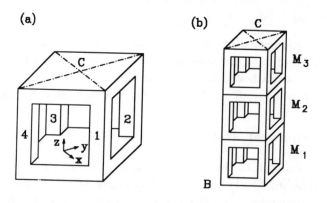

Figure 6.   Example of a frame structure (a) and its serial 3-module expansion (b).

Examples of graphic representations of controlled deformation processes
of the single module structure and the serial 3-module structure are illus-

trated in Fig. 7 and Fig. 8, respectively.

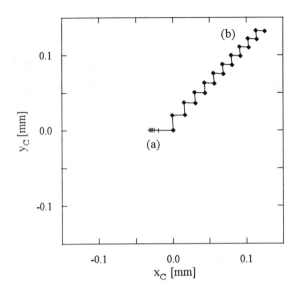

Figure 7.   Examples of laser control of the elementary structure.

In the process illustrated in Fig. 7a, laser pulses were applied to beam 3 (see Fig. 6a) in a sequence $(3)^5$ according to the introduced notation. Irradiations number 2 ... 5 yield significantly smaller displacement of the selected point of the upper table than the first irradiation. This nonlinearity is beneficial to accomplish fine positioning or adjustment. Fig. 7b represents the process $(2, 1)^{10}$. Consequtive deformations produced in the sequence of alternate irradiating of two neighbouring beams do not decrease as fast so in the case of applying laser pulses to a single beam. The presented characteristic illustrates dependence of the laser control process on the deformation history.

Examples of deformation of a 3-module structure shown in Fig. 6b are presented in Fig. 8, where (a), (b) and (c) correspond to processes $(1)^3(2)^3$ applied to module M_1, M_2 and M_3, respectively. Serial topology of the considered structure implies difference between the amplitudes of deformation when applying laser irradiation to different modules.

To illustrate possibilities of application of the considered multimodule structure in micropositioning, the set of achievable outputs $Y = [x_C, y_C]^T$ has been determined in the simplest case, when at most one laser pulse is applied to each module (Fig. 9). The number of achievable states of the structure is equal to $[N(4, 1, 1)]^3 = 5^3 = 125$ in this case.

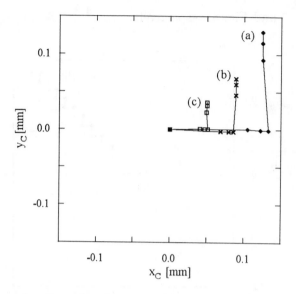

Figure 8.   Examples of laser control of the 3-module structure.

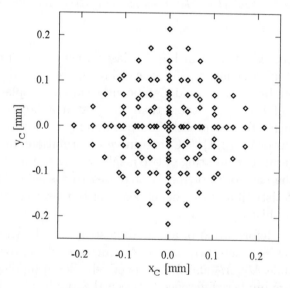

Figure 9.   Achieveble output points of the 3-module structure.

The optimization of the geometric parameters of the structure enables us to obtain more uniform distribution of the achievable points, as is shown in Fig. 10, where coordinates x_C, y_C are expressed in nondimensional units.

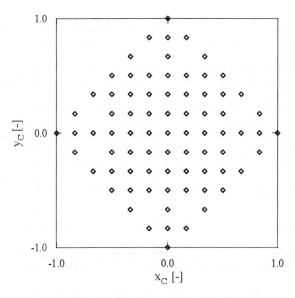

Figure 10. Achievable output points of the 3-module optimized structure.

5. Conclusions

The general problem of laser control of frame microstructures has been formulated in this paper. The results of experimental studies indicate the possibility of precise laser control. The controlled multimodule microstructures of the proposed topology seem to have a great potential of technological applications, for example in micromechanics (contactless positioning of sub-assemblies and alignment of micro-components).

References

1. Namba, Y. (1987) Laser Forming of Metals and Alloys, in *Proc. of LAMP'87*, Osaka, pp. 601–606.
2. Fr/ackiewicz, H., Mucha, Z., Tr/ampczy/nski, W., Baranowski, A., Cybulski, A. (1988) A method of bending metal objects, *European Patent Specification 0 317 830*.
3. Geiger, M., Vollertsen, F., Amon, S. (1991) Flexible Blechumformung mit Laserstrahlung – Laserstrahlbiegen, *Blech Rohre Profile*, **38**, **11**, pp. 856–861.
4. König, W., Weck, M., Herfurth, H. J., Ostendarp, H., /Zaboklicki, A. K. (1993) Formgebung mit Laserstrahlung - Neue Wege der Blechumformung in der Prototypenfertigung, *VDI-Z*, **135**, **4**, pp. 14–17.
5. Kittel, S., Küpper, F. et al. (1993) Für Einzelteilen und Kleinserien: Laserstrahlumformen von Blechen, *Bänder, Bleche, Rohre*, **3**, pp. 54–62.
6. Sprenger, A., Vollertsen, F., Steen, W., M., Watkins, K., (1994) Influence of strain hardening on laser bending, in M. Geiger and F. Vollertsen (eds.), *Laser-Assisted Net Shape Engineering, Proc. of the LANE'94*, Meisenbach Bamberg, pp. 361–370.
7. Fratini, L., Lo Nigro, G., (1995) Neural network application in the laser bending process: direct and inverse approaches, in *Proc. of The II Meeting of AITEM, Padova*,

pp. 11–20.

8. Pridham, M. S., Thomson, G. A. (1995) An investigation of laser forming using empirical methods and finite element analysis, *Journal of Design and Manufacturing*, **5**, pp. 203–211.

9. Vollertsen, F. (1995) *Laserstrahlumformen – Lasergestützte Formgebung: Verfahren, Mechanismen, Modellierung*, Meisenbach Bamberg, Friedrich-Alexander-Universität Erlangen-Nürnberg.

10. Holzer, S. (1996) *Berührungslose Formgebung mit Laserstrahlung,* Reihe Fertigungstechnik, Bd. 57, Meisenbach Bamberg, Erlangen.

11. Mucha, Z., Hoffman, J., Kalita, W., Mucha, S. (1997) Laser Forming of Thick Free Plates, in M. Geiger and F. Vollertsen (eds.), *Laser Assisted Net Shape Engineering 2, Proc. of the LANE'97*, Meisenbach Bamberg, pp. 383–392.

12. Wid/laszewski, J. (1997) Precise laser bending, in M. Geiger and F. Vollertsen (eds.), *Laser Assisted Net Shape Engineering 2, Proc. of the LANE'97*, Meisenbach Bamberg, pp. 393–398.

13. Wid/laszewski, J. (1997) Laser microbending, in *Laser Technology in Materials Surface Engineering* (in Polish), Military Academy of Technology, Warsaw, pp. 141–148.

14. Dovč, M., Možina, J., Kosel, F., (1998) An analytical model for time-dependent deformation of a circular plate illuminated by a laser pulse, *J. Phys. D: Appl. Phys.*, **31**, pp. 65–73

15. Steiger, E. (1984) Führungsloses Justieren der Mittelkontaktfedern des Kleinrelais D2 in einem Pulslasersystem, *Siemens Components*, **22, 3**, pp. 135–137.

16. Hanebuth, H., Hamann, Chr. (1997) Suitability of CuCoBe-alloys for laser beam bending, in M. Geiger and F. Vollertsen (eds.), *Laser Assisted Net Shape Engineering 2, Proc. of the LANE'97*, Meisenbach Bamberg, pp. 367–374.

17. Vollertsen, F., Komel, I., Kals, R. (1995) The laser bending of steel foils for microparts by the buckling mechanism – a model, *Modelling Simul. Mater. Sci. Eng.*, **3**, pp. 107–119.

18. Arnet, H., Vollertsen, F. (1995) Extending Laser Bending for the Generation of Convex Shapes, *Proc. Inst. Mech. Engrs, Part B: Journal of Engineering Manufacture*, **209**, pp. 433–442.

19. Huber, A., Müller, B., Vollertsen, F. (1997) A measuring device for small angles and displacements, in M. Geiger and F. Vollertsen (eds.), *Laser Assisted Net Shape Engineering 2, Proc. of the LANE'97*, Meisenbach Bamberg, pp. 325–330.

20. Hoving W. (1997) Laser applications in micro technology, in M. Geiger and F. Vollertsen (eds.), *Laser Assisted Net Shape Engineering 2, Proc. of the LANE'97*, Meisenbach Bamberg, pp. 69–88.

FINITE ELEMENT ANALYSIS OF THE DYNAMIC RESPONSE OF COMPOSITE PLATES WITH EMBEDDED SHAPE-MEMORY ALLOY FIBERS

W. OSTACHOWICZ, M. KRAWCZUK, A. ŻAK
Institute of Fluid Flow Machinery PAS
ul. Gen. J. Fiszera, 80–952 Gdańsk, Poland

1. Introduction

This paper illustrates the stress–strain relationships for a single composite layer with embedded SMA (Shape-Memory Alloy) fibers and also their influence upon changes in natural frequencies of a composite multilayer plate. Governing equations based on the finite element method are presented in the paper. The plate is modeled by plate finite elements. These elements have eight nodes with five degrees of freedom at each node (i.e. three axial displacements and two independent rotations). For both axial displacements and independent rotations biquadratic shape functions are used.

A limited number of papers concerned with natural vibration of SMA fiber–reinforced composite plates can be found in the present literature. Zhong, Chen, and Mei [8], analyzed thermal buckling and postbuckling of composite plates with embedded SMA fibers using a finite element technique. Ro and Baz [4] introduced fundamental equations governing dynamic behavior of Nitinol–reinforced plates. Applying the finite element method they studied dynamic characteristics of such structures.

In the present work a more general description of this problem is introduced. The finite element analysis results presented in this work are compared to those obtained from an analytical continuum solution.

2. Mechanical and physical properties of shape-memory alloys

Mechanical and physical properties of SMA strongly depend on temperature and initial stresses [2, 3]. Changes in temperature and initial stresses involve changes in the volume fraction of martensite in the alloys. During the martensite transformation recovery stresses appear. These recovery stresses are not only a function of alloys temperature but also depend on initial strains ε_0. In Figure 1 SMA recovery stresses versus SMA temperature for four different initial strains are presented.

J. Holnicki-Szulc and J. Rodellar (eds.), Smart Structures, 265–274.

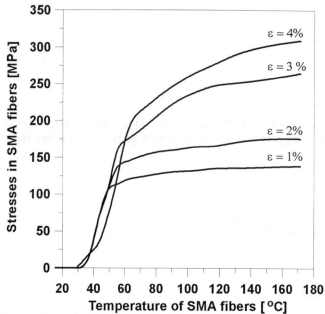

Figure 1. Stresses in SMA fibers as a function of temperature and initial strains.

It can be easily noticed that an increase in the initial strains involves higher recovery stresses in SMA wires and simultaneously, changes in the temperatures of phase transformation are also observed.

Material properties of SMA (Nitinol) wires used in this paper (See also [5]) are presented in Table 1.

TABLE 1. Material properties of Nitinol wires.

Austenite start temperature T_s	37.8°C
Austenite finish temperature T_f	62.8°C
Young modulus for $T < T_s$	27.50 GPa
Young modulus for $T > T_s$	82.60 GPa
Poisson ratio	0.3
Density	6450 kg/m^3
Thermal expansion coeff.	$10.26 \cdot 10^{-6}$ 1/$^\circ$C

3. Problem formulation

A SMA fiber–reinforced laminated composite plate which contains embedded SMA wires is shown in Figure 2. A single composite layer of the plate has arbitrary orientation of graphite fibers. SMA fibers are placed in the neutral plane of the plate.

The displacements field within a single layer of the plate is assumed due to the first order shear deformation theory of Reissner–Mindlin. Using standard finite element techniques, the displacements field can be easily calculated from the following relations:

$$\begin{cases} u(x, y, z) = u_0(x, y) + z \cdot \alpha(x, y), \\ v(x, y, z) = v_0(x, y) + z \cdot \beta(x, y), \\ w(x, y, z) = w_0(x, y), \end{cases} \tag{1}$$

where u_0, v_0 and w_0 are middle surface displacements while α and β are related to rotations.

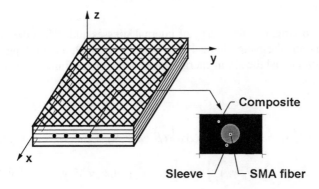

Figure 2. A SMA fiber–reinforced multilayer composite plate.

In this analysis plate–bending quadrilateral finite elements, called PBQ8 [7], were used. The applied element has eight nodes and five degrees of freedom at each node. In order to approximate the displacement field in the element, biquadratic shape functions were used.

Considering a composite plate composed of M layers, for k–th layer of the plate the constitutive relations can be written in the form:

$$col[\sigma_x, \sigma_y, \sigma_{yz}, \sigma_{xz}, \sigma_{xy}] = [Q_k] \begin{Bmatrix} \varepsilon_{x_0} + z\kappa_x - \alpha_x \Delta T \\ \varepsilon_{y_0} + z\kappa_y - \alpha_y \Delta T \\ \varepsilon_{yz} \\ \varepsilon_{xz} \\ \varepsilon_{xy_0} + z\kappa_{xy} - \frac{1}{2}\alpha_{xy} \Delta T \end{Bmatrix}. \tag{2}$$

Symbols α_x, α_y and α_{xy} denote the coefficients of thermal expansion while strains in the single layer of the plate can be evaluated from the following relations:

$$\begin{cases} \varepsilon_{x_0} = \partial_x u_0, & \kappa_x = \partial_x \alpha, & \varepsilon_{xz} = \tfrac{1}{2}(\alpha + \partial_x w), \\ \varepsilon_{y_0} = \partial_y v_0, & \kappa_y = \partial_y \beta, & \varepsilon_{yz} = \tfrac{1}{2}(\beta + \partial_y w), \\ \varepsilon_{xy_0} = \tfrac{1}{2}(\partial_y u_0 + \partial_x v_0), & \kappa_{xy} = \tfrac{1}{2}(\partial_y \alpha + \partial_x \beta). \end{cases} \tag{3}$$

Matrix $[\mathbf{Q}_k]$ in Equation 2 has a well–known structure [6]. For SMA fiber–reinforced hybrid composite plates subjected to a combined external, thermal and recovery stresses load governing equations of motion can be obtained through the principle of virtual work:

$$\delta U_e = \delta V_e = \delta(V_e' + V_e''), \tag{4}$$

where δU_e is the virtual strain energy of internal stresses and δV_e is the virtual work of external actions on the plate element. For the plates which are mid–plane symmetric there is no coupling and the strain energy stored in the plates finite element is given by

$$\begin{aligned} U_e = \iint_A \Big\{ &\tfrac{1}{2} A_{11}(\partial_x u_o)^2 + A_{12} \partial_x u_o \partial_y v_o + \tfrac{1}{2} A_{22}(\partial_y v_o) + \tfrac{1}{2} D_{11}(\partial_x \alpha)^2 + \\ &+ A_{44}\Big[\tfrac{1}{2}\beta^2 + \beta \partial_y w_0 + \tfrac{1}{2}(\partial_y w_0)^2\Big] + A_{45}\Big[\alpha\beta + \alpha \partial_y w_0 + \beta \partial_x w_0 + \partial_x w_0 \partial_y w_0\Big] + \\ &+ A_{55}\Big[\tfrac{1}{2}\alpha^2 + \alpha \partial_x w_0 + \tfrac{1}{2}(\partial_x w_0)^2\Big] + A_{66}\Big[\tfrac{1}{2}(\partial_y u_0)^2 + \partial_y u_0 \partial_x v_0 + \tfrac{1}{2}(\partial_x v_0)^2\Big] + \\ &+ D_{12} \partial_x \alpha \partial_y \beta + \tfrac{1}{2} D_{22}(\partial_y \beta)^2 + D_{66}\Big[\tfrac{1}{2}(\partial_y \alpha)^2 + \partial_y \alpha \partial_x \beta + \tfrac{1}{2}(\partial_x \beta)^2\Big] \Big\} dA \end{aligned} \tag{5}$$

where:

$$A_{ij} = \sum_k (Q_{ij})_k (h_k - h_{k-1}) \quad \text{for } i,j=1,\,2,\,6, \tag{6}$$

$$A_{ij} = \tfrac{5}{4}\sum_k (Q_{ij})_k \Big[h_k - h_{k-1} - \tfrac{4}{3}(h_k^3 - h_{k-1}^3)/h^2\Big] \quad \text{for } i,j=4,\,5, \tag{7}$$

and

$$D_{ij} = \tfrac{1}{3}\sum_k (Q_{ij})_k (h_k^3 - h_{k-1}^3) \quad \text{for } i,j=1,\,2,\,6. \tag{8}$$

Work done by the in–plane forces which includes the recovery forces of the Nitinol fibers, the thermal loads induced by heating the Nitinol fibers and the external in–plane loads can be written in the following form [4]:

$$\delta V_e' = \tfrac{1}{2} h \iint_A \delta w \Big[N_x(\partial_x w)^2 + N_y(\partial_y w)^2 + 2 N_{xy}(\partial_x w)(\partial_y w)\Big] dA, \tag{9}$$

where: $N_x = P_{m,x} + P_{t,x} - F_{w,x}$, $N_y = P_{m,y} + P_{t,y}$, $N_{xy} = P_{m,xy} + P_{t,xy}$, with $P_{m,x,y,xy}$ and $P_{t,x,y,xy}$ denoting the compressive in–plane mechanical and thermal loads in x, y and xy directions. Symbol $F_{w,x}$ denotes the total tension developed in the Nitinol fibers:

$$F_{w,x} = \iint_{A_w} [\varepsilon_0 - \alpha_r (T - T_0)] E_r \, dA_w , \tag{10}$$

where α_r, E_r and A_w are the thermal expansion coefficient of SMA wires, their modulus of elasticity and their cross–sectional area, respectively. The thermal loads P_t are given by the relations:

$$col[P_{t,x}, P_{t,y}, P_{t,xy}] = \sum_k \int_{h_{k+1}}^{h_k} [Q_k] col[\alpha_x, \alpha_y, \alpha_{xy}] \Delta T \, dz \tag{11}$$

in which $\alpha_x = \alpha_1 m^2 + \alpha_2 n^2$, $\alpha_y = \alpha_2 m^2 + \alpha_1 n^2$ and $\alpha_{xy} = (\alpha_1 - \alpha_2)mn$. The principal material axes are labeled 1 and 2. Symbol 1 denotes the direction parallel to the fibers direction and symbol 2 the direction normal to them. The angle of the fibers orientation is denoted by symbol θ (symbols m and n denote $\cos(\theta)$ and $\sin(\theta)$, respectively).

The thermal loads are generated by changes in temperature ΔT of the element and are caused by both activation and deactivation processes of the Nitinol fibers. If dynamic effects have to be included in the formulation, then Equation 4 must be supplemented with $p_z = -\rho \ddot{w}$:

$$\delta V_e'' = -\iint_A \delta w h \rho \ddot{w} \, dA . \tag{12}$$

Using standard finite element formulae, the inertia and stiffness matrices of the plate can be now easily determined.
The governing equation of motion takes the following form:

$$[M]\{\ddot{q}\} + ([K_s] + [K_r] - \Delta T \cdot [K_T])\{q\} = \{F\} \tag{13}$$

where symbols $[M]$ and $[K_s]$ denote the system mass and stiffness matrices, $[K_r]$ and $[K_T]$ are the geometric stiffness matrices due to recovery and thermal stresses, respectively.

The natural frequencies ω and mode shapes of vibration $\{q^0\}$ can be obtained by solving the following eigenvalue problem:

$$([K_s] + [K_r] - \Delta T \cdot [K_T] - \omega^2 [M])\{q^0\} = \{0\} . \tag{14}$$

4. Numerical calculation

As an example, a simply supported, square plate is investigated (length 600 mm, width 600 mm, thickness 8 mm). The analyzed plate possesses eight layers of graphite–epoxy composite material with the following orientation of graphite fibers: (0/90/90/0/0/90/90/0). Material properties for both components of the composite material are presented in Table 2.

TABLE 2. Material properties of graphite–expoy composite components.

	Matrix/Epoxy	Fiber/Graphite
Young modulus	3.43 GPa	275.6 GPa
Shear modulus	1.27 GPa	114.8 GPa
Poisson ratio	0.35	0.2
Density	1250 kg/m^3	1900 kg/m^3
Thermal expansion coeff.	$64.8 \cdot 10^{-6}$ 1/$^\circ$C	$24.4 \cdot 10^{-6}$ 1/$^\circ$C

The plate is modeled with 64 PBQ8 plate elements (at mesh size 8×8). Assumed volume fraction of the graphite fibers was 0.2 in each layer. In the present study, SMA Nitinol fibers cover only 5% of the cross–sectional area of the plate. They were embedded inside the neutral plane of the plate. It was also assumed that the material properties (except density, the angle of the graphite fibers and the coefficients of thermal expansion) are functions of temperature, as it is presented in Figure 4 (See also [1]). In this study, only uniform temperature distribution in the plate is considered (See Figure 3).

In order to determine the influence of the mesh size on convergence of the numerical results to analytical values, the numerical calculations for different mesh sizes were carried out. The natural frequencies calculated numerically were compared to those obtained from the analytical formula given in [6] and relative errors were determined (as shown in Table 3). Taking into account the results presented in Table 3, further calculations were carried out for 10x10 mesh size.

TABLE 3. Relative errors of natural frequencies for different mesh sizes.

Frequency Number	Mesh size								
	2x2	3x3	4x4	5x5	6x6	7x7	8x8	9x9	10x10
ω_{01}	25.90	2.10	0.41	0.01	0.15	0.23	0.27	0.30	0.32
ω_{02}	84.06	14.13	3.84	1.43	0.46	0.02	0.29	0.46	0.56
ω_{03}	84.06	14.13	3.84	1.43	0.46	0.02	0.29	0.46	0.56
ω_{04}	95.58	25.97	7.31	1.19	0.23	0.74	0.99	1.12	1.20
ω_{05}	249.07	16.63	13.04	5.85	2.73	1.14	0.25	0.28	0.62
ω_{06}	249.07	38.37	13.04	5.85	2.73	1.14	0.25	0.28	0.62
ω_{07}	159.02	21.02	20.39	4.93	0.91	0.47	1.11	1.45	1.66
ω_{08}	164.36	94.15	21.19	4.93	0.91	0.47	1.11	1.45	1.66
ω_{09}	209.82	73.19	8.11	7.24	6.66	3.31	1.41	0.25	0.47
ω_{10}	220.30	114.13	8.47	12.60	6.66	3.31	1.40	0.25	0.47

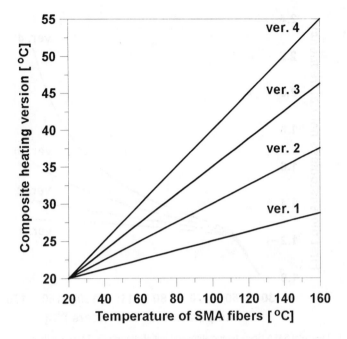

Figure 3. Temperature of the plate vs. temperature of SMA wires.

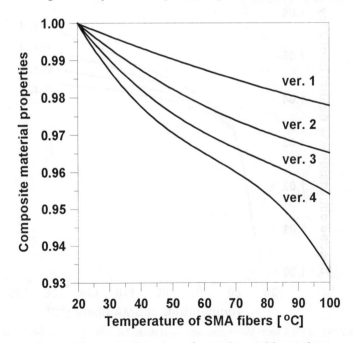

Figure 4. Temperature dependence of composite material properties.

272

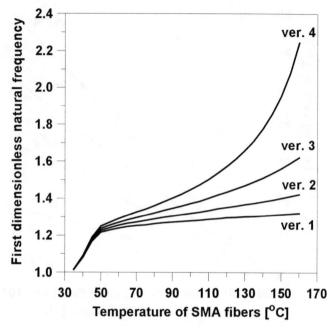

Figure 5. Influence of SMA fibers temperature and initial strains (ε_0=1%) upon the relative change of the first natural frequency.

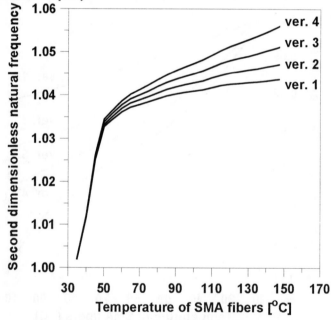

Figure 6. Influence of SMA fibers temperature and initial strains (ε_0=1%) upon the relative change of the second natural frequency.

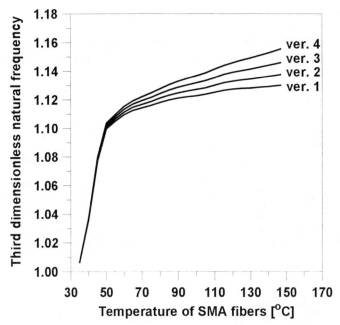

Figure 7. Influence of SMA fibers temperature and initial strains ($\mathcal{E}_0$=1%) upon the relative change of the third natural frequency.

The results of numerical investigations illustrate influence of SMA fibers temperature upon the changes in bending of the natural frequencies of the analyzed plate. In Figures 5–7 changes in the three first natural bending frequencies are presented. The results of the calculations are related to those obtained for the same plate without SMA wires.

The results obtained in this paper demonstrate the potential effectiveness of SMA fiber–reinforcement in composite structural elements in the process of controlling the vibration. The effect of SMA fibers activation on the amplitude of vibration normalized with respect to the amplitude of the uncontrolled vibration can also be analyzed.

It can be clearly observed that the activation process of SMA wires involves an increase in the natural bending frequencies. This effect rises when both the temperature and initial strains are higher.

5. Conclusions

Applications of SMA Nitinol fibers in the natural frequency analysis of composite plates have been successfully demonstrated. The fundamental equations governing the behavior of SMA fiber–reinforced multilayer composite plates have also been introduced in this work. The stress–strain relationships for a single composite layer with embedded SMA fibers has been presented. The finite element formulation to predict changes in natural frequencies and modes of vibration of composite laminates with SMA fibers has also been shown in the paper.

274

Acknowledgments

This work has been made possible through the financial support and sponsorship of the Polish State Committee for Scientific Research, Grant Number 1035/T07/96/11.

References

1. Baz, A., Poh, S., Ro, J. and Gilheany, J. (1995) Control of the natural frequencies of Nitinol–reinforced composite beams, *Journal of Sound and Vibration* **185**, 171–185.
2. Birman, V., Saravanos, D.A. and Hopkins, D.A. (1996) Micromechanics of composites with shape memory alloy fibers in uniform thermal fields, *AIAA Journal* **34**, 1905–1912.
3. Epps, J.J. and Chopra, I. (1997) Comparative evaluation of shape memory alloy constitutive models with test data, *Proceedings of 38th AIAA Structures, Structural Dynamics and Materials Conference and Adaptive Structures Forum*, Kissimmee, Florida, April 7–10.
4. Ro, J. and Baz, A. (1995) Nitinol–reinforced plates, Part I–III, *Composites Engineering* **5**, 61–106.
5. Turner, T.L., Zhong, Z.W. and Mei, C. (1994) Finite element analysis of the random response suppression of composite panels at elevated temperatures using shape memory alloy fibers, *AIAA–94–1324–CP*, 136–146.
6. Vinson, J.R. and Sierakowski, R.L. (1981) *Behaviour of Structures Composed of Composite Materials*, Martinus Nijhoff, London.
7. Weaver, W. and Johnston, P.R. (1984) *Finite Element for Structural Analysis*, Prentice Hall Inc., Englewood Cliffs, New Jersey.
8. Zhong, Z.W., Chen, R.R. and Mei, C. (1994) Buckling and postbuckling of shape memory alloy fiber reinforced composite plates, *Buckling and Postbuckling of Composite Structures, ASME.AD–Vol. 41/PVP–Vol. 233*.

APPLICATION OF ACTIVE TENDONS TO THE DAMPING OF AEROSPACE AND CIVIL ENGINEERING STRUCTURES

ANDRÉ PREUMONT & FRÉDÉRIC BOSSENS

Active Structures Laboratory
Université Libre de Bruxelles,
Brussels, Belgium

1. Introduction

In recent years, cable structures have had spectacular applications in large cable-stayed bridges. However, cable and deck vibrations have become a major design issue, because the ever increasing span of the bridges makes them more sensitive to flutter instability as well as to wind and traffic induced vibrations. This justifies the development of active damping devices for future large bridges.

Cable structures have also potential applications in the future large space structures, where they offer a light and effective way of stiffening large trusses. Many of these future large space structures, particularly those used for astrometric measurements, will require a very high accuracy, in the nanometer range, which can only be achieved with active vibration suppression devices.

The two types of structures described above constitute the driving force for our studies on robust decentralized active damping of cable structures using collocated force sensor and displacement actuator. This paper is a follow-up to several papers where the theory was gradually developed [1, 2, 3, 4, 5]; its main goal is to confirm previous results on structures more representative of realistic situations. It is organized as follows: section 2 summarizes the control strategy for active damping of cable structures; section 3 states without proof approximate analytical results for the closed-loop poles; the experimental results for the guyed truss and the cable-stayed bridge are described in section 4 and 5, respectively. Section 6 gives some concluding remarks.

J. Holnicki-Szulc and J. Rodellar (eds.), Smart Structures, 275–284.

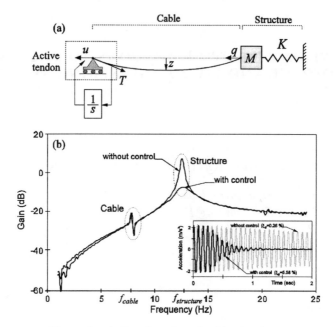

Figure 1. Active damping of cable structures.

2. Control of Cable Structures

2.1. CONTROL STRATEGY

It is widely accepted that the active damping of linear structures is much simplified if one uses collocated actuator-sensor pairs [3]; for nonlinear systems, this configuration is still quite attractive, because there exist control laws that are guaranteed to remove energy from the structure. The direct velocity feedback is an example of such "energy absorbing" control. When using a displacement actuator (active tendon) and a force sensor, the (positive) Integral Force Feedback

$$u = g \int T \, dt \tag{1}$$

(refer to Fig.1.a for notations) also belongs to this class, because the power flow from the control system is $W = -T\dot{u} = -gT^2$. This control law was already applied to the active damping of a truss structure in [6], and it is quite remarkable that it also applies to nonlinear structures; all the states that are controllable and observable are asymptotically stable for any value of g (infinite gain margin).

2.2. EXPERIMENT

The foregoing theoretical results have been confirmed experimentally with a laboratory scale cable structure similar to that represented schematically in Fig.1.a, where the active tendon consisted of a piezoelectric actuator [2]. Figure 1.b shows the experimental frequency response between a force applied to the structure and its acceleration; also shown in the figure is the free response of the structure with and without control. We see that the control system brings a substantial amount of damping in the structure, without destabilizing the cable (theoretically, the control system does indeed bring a small amount of damping to the cable, which depends on the sag); this behaviour is maintained at the parametric resonance, when the natural frequency of the structure is twice that of the cable.

2.3. DECENTRALIZED CONTROL

The foregoing approach can readily be extended to the decentralized control of a structure with several active cables, each tendon working for itself with a local feedback following Equ.(1). This statement was verified experimentally on a T structure controlled with two cables [4]. It is important to point out that the concept of active tendon control of cable structures does not require that all the cables be active; on the contrary, the control system would normally involve only a small set of cables judiciously selected. Next section summarizes the main results of an approximate linear theory to predict the performance of the control system and provides design guidelines to select the active cables.

3. Closed-Loop Poles

If we assume that the dynamics of the active cables can be neglected and that their interaction with the structure is restricted to the tension in the cables, it is possible to develop an approximate linear theory of the closed-loop system.

For a decentralized feedback control law

$$\delta = \frac{g}{s} K_c^{-1} T \tag{2}$$

where T is the local force measurement, δ is the active tendon displacement, K_c is the stiffness of the active cable ($K_c^{-1}T$ represents the elastic extension of the active cable) and g is the control gain (the same for all control elements), the following results have been established in earlier studies [2]:

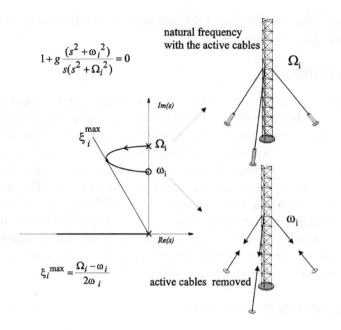

$$1+g\frac{(s^2+\omega_i{}^2)}{s(s^2+\Omega_i{}^2)}=0$$

$$\xi_i^{max}=\frac{\Omega_i-\omega_i}{2\omega_i}$$

Figure 2. Root locus of the closed-loop poles.

1. If we assume no structural damping, the open-loop zeros are $\pm j\omega_i$ where ω_i are the natural frequencies of the structure where the active cables have been removed.

2. The open-loop poles are $\pm j\Omega_i$ where Ω_i are the natural frequencies of the structure including the active cables.

3. As g goes from 0 to ∞, the closed-loop poles follow the root locus corresponding to the open-loop transfer function (Fig.2)

$$G(s) = g\frac{(s^2 + \omega_i^2)}{s\left(s^2 + \Omega_i^2\right)} \tag{3}$$

Thus, the closed-loop poles go from the open-loop poles at $\pm j\Omega_i$ for $g = 0$ to the open-loop zeros at $\pm j\omega_i$ for $g \to \infty$.

4. The depth of the loop in the left half plane depends on the frequency difference $\Omega_i - \omega_i$ and the maximum damping, obtained for $g = \Omega_i\sqrt{\Omega_i/\omega_i}$, is

$$\xi_i^{max} = \frac{\Omega_i - \omega_i}{2\omega_i} \tag{4}$$

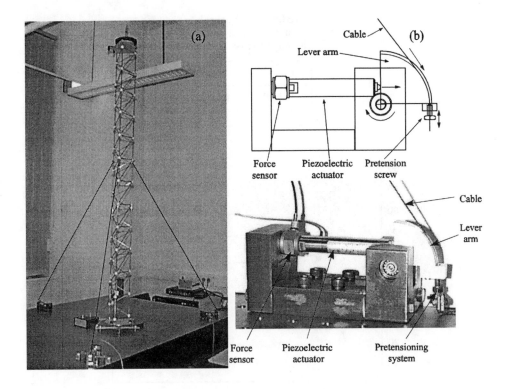

Figure 3. (a) Guyed truss. (b) Design of the active tendon.

5. For small gains, the modal damping ratio resulting from the active tendon control is given by

$$\xi_i \approx \frac{g\nu_i}{2\Omega_i} \tag{5}$$

where $\nu_i = (\Omega_i^2 - \omega_i^2)/\Omega_i^2$ is the *modal fraction of strain energy* in the active cables.

Equations (4) and (5) can be used very conveniently in the design of actively controlled cable structures.

4. Guyed Truss Experiment

The first experiment concerns the guyed truss of Fig.3.a; the guy cables are made of a synthetic fiber "Dynema" of 1 mm diameter; their tension is adjusted in order to achieve a cable frequency larger than 500 rad/s, well above the first two bending modes of the truss.

TABLE 1. Natural
frequencies (rad/s) of
the guyed truss.

i	ω_i	Ω_i
1	53.8	67.9
2	66	78.9

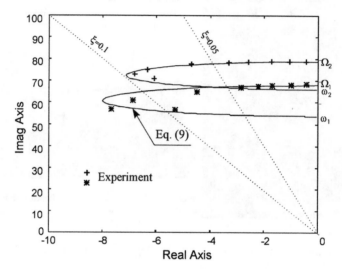

Figure 4. Experimental vs. analytical closed-loop poles.

The design of the active tendon is shown in Fig.3.b; the amplification ratio of the lever arm is 3, leading to a maximum stroke of about 150μ. The natural frequencies with and without the active cables (respectively Ω_i and ω_i) are given in Table 1. Figure 4 shows the root locus predicted by the linear model together with the experimental results for various values of the gain; only the upper part of the loops is available experimentally because the control gain is limited by the saturation due to the finite stroke of the actuators. The agreement between the experimental results and the linear predictions is quite good. The guyed truss is actually the same as the one used in a previous study on active damping of trusses [6]; it is provided with two active struts near its base (visible on Fig.3.a); each active strut consists of a piezoelectric actuator collocated with a force sensor. Figure 5 displays a typical frequency response function for various operating conditions: **(a)** open-loop (no control), **(b)** active truss alone, **(c)** active cables, **(d)** active truss + active cables. Figure 5.d shows that the combined use of an active

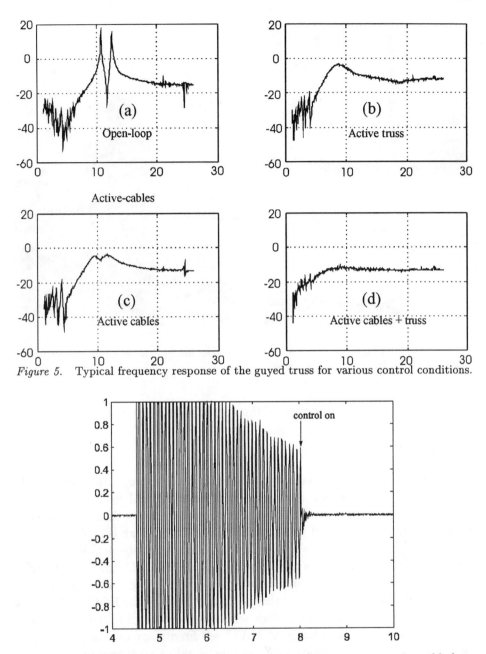

Figure 5. Typical frequency response of the guyed truss for various control conditions.

Figure 6. Time response with and without control (active truss + active cables).

truss and the active cables eliminates completely the resonant peaks of the frequency response. Figure 6 shows a typical time response with and without control, for case (d), after an impulsive load (the control is turned

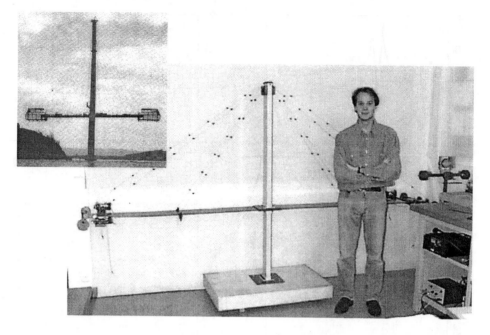

Figure 7. Experimental set-up for the cable-stayed bridge (the small picture shows an actual bridge in its construction phase).

on after 8 sec). The decay rate of the controlled structure is spectacular.

5. Cable-Stayed Bridge

The test structure is a laboratory model of a cable-stayed bridge during its construction phase, which is amongst the most critical from the point of view of the wind response. The structure consists of two half decks mounted symmetrically with respect to a central column of 2 m high (Fig.7); each side is supported by 4 cables, two of which are equipped with active piezoelectric tendons (Fig.3.b). This experiment is part of an EC funded industrial project aimed at demonstrating the applications of active control techniques in civil engineering [7]. Figure 8 shows the evolution of the first bending and torsion closed-loop poles of the deck when the control gain increases (these poles have been obtained with the MATLAB Frequency Domain Identification Toolbox from frequency response functions). Also shown on the figure are the predictions of Equ.(3); the agreement is good for moderate values of the gain. For larger gains, when the modal damping exceeds 20%, the discrepancy between experimental and analytical results increases, but this seems to be essentially related to the identification algorithm which cannot distinguish oscillating poles from highly damped frequency response

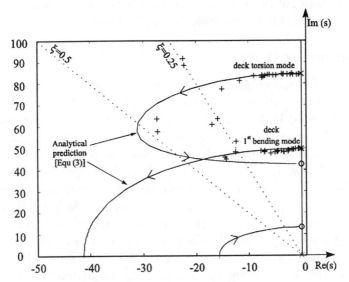

Figure 8. Evolution of the first bending and torsion poles of the deck with the control gain.

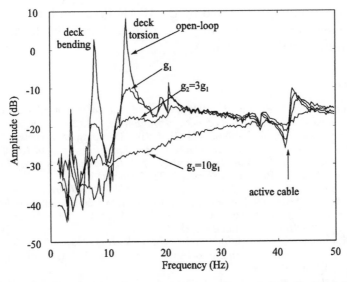

Figure 9. Frequency response between an active tendon and the collocated force sensor.

(Fig.9). Note that damping ratios significantly larger than 20% can be obtained. Figure 9 shows a typical frequency response between the voltage applied to one of the active tendons and its collocated force sensor, for several values of the gains (respectively g_1, $3g_1$,$10g_1$). We note that for large gains the resonant peaks of the structure have totally disappeared and that the one of the cable is also considerably reduced.

6. Concluding Remarks

The use of tension cables for active damping of cable structures has been investigated theoretically and experimentally. The decentralized control approach proposed is simple, robust and easy to implement with analog electronics (advantage for space applications); in case of a sensor or actuator failure in an active tendon, the corresponding control loop simply returns to its passive state. Simple formulae have been developed for predicting the closed-loop poles; these formulae have been confirmed experimentally and they can be used very conveniently in the design, for selecting the number and location of the active tendons. Both test structures considered in this study use piezoelectric active tendons; those are well adapted for space applications, but are not suitable for civil engineering. For bridges and other civil engineering applications, special hydraulic divices and giant magnetostrictive actuators are being developed within the ACE project [7].

Acknowledgment

This study was partly supported by the IUAP IV-24 on Intelligent Mechatronics Systems and by the Brite-Euram project n^o BE96-3334: ACE (Active Control in Civil Engineering).

References

1. Y. Achkire and A. Preumont, (1996) Active tendon control of cable-stayed bridges, *Earthquake Engineering and Structural Dynamics*, Vol. **25**, No. **6**,585-597.
2. A. Preumont and Y. Achkire, (1997) Active Damping of Structures with Guy Cables, *AIAA, J. of Guidance, Control, and Dynamics*, Vol. **20**, No. **2**, 320-326
3. A. Preumont, (1997) *Vibration Control of Active Structures : An Introduction*, Kluwer Academic Publishers.Dordrecht, The Netherlands
4. Y. Achkire,(1997) *Active Tendon Control of Cable-Stayed Bridges*, Ph.D. dissertation, Active Structures Laboratory, Université Libre de Bruxelles, Belgium.
5. A. Preumont and Y. Achkire and F. Bossens, (1998) Active tendon control of large trusses, *39th SDM conf.*, Long Beach, CA,
6. A. Preumont and J.P. Dufour and Ch. Malekian, (1992) Active Damping by a Local Force Feedback with Piezoelectric Actuators, *AIAA, J. of Guidance*, Vol. **15**, No **2**,390-395
7. ACE, Active Control in Civil Engineering, *Brite Euram project n^o BE96-3334,*

CONTROL THEORY SOURCES IN ACTIVE CONTROL OF CIVIL ENGINEERING STRUCTURES

J. RODELLAR
Dept. of Applied Mathematics III, School of Civil Engineering,
Technical University of Catalunya, 08034 Barcelona, SPAIN.

1. Introduction

Research in system analysis and control theory has lead to a broad variety of methods over the last four decades. Many of those have been transferred to (and even motivated by) different areas of science and technology. Within this scenario, feedback control has grown in the field of structural dynamics as an innovative mean to reduce vibrations produced by dynamic loads. Control systems in this field have been referred to as active control systems as counteracting to the passive systems based on absorbing or dissipating energy, both classes of systems, plus combinations such as hybrid and semiactive systems, delimiting the field labeled to as structural control. Structural control had origins in flexible space structures and rapidly extended to civil engineering structures [1] as a promising tool to protect bridges, buildings and other structures against loads induced by earthquakes, wind, etc. It is an always incomplete task to review the research effort in structural control, which nowadays involves a variety of disciplines and technologies such as mechanics, structures, materials, sensors, actuators, mathematics, computation and control theory, among others. We can mention a number of books and proceedings of structural control conferences [2-8] and recent review papers [9,10]. This paper will focus on control theory issues outlining some of the most relevant methods adopted to cope with specific features of civil engineering active structural control.

2. Essential issues in feedback control systems

A feedback controller is designed with the objective of making a set of output variables to exhibit some kind of desired behaviour. The desired behaviour is

J. Holnicki-Szulc and J. Rodellar (eds.), Smart Structures, 285–294.

translated into a set of specifications or performance criteria in terms of transient response, frequency response, steady state response, control effort, robustness in the presence of uncertainties in modelling or disturbances, reliability, etc. The control design is caste within a theoretical framework in which two essential elements interplay: a model of the closed control loop and a control strategy. The issue of modelling for control purposes is very crucial since the controller design is performed to guarantee a desired performance of the system, but "system" is usually synonymous of "model" in this design. Most of the control strategies are associated to models described in the form of differential equations (in the case of continuous time approaches) or difference equations (in the case of discrete time formulations). Alternatively to "model based" strategies, there are approaches that can be referred to as "model-free". In this context we can frame the classical control techniques, first proposed in the 40's, such as PID controllers which are designed through engineering tuning rules. In recent years, model-free approaches have been proposed based on neural networks [11,12] or fuzzy logic [12,13] which rely on building relations between inputs, outputs and environmental loads with little a priori knowledge combined with rules with a highly qualitative contents. Some other model-free approaches can be mentioned, as the one proposed in [14] based on displaying eigenvalue orbits of a matrix that relates real time outputs given by sensors on a moving time window.

3. Specific features of active structural control

As mentioned above, the controller aims to actively induce a desired behavior to the system. For a structure, in physical terms, the desired behaviour [9] can be stated in keeping stresses, strains, accelerations and displacements at specified locations below given bounds (maximum values, root mean square, etc.) in the presence of disturbances (earthquakes, wind, etc.) that are below certain bounds measured by peaks, energy, specific scales, etc. Some of these variables are relevant to safety (stresses, displacements) and others may be important to keep comfort within certain specifications (accelerations). Clearly the input generated by the controller is a signal that makes the actuators apply appropriate actions to the structure (forces, torques). Some specific characteristics of structures pose crucial problems to achieve this objective, which we may associate to physical constrains and modeling aspects.

From a physical side, structures are parameter distributed systems, are massive and have large dimensions. This poses practical problems to deal with the selection of the number of sensors and actuators and where is their best location and with the design of appropriate devices to apply large and fast enough actions to the structure. From a modeling view, two key issues are particularly significant: the model class and the uncertainties to describe reality. Models can be built in terms of equations derived from first principles or by identifying input/output relations on an experimental/empirical basis. The first case leads to partial differential equations (infinite dimension models) which, by appropriate discretization tools, can be approximated by ordinary differential equations. It can be expected that the order of these finite dimensional models be high, so that structures are viewed as large scale systems. The second case can lead to identify transfer functions (or equivalent differential equations), identify discrete time equations, build neural networks, etc.

In any case, models are "ideal" descriptions with doses of discrepancies with reality. Uncertainties sources in an active control may come from the order model in the discretization and dimension reduction, from erroneous selection or identification of the model parameters, from non linear components and actuator dynamics (difficult to model) and from the lack of knowledge or prediction of the environmental excitations.

4. Some Active Control Approaches

This section outlines some relevant approaches known in the literature on active control of civil engineering structures associated to the issues discussed in the previous sections. It is organized in two subsections corresponding to a design of the control based on mathematical models and to a model-free design, respectively. When considering mathematical models, we can distinguish the two classes of (1) distributed parameter systems, infinite dimensional, partial differential equation models and (2) lumped parameter systems, finite dimensional, ordinary differential (or difference) equation models. A mathematical body of control theory for distributed parameter systems exists [15] and rigorous approaches for structures such as beams and plates [16-19] have been proposed in the literature, mainly in aerospace or mechanical engineering scenarios. Although it is theoretically appealing and existence of distributed controllers is rigorously proved, distributed control is difficult to implement. In fact, analytical treatment of the model in practical cases is not always possible due, for instance, to the boundary conditions. Moreover, a distributed control loop would require a continuous distribution of sensors and actuators. To overcome these problems, sensors and point controllers are considered at discrete locations and, often, numerical approximations of the equation solutions are considered. In this case, the control problem resembles to a finite dimension model based control design and implementation. Here we will concentrate on finite dimensional approaches.

4.1. MODEL BASED APPROACHES

The conducting line that we will follow to discuss control approaches is to relate a control principle or objective that states the desired behaviour of the structure based on the mathematical model available for the design. In this sense we will discuss optimal and predictive control approaches and a class of control approaches deeply oriented to guarantee structural properties in spite of uncertainties.

One way of characterizing the desired behaviour is in terms of a performance criterion. Optimal control refers to a broad methodology characterized by finding the control that minimizes a performance criterion. This minimization is compatible with the model relating inputs and outputs and possible constraints on the variables. Optimal control is a well known and established body of theory [20] and many uses of it have been reported in the structural control literature, particularly for the case of linear time invariant models with linear quadratic (LQ) cost functions [3,4,21-23]. LQ control is defined by the minimization of an integral in which weighted squares of the response and the control are explicitly formulated [24]. Minimization of such a performance criterion, when no restrictions on the variables are imposed, is analytically solved through a matrix Riccati differential equation when the integral is extended over a finite time. In practice, the cost function integral is extended over an infinite time horizon, what leads to an algebraic Riccati equation and a linear controller with a constant gain matrix. The most stringent issue in LQ optimal control implementation is the selection of the weighting matrices. Some practical issues associated to the choice of weights in optimal structural control are described in [25-27]. Cost functions can be formulated in a discrete time setting with summations instead of integrals in combination with discrete time models [28].

In the above description, uncertainties and disturbances on the system are ignored so that "optimal control" means the "best" possible control using the "best" model. Actual performance deteriorates due to the presence of external loads and model discrepancies, but LQ optimal control has some degree of robustness. In the specific case of assuming random disturbances, the system becomes a stochastic process and the deterministic cost function is replaced by its expectation. The result is the well known linear quadratic Gaussian (LQG) control which is combined with an observer that performs an optimal estimation of the state variables [20]. State estimation and LQG control methods have been invoked by several authors for civil engineering structures, with the main motivation in that states, like displacements and velocities of the structural degrees of freedom, may not be all of them available and even some states can be immeasurable, like states associated to actuator dynamics [29-32].

Some variants of stochastic optimal control have been proposed, such as the so called covariance control [33-34]. This approach considers a linear time invariant state model and aims to design a linear feedback control to ensure that the controlled system is stable in the mean square, that is that the covariance matrix of the state describing the closed loop is positive definite. Since stochastic control is motivated mainly by the randomness of disturbances, some authors have considered the problem of designing deterministic LQ active controllers and assessing its robustness against stochastic disturbances in terms of probability of destabilizing the control loop [35,36].

Predictive control is a control methodology that relies on the prediction of the system output by a model on a finite time horizon and the computation of the control that makes this predicted output to exhibit a desired behaviour. Predictive control is usually formulated in a discrete time setting by using an input/output, state representation or impulse response model. At each sampling instant, this model is used to predict the system response on a prediction horizon and then the control sequence that gives a predicted trajectory minimizing a performance criterion is computed. With a receding horizon strategy, only the first value of the control sequence is applied to the system and the procedure is redefined at each sampling time to account for the actual system inputs and outputs. Predictive control is quite popular nowadays in several domains of control engineering. The concepts underlying the methodology and applications may be found in [37]. Application of predictive control in civil engineering structural control has been reported in papers like those in [38-41].

The above outlined methods of optimal and predictive control have in common to rely on a control objective based on performance criteria. There is a broad family of methods, which can be labeled as robust control, whose objective is guarantee some system properties in spite of the presence of uncertainties which are unknown but characterized in some way. If uncertainties, in the form of unknown disturbances (excitations), are described probabilisticaly, the above discussed LQG control and variants can be considered within this category. Inability of LQG to cope with uncertainties in system models was the motivating factor for the development of the so called H-infinity control method (see for instance [42] for a general reference). The basic background of this method is to find a linear output feedback control that minimizes the infinity norm of the transfer matrix that relates the external excitations with the responses of the system that one wants to control. Solution of this problem resembles the classical problem of LQG control in that a Riccati equation is needed to be solved. The infinity norm is interpreted as a measure of the worst case response for a class of exogenous disturbances. References [30, 31, 43-49] present applications of this methodology to civil engineering structural control.

Another approach to robust control is characterized in terms of stability. As clearly proposed in [50], the idea is to find a control law (linear, non linear of a combination) that guarantees a form of stability of the controlled system for any realization of the uncertainties that belongs to a prescribed class. In this approach uncertainties are deterministic, bounded in some assumed form and correspond to model errors and unknown external excitations. Applications of this methodology in active control of civil engineering structures can be found in [51-55]. In a similar stability vein, sliding mode control aims to drive the state of the system to a specified surface in the state space and keep the closed loop robust against uncertainties. [56] is a good general fundamental reference on sliding mode control. Applications to civil engineering structures have been reported in [57-60].

4.2. MODEL–FREE APPROACHES

Approaches discussed in the previous subsection have in common the use of a standard (let us say) mathematical model describing the system to be controlled. For cases in which the knowledge involved in model building is highly incomplete and with qualitative components the use of so called "soft" or "intelligent" techniques have been proposed, neural networks and fuzzy logic being the two main methodologies in this context [11-13]. Applications of neural networks in active control of civil structures can be found in [61-64], while references [65-67] describe fuzzy control applications.

5. References

1. Yao, J.T.P. (1972) Concept of structural control, *Journal of Structural Division ASCE*, **98**, 1567-1574.

2. Leipholz, H.H.E., ed. (1987) *Structural Control*, Martinus Nijhoff Publ., Dordrecht, The Netherlands.

3. Leipholz, H.H.E. and Abdel-Rohman, M. (1986) *Control of Structures*, Martinus Nijhoff Publ., Dordrecht, The Netherlands.

4. Soong, T.T. (1990) *Active Structural Control*, Longman Scientific and Technical, New York, USA.

5. Constantinou, M.C. and Soong, T.T., eds. (1994) *Passive and Active Structural Vibration Control in Civil Engineering*, Springer Verlag, Wien, Austria.

6. Housner, G.W., Masri, S.F. and Chassiakos, A.G., eds. (1994) *Proc. First World Conference on Structural Control*, Pasadena, USA.

7. Baratta, A. and Rodellar, J., eds. (1996) *Proc. First European Conference on Structural Control*, Barcelona, Spain, World Scientific, Singapore.

8. Chen, J.C., chairman (1996) *Second International Workshop on Structural Control*, Hong Kong.

9. Housner, G.W., Bergman, L.G., Caughey, T.K. et al. (1997), Structural control: past, present and future, *Journal of Engineering Mechanics ASCE*, **123(9)**, 897-971.

10. Spencer Jr., B.F. and Sain, M.K. (1997) Controlling buildings: a new frontier in feedback, *IEEE Control Systems Magazine: Special Issue on Emerging Technologies*, 19-35.

11. Special issue on intelligent control, *IEEE Control System Magazine*, June, 1993.

12. Brown, M. and Harris, C. (1994) *Neurofuzzy Adaptive Modelling and Control*, Prentice Hall, Englewood Cliffs, USA.

13. De Silva, C.W. (1995) *Intelligent Control: Fuzzy Logic Applications*, CRC Press, Boca Raton, USA.

14. Marczyk, J., Rodellar, J. and Barbat, A.H. (1996) Semiactive control of nonlinear systems via eigenvalue orbits, *Proc. First European Conference on Structural Control*, World Scientific, Singapore, pp. 451-458.

15. Omatu, S. and Seinfeld, J.H. (1989) *Distributed Parameter Systems. Theory and Applications*, Clarendon Press, Oxford, UK.

16. Langnese, J. and Lions, J.L. (1988) *Modelling, Analysis and Control of Thin Plates*, Collection RMA, 6, Masson, Paris, France.

17. Meirovitch, L. and Baruh, H. (1982) Control of self-adjoint distributed parameter systems, *AIAA Journal of Guidance and Control*, **5**, 60-66.

18. Lions, J.L. (1988) Exact controllability, stabilizability and perturbations for distributed systems, *SIAM Reviews*, **30**, 1-68.

19. Bourquin, F. A numerical controllability test for distributed systems, *Journal of Structural Control*, **2(1)**, 5-23.

20. Kwakernaak, H. and Sivan, R. (1972) *Linear Optimal Control Systems*, Wiley, New York, USA.

21. Yang, J.N. (1975) Application of optimal control theory to civil engineering structures, *Journal of Engineering Mechanics ASCE*, **101**, 818-838.

22. Yang, J.N., Akbarpour, A. and Ghammaghami, P. (1987) New control algorithms for structural control, *Journal of Engineering Mechanics ASCE*, **113**, 1369-1386.

23. Chung, L.L., Reinhorn, A.M. and Soong, T.T. (1988) Experiments on active control of structures, *Journal of Engineering Mechanics ASCE*, **114**, 241-256.

24. Anderson, B.D.O. and Moore, J.B. (1990) *Optimal Control, Linear Quadratic Methods*, Prentice Hall, New Jersey, USA.

25. Seto, K. (1994) Vibration control method for flexible structures arranged in parallel, *Proc. First World Conference on Structural Control*, Pasadena, USA, **3(FP3)**, 62-71.

26. Kubo, T. and Furuta, E. (1994) Seismic response and its stability of an AMD controller building considering control signal delay and control force saturation, *Proc. First World Conference on Structural Control*, Pasadena, USA, **3(FP4)**, 12-21.

27. Shibata, H., Shigeta, T., Nakakoji, Y. et al. (1994) Direct control structure against long-period ground motion, *Proc. First World Conference on Structural Control*, Pasadena, USA, **3(FP2)**, 3-10.

28. Ogata, K. (1987) *Discrete Time Control Systems*, Prentice Hall, New Jersey, USA.

29. Dyke, S.J., Spencer Jr., B.F., Quast, P. and Sain, M.K. (1995) The role of control-structure interaction in protective system design, *Journal of Engineering Mechanics ASCE*, **121**, 322-338.

30. Suhardjo, J., Spencer Jr., B.F. and Kareem, A. (1992) Frequency domain optimal control of wind excited buildings, *Journal of Engineering Mechanics ASCE*, **118(12)**, 2463-2481.

31. Yoshida, K., Kang, S. and Kim, T. (1994) LQG and Hinf control of vibration isolation for multi-degree-of-freedom systems, *Proc. First World Conference on Structural Control*, Pasadena, USA, **TP4**, 53-62.

32. Dyke, S.J., Spencer Jr., B.F., Quast, P., Sain, M.K. and Kaspari Jr., D.C. (1986) Acceleration feedback control of MDOF structures, *Journal of Engineering Mechanics ASCE*, **122(9)**, 907-918.

33. Skelton, R., Iwasaki, T. and Grigoriadis, K. (1996) *A Unified Algebraic Approach to Control Design*, Taylor and Francis, London, UK.

34. Zhu, G., Rotea, M. and Skelton, R. (1997) A convergent algorithm for the output covariance constraint problem, *SIAM Journal of Control and Optimization*, **35(1)**, 341.

35. Field, R.V., Voulgaris, P.G. and Bergman, L.A. (1996) Probabilistic stability robustness of structural systems, *Journal of Engineering Mechanics ASCE*, **122(10)**, 1012-1021.

36. Spencer Jr., B.F., Sain, M.K., Won, C.H., Kaspari Jr. D. and Sain, P. (1994) Reliability-based measures of structural control robustness, *Structural Safety*, **15**, 111-129.

37. Martín Sánchez, J.M. and Rodellar, J. (1995) *Adaptive Predictive Control. From the Concepts to Plant Optimization*, Prentice Hall, Englewood Cliffs, USA.

38. Rodellar, J., Barbat, A.H. and Martín Sánchez, J.M. (1987) Predictive control of structures, *Journal of Engineering Mechanics ASCE*, **113(6)**, 797-812.

39. Rodellar, J., Chung, L.L., Soong, T.T. and Reinhorn, A.M. (1989) Experimental digital control of structures, *Journal of Engineering Mechanics ASCE*, **115(6)**, 1245-1261.

40. Inaudi, J., López Almansa, F., Kelly, J.M. and Rodellar, J. (1992) Predictive control of base isolated structures, *Earthquake Engineering and Structural Dynamics*, **21**, 471-482.

41. López Almansa, F., Andrade, R., Rodellar, J. and Reinhorn, A.M. (1995) Modal predictive control of structures. Part I: Formulation. Part II: Implementation, *Journal of Engineering Mechanics ASCE*, **120(8)**, 1743-1772.

42. Stoorvogel, A. (1992) *The H-infinity Control Problem. A State Space Approach*, Prentice Hall, Englewood Cliffs, USA.

43. Spencer Jr., B.F., Suhardjo, J. and Sain, M. (1994) Frequency domain optimal control strategies for seismic protection, *Journal of Engineering Mechanics ASCE*, **120(1)**, 135-158.

44. Schmitendorf, W.E., Jabbari, F. and Yang, J.N. (1994) Robust control techniques for buildings under earthquake excitations, *Earthquake Engineering and Structural Dynamics*, **23**, 539-552.

45. Miyata, T., Yamada, H., Dung, N.N. and Kazama, K. (1994) On active control and structural response control of the coupled flutter problem for long span bridges, *Proc. First World Conference on Structural Control*, Pasadena, USA, **WA4**, 13-22.

46. Nishitani, A. and Yamada, N. (1994) H-infinity structural response control with reduced-order controller, *Proc. First World Conference Structural Control*, Pasadena, USA, *TP4*, 110-119.

47. Chase, J.G. and Smith, H.A. (1996) Robust H-infinity control considering actuator saturation. Part 1: theory, *Journal of Engineering Mechanics ASCE*, **122(10)**, 976-983.

48. Chase, J.G., Smith, H.A. and Suzuki, T. (1996) Robust H-infinity control considering actuator saturation. Part 2: aplication, *Journal of Engineering Mechanics ASCE*, **122(10)**, 984-993.

49. Kozodoy, D.A. and Spanos, P.D. (1995) Robustly stable active structural control, *Journal of Structural Control*, **2(1)**, 65-77.

50. Corless, M. and Leitmann, G. (1981) Continuous state ffedback guaranteeing uniform ultimate boundedness for uncertain dynamic systems, *IEEE Transactions on Automatic Control*, **AC-26**, 1139-1144.

51. Kelly, J.M., Leitmann, G. and Soldatos, A.G. (1987) Robust control of base-isolated structures under earthquake excitation, *Journal of Optimization Theory and Applications*, **53(1)**, 159-180.

52. Rodellar, J., Leitmann, G. and Ryan, E.P. (1993) Output feedback control of uncertain coupled systems, *International Journal of Control*, **58(2)**, 445-457.

53. Leitmann, G. and Reithmeier, E. (1993) Semi-active control of a vibrating system by means of electro-rheological fluids, *Dynamics and Control*, **3(1)**, 7-33.

54. Magaña, M.E. and Rodellar, J. (1998) Nonlinear decentralized active tendon control of cable-stayed bridges, *Journal of Structural Control*, **5(1)**, 45-62.

55. Barbat, A.H., Rodellar, J., Ryan, E.P. and Molinares, N. (1995) Active control of non-linear base-isolated buildings, *Journal of Engineering Mechanics ASCE*, **121(6)**, 676-684.

56. Utkin, V.I. (1992) *Sliding Modes in Control and Optimization*, Springer Verlag, Berlin, Germany.

57. Yang, J.N., Wu, J.C. and Agrawal, A.K. (1995) Sliding mode control for nonlinear and hysteretic structures, *Journal of Engineering Mechanics ASCE*, **121(12)**, 1330-1339.

58. Adhikari, R. and Yamaguchi, H. (1994) Adaptive control of nonstationary wind-induced vibration of tall buildings, *Proc. First World Conference on Structural Control*, Pasadena, USA, **TA4**, 43-52.

59. Singh, M.P. and Matheu, E.E. (1994) Sliding mode control of civil structures with generalized sliding surface, *Proc. First European Conference on Structural Control*, World Scientific, Singapore, pp. 551-558.

60. Luo, N., Rodellar, J. and de la Sen, M. (1998) Composite robust active control of seismically excited structures with actuator dynamics, *Earthquake Engineering and Structural Dynamics*, **27**, 301-311.

61. Faravelli, L. and Venini, P. (1994) Active structural control by neural networks, *Journal of Structural Control*, **1(1-2)**, 79-101.

62. Ghaboussi, J. and Joghataie, A. (1991) Active control of structures using neural networks, *Journal of Engineering Mechanics ASCE*, **117(1)**, 132-153.

63. Nikzad, K., Ghaboussi, J. and Paul, S. (1996) Actuator dynamics and delay compensation using neurocontrollers, *Journal of Engineering Mechanics ASCE*, **122**, 966-975.

64. Masri, S.F., Nakamura, M., Chassiakos, A.G. and Caughey, T.K. (1996) A neural network approach to the detection of changes in structural parameters, *Journal of Engineering Mechanics ASCE*, **122(5)**.

65. Iiba, M., Fujitani, H., Kitagawa, Y. et al. (1994) Shaking table test on sesmic response control system by fuzzy optimal logic, *Proc. First World Conference on Structural Control*, Pasadena, USA, **WP1**, 69-77.

66. Nagarajaiah, S. (1994) Fuzzy controller for structures with hybrid isolation systems, *Proc. First World Conference on Structural Control*, Pasadena, USA, **TA2**, 67-76.

67. Sun, L. and Goto, Y. (1994) Applications of fuzzy theory to variable dampers for bridge vibration control, *Proc. First World Conference on Structural Control*, Pasadena, USA, **WP1**, 31-40.

ANALYSIS AND OPTIMIZATION OF ENERGY CONVERSION EFFICIENCY FOR PIEZOELECTRIC TRANSDUCERS

N.N.ROGACHEVA
Institute for Problems in Mechanics
Russian Academy of Sciences, Moscow, Russia
prospekt Vernadskogo 101, Moscow 117526, Russia

1. Introduction

Different formulas for electromechanical coupling coefficient (EMCC) characterizing the energy conversion of piezoelectrical structures are analyzed. As a result, the fields of application for each EMCC formula are established. A general-purpose method for the EMCC determination is worked out. The dependence of the EMCC on the size, shape of electrodes, and thicknesses of elastic and piezoelectric layers is analyzed. As an example, the EMCC for cylindrical piezoelectric transducer and a three-layered beam are calculated and optimized as a function of different parameters. It is widely believed that the EMCC for dynamic state of any structure is less than the EMCC for its static state. In this paper we show the for first time that the EMCC for a dynamic behavior can be extended to a maximum static value.

2. Different Ways of Determination of the EMCC

An important characteristic of the performance of piezoceramic elements is the electromechanical coupling coefficient. Most of the scientists introduce the concept of electromechanical coupling coefficient as follows: k^2 is the ratio of electrical (mechanical) energy stored in the volume of a piezoceramic body and capable of conversion, to the total mechanical (electrical) energy supplied to the body. This determination is the most complete one, but it is difficult to understand what means "capable of conversion".

There are different formulas for calculating the EMCC. In this paper we discuss three of them:

J. Holnicki-Szulc and J. Rodellar (eds.), Smart Structures, 295–302.

The simple and popular formula [1],[2] defines the EMCC (we denote it by k_s) as the ratio of the mutual elastic and electrical energy U_m to the geometric mean of the elastic U_e and electrical U_d energy densities. They derived the formula for k_s in the following formal way:

The formula for internal energy can be written in short index notation as

$$U = \frac{1}{2}(\sigma_i \epsilon_i + E_m D_m), \quad i = 1, 2, ..., 6, \quad m = 1, 2, 3. \tag{1}$$

Here and below we use the same notations as in [3].

Using the constitutive equations,

$$\epsilon_i = s_{ij}^E \sigma_i + d_{ki} E_k, \quad D_m = d_{mn} \sigma_n + \epsilon_{mk} E_k, \tag{2}$$

the expression for the internal energy can be rearranged as follows:

$$U = U_e + 2U_m + U_d \tag{3}$$

where

$U_e = \frac{1}{2}\sigma_i s_{ij}^E \sigma_i$ is the elastic energy,

$U_m = \frac{1}{2}\sigma_i d_{ki} E_k$ is the mutual energy,

$U_d = \frac{1}{2}E_m \varepsilon_{mk}^T E_k$ is the electrical energy.

The electromechanical coupling coefficient is determined according to the following formula:

$$k_s = \frac{U_m}{\sqrt{U_e U_d}}. \tag{4}$$

Formula (4) is used sometimes as a general formula for the EMCC suitable for both statics and dynamics.

Another popular formula for the electromechanical coupling factor was suggested by Mason (we call it as a dynamic electromechanical coupling coefficient and denote by k_d):

$$k_d = \frac{\sqrt{\omega_a^2 - \omega_r^2}}{\omega_a}, \tag{5}$$

where ω_a is the resonance frequency and ω_r is the antiresonance frequency nearest to the resonance frequency ω_a.

Note, that the formula (5) allows to determine the efficiency near resonance - regime operation only. This formula is widely used to define EMCC experimentally by measuring the resonance and antiresonance frequencies.

Another formula for efficiency of piezoelectric structures was discussed in [4-6]. This method was developed and used for shells and plates in [3]. We call it the energy method and denote this EMCC by k_e.

The energy formula for computing the efficiency has the form

$$k_e = \sqrt{\frac{U^{(d)} - U^{(sh)}}{U^{(d)}}}, \qquad (6)$$

where $U^{(d)}$ is the internal energy of the body when the electrodes are disconnected and $U^{(sh)}$ is the internal energy for short-circuited electrodes. In order to find $U^{(d)}$ and $U^{(sh)}$, we should first solve our initial problem and then the two additional problems. When we solve these two problems we suppose that the strains are known from the solution of the initial problem.

The formula (6) is not so popular as (4), (5). The reason is that

1) the existence of very simple formula (4) for coupling coefficient, and

2) the use of the formula (6), is a cumbersome procedure since in order to find k_e, we must find the solutions of three problems.

Our earlier investigation [3] shown that the formula (4) is true only for a static uniform state with constant values of stresses, strains, and electrical quantities over the whole piezoelectric element, that is, for the electroelastic state independent of the coordinates and time, and for the body of the simplest geometry. It can be used to determine the EMCC as material properties only. Moreover, it has been shown that

the formula (5) is true only for vibration with a frequency what is equal to the arithmetic mean of the resonance and antiresonance frequency;

the energy formula (6) is general, it is true for any static and dynamic state, for any electroelastic structure.

3. The EMCC for a Three-Layered Piezoelectric Beam

We will study electroelastic states of a three-layered beam. The upper and lower piezoceramic layers are located symmetrically with respect to the middle elastic layer. Piezoceramic layers are polarized in the direction of thickness. The elastic layer's thickness is $2h_0$, the thickness of each piezoceramic layer is h_1. The length of the beam is $2l$ for the beam with both edges free or both edges rigidly clamped, and l for the cantilever beam. The faces of piezoceramic layers are covered by electrodes totally or partially.

To solve the problem we use the equations presented in [3].

We suppose that the electrical loading is that which produces axial expansion-contraction or pure bending states.

3.1 EXPANSION-CONTRACTION STATE OF A BEAM

The faces of piezoceramic layers are covered by electrodes completely. We suppose that the electrical loading is such that it causes an axial expansion-contraction only.

3.1.1 *Static problem.*

Determining the EMCC k_s and k_e for the beam with free edges by formulas (4) and (6), we find

$$k_s^2 = \frac{k_{31}^2 \alpha}{1 + \alpha}, \qquad k_e^2 = \frac{k_{31}^2}{k_{31}^2 + (1 - k_{31}^2)(1 + \alpha)}, \qquad \alpha = \frac{h_0 E s_{11}^E}{h_1}, \qquad (7)$$

where E is the Young modulus for the elastic layer. Note that if h_0 is equal to zero, the layered beam becomes uniformly piezoelectric with the EMCC equal k_{31}. Clearly the first formula for k_s is in error, the second one for k_e is true: as the thickness of elastic layer approaches zero, the k_s also approaches zero and k_e approaches k_{31}.

If the edges of the beam are rigidly fixed, the EMCC values k_s and k_e are defined by formulas (4) and (6), and $k_s = k_{31}, k_e = 0$.

In this case all strains are zero. Evidently the EMCC is equal to zero. The formula (4) leads to a wrong result.

3.1.2. *Harmonic vibration.*

The EMCC k_e for longitudinal harmonic vibrations in the axial direction is defined by the following expression (the faces of the piezoelectric layers are covered by electrodes completely):

$$k_e^2 = \frac{2k_{31}^2 \sin^2 \lambda}{2k_{31}^2 \sin^2 \lambda + (1 - k_{31}^2)(1 + \alpha)\lambda(2\lambda + \sin 2\lambda)}, \qquad \lambda^2 = \frac{l^2 \omega^2 \rho_1 s_{11}^E}{2h_1(1 + \alpha)}, \qquad (8)$$

$$\rho_1 = 2h_0\rho_e + 2h_1\rho_p,$$

where ρ_e and ρ_p are the densities of the elastic layer and piezoceramics respectively, ω is the angular frequency of the vibrations, λ is a dimensionless frequency parameter.

Our computation lends support to a good agreement between k_e and k_d for the frequency of vibration in the middle of the interval $[\lambda_r, \lambda_a]$, where λ_r, λ_a are the frequency parameter at the resonance and antiresonance respectively.

To optimize the EMCC we change the size and the position of the electrodes.

As an example we will perform the calculations for a beam made from PZT-5 and aluminium with $h_0 = h_1$. If all the piezoceramic layer's faces are completely covered with electrodes, the EMCC k_e for statics calculated from equation (7) equals 0.242, and the EMCC k_e at the first and second resonance frequencies, calculated from (8), are 0.219 and 0.075, respectively.

Our calculation shows that the EMCC values can be increased if we place some electrodes in the regions of the maximum strain.

If we place electrodes in the area $0 \leq x \leq 0.75, l$ the EMCC rises to 0.233 at the first resonance (Fig.1(a)) (there is one electrode on each face, each face of piezoceramic layer is covered by electrode partially). For another shape of electrodes shown in Fig.1(b), the EMCC at the first resonance equals 0.216 exactly as for the electrodes completely covered the beam faces. If we use the electrodes in the form shown in Fig.1(c), the EMCC peaks at about maximum static value 0.242.

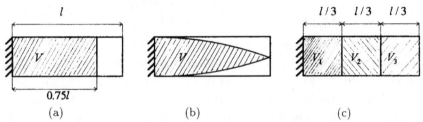

(a)   (b)   (c)

Figure 1. Electrical loading and geometry of cantilever beam's electrodes at the first resonance ($V_2 = 0.732V_1$, $V_3 = 0.268V_1$)

The shaded regions in the figures present the electrode-covered part of the faces of piezoelectric layer. Different shades correspond to different values of electrical potential V.

Fig.2 shows that at the second resonance for the cases presented in Fig.2(a),(b),(c), the EMCC is equal to 0.233, 0.216, 0.242, respectively.

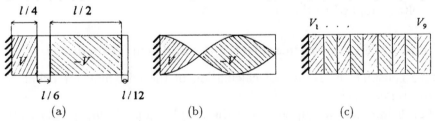

(a)   (b)   (c)

Figure 2. Electrical loading and geometry of cantilever beam's electrodes at the second resonance ($V_1 = -V_6 = -V_7$, $V_2 = -V_5 = -V_8 = 0.732V_1$, $V_3 = -V_4 = -V_9 = 0.268V_1$)

3.2 BENDING STATE OF A BEAM

Let us consider pure bending harmonic vibrations of the same three-layered cantilever beam under special electrical loading [3]. The EMCC value k_e for bending harmonic vibrations is defined by the following expression (the faces of the piezoelectric layers are covered by electrodes completely):

$$k_e^2 = \frac{AS^2}{AS^2 + m\mu CS}$$

where

$$S = \cos\mu \sinh\mu + \cosh\mu \sin\mu, \quad C = (\mu(\cos\mu + \cosh\mu)^2 + 3(1 + \cos\mu \cosh\mu),$$

$$A = \frac{2k_{31}^2}{(1 - k_{31}^2)} \frac{(2h_0 + h_1)^2 h_1}{h^3 - h_0^3}, \qquad m = \frac{2}{3}(\beta + Es_{11}^E \frac{h_0^3}{h^3 - h_0^3}),$$

$$\beta = 1 + \frac{k_{31}^2}{4(1 - k_{31}^2)} \frac{h_1^3}{h^3 - h_0^3}, \qquad \mu^4 = \frac{l^4 \omega^2 \rho_1 s_{11}^E}{m(h^3 - h_0^3)}, \qquad h = h_1 + h_0.$$

Here $2h$ is the beam thickness, l is the length of the beam. We studied k_e as a function of the ratio h_0/h on condition that the thickness h is constant. We obtained that the EMCC peaks at definite ratio h_0/h are independent of the frequency of vibration. For example, maximum of the EMCC for three-layered beam made from PZT-5 and aluminium, is achieved for $h_0/h = 0.3$. To optimize the EMCC, we change the size and the position of the electrodes. For example, the EMCC k_e for a beam made from PZT-5 and aluminium with $h_0/h = 0.3$ at the first and second resonance frequencies are 0.26 and 0.143, respectively. The EMCC values can be increased to 0.309 if we place the electrodes in the area $0 \le x \le 0.5l$ at the first resonance (one electrode on each face), and in area $0 \le x \le 0.15l$, $0.29l \le x \le 0.81l$ at the second resonance (two electrodes on each face with opposite signs of the voltage applied to them).

4. The EMCC for a Piezoelectric Cylindrical Transducer

Let us consider the results of calculation of the efficiency for cylindrical piezoceramic shell.

For definiteness we suppose that the shell has a preliminary polarization in the direction of thickness, the length of the shell $2L$ is equal to its diameter $2R$, piezoceramic material is PZT-5.

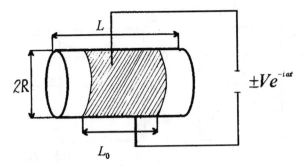

Figure 3. Cylindrical shell under the action of electrical loading

In order to solve the problem we use the theory of the piezoceramic shells, presented in [3].

If the cylinder edges are rigidly fixed, electrodes cover the faces completely, the value of k_s and k_e for the axially-symmetrical static state of the cylindrical shell are determined by the following formulas:

$$k_s^2 = \frac{2k_{31}^2}{1 - \nu}, \qquad k_e = 0.$$

As before, in this case formula (4) for k_s is not suitable, since all strains and all displacements are equal zero in the whole shell.

Let us compare the EMCC k_e and k_d near to the first and second resonance frequencies:

near to the first resonance frequency
$\lambda_r = 1.628$, $\lambda_a = 1.904$, $\lambda_m = (\lambda_r + \lambda_a)/2 = 1.766$, $k_d = 0.518$,
$k_e = 0.518$ at $\lambda = 1.766$;

near to the second resonance frequency
$\lambda_r = 4.726$, $\lambda_a = 4.810$, $\lambda_m = (\lambda_r + \lambda_a)/2 = 4.768$, $k_d = 0.186$,
$k_e = 0.186$ at $\lambda = 4.768$.

Here λ is the dimensionless frequency parameter

$$\lambda^2 = \rho\omega^2 R^2 s_{11}^E (1 - \nu^2),$$

λ_r, λ_a correspond to the resonance and antiresonance frequencies.

We emphasize that the values k_s and k_e coincide exactly.

To optimize the EMCC at the first resonance, we consider the case when electrodes cover only the central part of the shell $x \leq L_0 \leq L$.

We found that the EMCC peaks at $L_0 = 0.3R$ for the shell with rigidly fixed edges, and at $L_0 = 0.7R$ for the shell with free edges at the first resonance frequency (Fig.4).

302

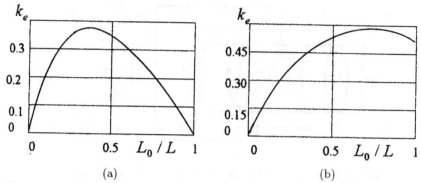

Figure 4. The EMCC k_e of the cylindrical shell at the first resonance frequency as a function of the ratio L_0/L for the shell with (a) rigidly fixed edges, (b) free edges.

5.Conclusions

Analysis of the EMCC given by different formulas (4), (5), (6) shows that the energy formula (6) is valid for all static or dynamic problems. It is shown that for any electroelastic structure it is possible to increase the EMCC to the maximum static value by choosing a special shape and size of electrodes and a special electrical loading. Choosing the optimal shape, the size electrodes and electrical loading depend on structure's geometry, boundary conditions, material properties, and kind of vibrations.

Acknowledgement

This work was supported by the Russian Fundamental Research Foundation (grant No. 96-01-01098).

References

1. Berlincourt, D.A, Curran, D.R., and H. Jaffe, H. (1964) Piezoelectric and piezomagnetic materials and their function as transducers, W.P.Mason (eds.), *Physical Acoustics, 1A*, Academic Press, New York, pp.204-326.
2. IEEE Standart on Piezoelectricity, ANSI-IEEE Std. 176 (1987), IEEE, New York.
3. Rogacheva, N. N.(1994) *The Theory of Piezoelectric Shells and Plates*, CRC Press, Boca Raton.
4. Toulis, W.J. (1963) Electromechanical coupling and composite trans-ducers, *J. Acoust. Soc. Am.* **35**, 74-80.
5. Woollett, R.S. (1963) Comments on "Electromechanical coupling and composite transducers",*J. Acoust. Soc. Am.*,**35**, 1837-1838.
6. Grinchenko, V.T., Ulitko, A.F., and Shulga, N.A. (1989) *Electroelas-ticity*, Naukova Dumka, Kiev.

NUMERICAL AND EXPERIMENTAL BEHAVIOUR OF THE CORGE RAILWAY BRIDGE WITH ACTIVE RAILWAY SUPPORT

LUÍS A. P. SIMÕES DA SILVA AND SANDRA F. S. J. ALVES
Department of Civil Engineering, University of Coimbra
3049 Coimbra Codex, Portugal

RYSZARD KOWALCZYK
Department of Civil Engineering, University of Beira Interior,
6200 Covilhã, Portugal

1. Introduction

1.1. PORTUGUESE RAILWAY BRIDGES

Of all structural types, bridges are probably among those that stay in service well beyond their predicted service life. In fact, in Portugal alone, a reasonable number of roman bridges are still in use. Extending the life-span of existing structures is a current concern of all governments and the scientific community, and railway bridges around the world clearly provide the best possible example to this problem.

The backbone of all railway networks around Europe and the USA were planned and built in the second half of the XIXth century and early years of the 20th century. Again, focussing on the Portuguese example [1], over 80% of the Portuguese railway system was finished by 1910. In terms of railway bridges, corresponding numbers may be referred, the majority of which are still in service. These numbers would be even higher, had some lines not been closed for economical reasons.

Compared to road bridges, railway bridges are probably unique in that they usually provide an economical alternative to meeting current railway criteria and specifications through rehabilitation. In fact, while present road traffic specifications are not comparable to those prevailing 100 years ago (road geometry, speed, loads), railway bridges specifications have had a more modest increase, with the added advantage of maintaining a fixed width.

Certainly, the recent introduction of high-speed trains (over the past 15 years in Europe) has forced the construction of completely new lines, but the extremely high

J. Holnicki-Szulc and J. Rodellar (eds.), Smart Structures, 303–312.

304

costs associated these projects limt such new infrastructures to a few major routes where high passenger traffic can yield a profitable return to such investments.

Current strengthening and rehabilitation techniques for railway bridges provide an effective means to meet current load requirements at a price representing between 35% to 50% of the cost of a new replacement bridge, while achieving a compromise in terms of speed.

Achieving speed enhancements through rehabilitation of older bridges has stirred the imagination to try new concepts and technologies. The application of active control strategies to improve the behaviour of railway bridges stood out as a possibility and led to the feasibility study within a Copernicus project next described.

1.2. THE ACTIVE RAILWAY TRACK SUPPORT CONCEPT

Achieving speed improvements requires that both structural criteria for the bridge and operational and confort criteria for the train and passengers are met. From the

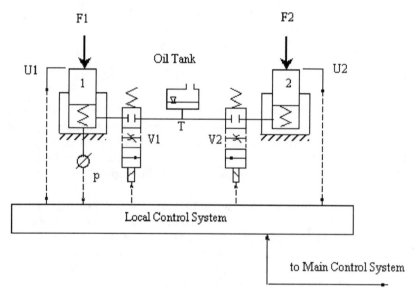

Figure 1. Smart sleeper

operation of the train point of view, the optimum situation would be attained were the bridge to have no effect on the train. This, of course, is an impossible result, since any bridge has a finite stiffness and hence deforms and vibrates under a passing train.

Minimisation of deformation and vibration using actively controlled devices was thus proposed as a way to achieve speed improvements on existing bridges. Although a detailed description of the proposed system lies outside the scope of this paper and can be found elsewhere [2], it is useful to briefly describe the concept.

Given that the objective was to maintain the track in a straight and undeformed configuration as the train passed, and since the bridge superstructure would simultaneously and inevitably deform and vibrate, it was proposed to manufacture special sleepers, composed of hydraulic actuators, valves and an active control system, as shown in figure 1, that would impose a relative movement between the track and the bridge. Assuming that an optimum control methodology could be found, the train would pass the bridge with reduced track deflections and vibrations and thus its velocity could be safely increased. The process is illustrated schematically in figure 2, which shows the various positions of the train with and without active railway track support.

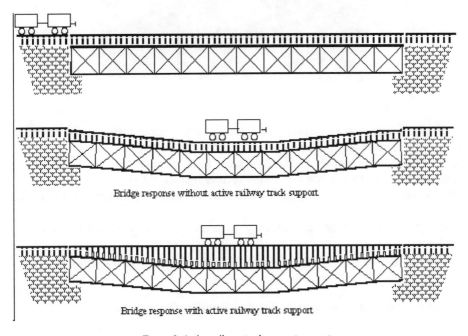

Bridge response without active railway track support

Bridge response with active railway track support

Figure 2: Active railway track support concept

1.3. THE TEST CASE: CORGE BRIDGE

The concept described in the previous section seemed very interesting but it remained to be proven that, assuming it was possible to manufacture such a system, it would really produce an improvement in terms of speed and an overall benefit considering all the implications and side effects.

In order to clarify this issue, it was decided to perform an extensive study of a real railway bridge, the Corge bridge. The bridge is located in Portugal and belongs to the railway network of the portuguese railway infrastructure operator, REFER. As seen in figure 3, it is a trussed four-span continuous bridge, total length of 176m, with four spans of 40-48-48-40m, supported on masonry columns, the track being supported on the upper chords and built in 1889. The truss is 3-meter high and is made of wrought iron plates riveted to form the various structural elements.

Figure 3. Corge bridge

The Corge bridge is still in service and was chosen for various reasons, next described:
Low traffic on the bridge, allowing easier experimental testing;
Speed limitation of 20 km/h, giving good scope for improvement;
Load limitation of 16 ton/axle and 5 ton/m (class A bridge);
Located very near to Covilhã, which University is one of the partners in the project, with easy access;
Multi-span continuous bridge, allowing to assess the effect of continuity over te supports;
A structural system that is easily and accurately translated into a structural model;

2. Structural modelling, experimental calibration and verification

An accurate model of the bridge was a pre-requisite for a meaningful assessment of the potential advantages of the active control system. Because it consists of relative movements imposed in time between the bridge and the track, it was necessary to implement a 3-D structural model to reproduce the track support system: track supported on sleepers, sleepers supported on secondary beams, secondary beams riveted to cross-beams, cross-beams attached to main trusses and trusses stabilised by bracing.

The structural model thus consists of a 3-D system with 2 noded beam elements with the usual six degrees of freedom per node and was implemented in the finite element package LUSAS, a detail of the structural model being shown in figure 4. In order to calibrate the numerical model, mechanical properties of the wrought iron were recovered from tests carried out on samples taken from the bridge by the portuguese railway operator, REFER, which revealed a brittle material with a uniaxial tensile stiffness of E = 182 GPa and a failure stress of $\square$ = 253 Mpa.

Secondly, statical in situ measurements of strains and deflections were carried out on the bridge in 1997, using a MLW Series 1570 locomotive, positioned at seven different locations. Deflections were measured at mid-span of each span and 40 extensometers

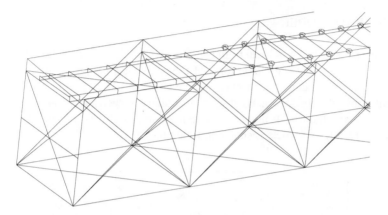

Figure 4. Finite element model of the bridge, including track and sleepers

were placed in a typical mid-span module. Measured results were in good agreement with the corresponding numerical calculations, as seen in the above figure, thus validating the numerical model for the dynamical analysis, next described. Full details of the experimental testing can be found in [4].

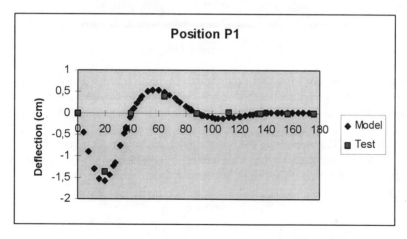

Figure 5. Measured and calculated values for ridge deflection at position 1

3. Train-bridge interaction

3.1.INTRODUCTION

Unlike road bridges, in railway bridges, particularly unballasted bridges, the weight of the train represents a considerable proportion of the total loading on the bridge (self-weight included). Under these conditions, the inertia forces corresponding to the moving train cannot be neglected. Additionally, the movement of the train requires dynamic equilibrium equations for all successive positions of the train on the bridge.

3.2. THE TRAIN MODEL

According to Frýba [5], a modern train consists of $1,..., q,..., Q$ vehicles which possess $1,..., n, ..., N$ axes, the static axle force of the qth vehicle being F_q. The body of each vehicle is considered elastic with mass m_{3q} and stiffness $E_{3q}I_{3q}$, while the boogies are represented by a secondary mass m_{2rq}, a secondary suspension of stiffness k_{2rq} and damping c_{2rq}, and the wheels and axles have mass m_{1sq}, stiffness k_{1sq} and damping c_{1sq}. Schematically, a train vehicle is shown in figure 6.

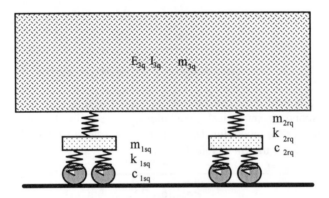

Figure 6. Structural model for the train

While the above model adequately represents a typical train, it is fairly complex. In terms of the global bridge response it is often sufficient to consider a simplified model consisting of an equivalent one degree of freedom spring mass system for each vehicle.

The movement of the train along the bridge may have an irregular velocity; the motion of the front axle in time is thus determined by the function u(t) where, in most situations, the train crosses the bridge at constant speed, giving

$$u(t) = ct \tag{1}$$

where c is the constant train speed.

3.3. EQUATIONS OF MOTION

Having established that the inertia forces of the train must be accounted for and the train itself behaves like an elastic body with stiffness and damping properties, the dynamic equilibrium equations must be derived for the structural assembly of train and bridge acting together, giving

$$\frac{d}{dt}\left(\left[\mathbf{m(t)}\right]\dot{\mathbf{r}}\right)+\left[\mathbf{c(t)}\right]\dot{\mathbf{r}}+\left[\mathbf{k(t)}\right]\mathbf{r}=\mathbf{F(t,s)}$$

(2)

written in a discrete format in space, where [m(t)] represents the mass matrix (variable in time), [c(t)] the damping matrix, [k(t)] the stiffness matrix, r, $\dot{\mathbf{r}}$ and $\ddot{\mathbf{r}}$ the displacement, velocity and acceleration vectors, respectively, and F(t,x) the loading vector representing the moving train, which varies in time and space along the longitudinal axes of the bridge denoted by the co-ordinate s.

Equation (2) highlights some of the difficulties to solve the problem, namely the dependence of the system properties m, c and k on time and the double variation of the load vector in time and space. Additionally, the interface between the train and the track represents a complicated contact problem with the possibility of separation.

Because the variable properties in time are only related to the train model for a fixed co-ordinate system linked to the bridge, separation of bodies between train and bridge yields considerable simplification. Assuming the bridge to be straight with its longitudinal axis lying in the x direction, equation. (2) yields two coupled equilibrium equations for bridge and train

$$\left[\mathbf{m}_b\right]\ddot{y}+\left[\mathbf{c}_b\right]\dot{y}+\left[\mathbf{k}_b\right]y = F(t,x)$$

$$\left[\mathbf{m}_t\right]\ddot{y}+\left[\mathbf{c}_t\right]\dot{y}+\left[\mathbf{k}_t\right]y = 0$$

(3)

shown above only with respect to its vertical direction y, where subscript b represents the bridge, subscript t the train. Although analytical solutions exist for the simple case of a single moving load on a beam [6], dealing with a real bridge structure and a realistic train requires the numerical solution of the multi degree-of-freedom coupled problem stated above.

3.4. ITERATIVE SOLUTION PROCEDURE

Solving equations (3) using a standard finite element code requires a double iterative procedure [7]. The outer loop corresponds to the successive train positions along the bridge, defined at adequate length increments.

310

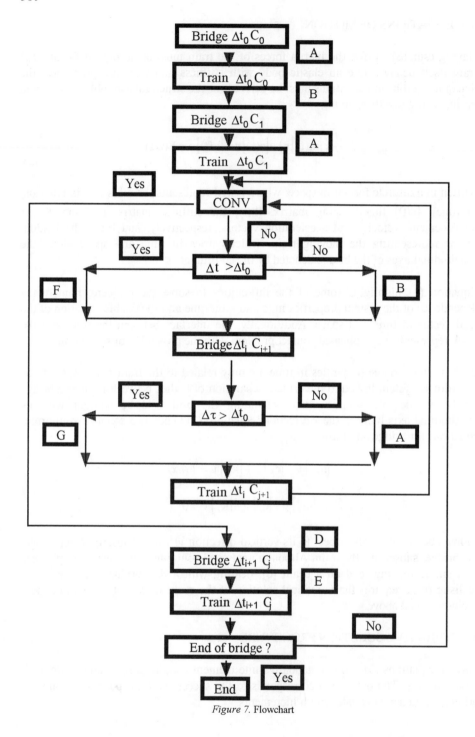

Figure 7. Flowchart

$$\Delta x = c\Delta t \qquad (4)$$

The inner loop ensures equilibrium between bridge and train through a common interactive force at the train wheels-bridge track contact. The iterative procedure thus consists of a sequence of finite element calculations and external control programs.

Schematically, the iterative procedure is illustrated in the following flowchart, which shows the external control programs and the finite element calculations

4. Active control

4.1. SMART SLEEPER CONTROL STRATEGY

Given the hardware described in figure 1, several control strategies may be implemented with increasing sophistication. Following [2], four different control schemes were considered, namely
passive sleepers, with y_0 = const., fixed for a particular bridge and an expected set of passing trains
passive adaptive sleepers, with y_0 = const., for a particular bridge and tailored for a particular, observed train
adaptive sleepers, for a particular bridge, tailored for a particular, observed train and moving in a controlled way, with $y_0 = y_0(t)$, $dy_0(t)/dt \leq 0$
fully actively controlled sleepers, including shifting up of the loaded track, with $y_0 = y_0(t)$

While the first two schemes correspond to global (i) or selective (ii) contraflexures applied to the track prior to the passing of a train, the latter two require controlled movement while the train is passing. The energy input for strategy (iii) is supplied by the weight of the passing train, whereas scheme (iv) requires an enormous external energy input which renders it of little practical use. Greater effort was thus focussed on scheme (iii), although the analysis that was implemented could equally deal with the fully actively controlled situation. Following the results of an experimental testing program on the response of the actuator under several loading strategies [7], a time-displacement curve was specified for each actuator, as shown in figure 8

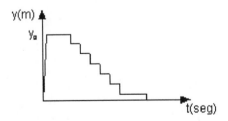

Figure 8. Optimal control strategy for each actuator

resulting from the minimisation of the mean square root measure of the resultant track deflection.

4.2 NUMERICAL IMPLEMENTATION

Introduction of an arbitrary time-displacement curve within the bridge model (equation (3b)) that could reproduce the actual movement of the actuator was implemented as a time-temperature load variation applied to each single smart sleeper. This strategy presented the added advantage of requiring no changes in the structural model of the bridge.

5. Results and conclusions

The concept of active railway track support through smart sleepers was tested in a real situation, the Corge bridge. While final results are still not available, preliminary results show that the system is technically viable and gains in terms of speed and control of stresses can be achieved. The economic feasibility of such a system remains the big challenge for the coming years.

References

1. Andrade Gil, J.M.D. (1997) Pontes metálicas ferroviárias, in A. Lamas, P. Cruz and L. Calado (eds.), Proceedings on I Conference on *Construção Metálica e Mista*, 20-21November 1997, Porto, Portugal, pp. 231-238.
2. Flont, P. and Holnicki-Szulc, J. (1997) Adaptive railway track with improved dynamic response, in W. Gutkowski and Z. Mróz (eds.), Proceedings of the *Second World Congress of Structural and Multidisciplinary Optimization*, 26-30 May 1997, Zakopane, Poland, pp. 339-344.
3. Simões da Silva, L.A.P, Alves, S.J. and Kowalczyk, R. (1998) Ensaios experimentais - Ponte do Corge, *Mecânica Experimental* 2, (in publication).
4. Frýba, L. (1996) *Dynamics of railway bridges*, T. Telford, London.
5. Inglis, C.E. (1934) *A mathematical treatise on vibrations in railway bridges*, Cambridge University Press.
6. Cruz, S. M. (1994) Comportamento dinâmico de pontes ferroviárias em vias de alta velocidade, MsC Thesis, FEUP, Porto.
7. Král, J. and Sutner, O. (1998) Tests on active support, COP 263/Czech Tech. University Report.

FLEXTENSIONAL ACTUATOR DESIGN USING TOPOLOGY OPTIMIZATION AND THE HOMOGENIZATION METHOD

Theory and Implementation

E. C. N. SILVA, S. NISHIWAKI, AND N. KIKUCHI
Department of Mechanical Engineering and Applied Mechanics
The University of Michigan,
Ann Arbor, 3005 EECS, MI,48109-2125, USA

1. Introduction

Flextensional actuators consist of a piezoceramic (or a stack of piezoceramics) connected to a flexible mechanical structure that converts and amplifies the output displacement of the piezoceramic. A well–known type of flextensional actuator is the "moonie" transducer, which consists of a piezoceramic disk sandwiched between two metal endcaps [1]. The performance of the flextensional actuator is measured in terms of output displacement and generative (or "blocking") force [2]. Generative force is the maximum force supported by the transducer without deforming, for a certain applied voltage.

Flextensional actuators have been developed by using simple analytical models and experimental techniques [1], and the finite element method [2]. However, the design is limited to the optimization of some dimension of a specific topology chosen for the coupling structure. These studies showed that the performance depend on the distribution of stiffness and flexibility in the coupling structure domain, which is related to the coupling structure topology. Therefore, the design of the coupling structure can be achieved by using topology optimization. By designing other types of flexible structures connected to the piezoceramic, we can obtain other types of flextensional actuators that produce high output displacements (or generative forces) in different directions, according to a specific application.

Based on this idea, in this work we propose a method for designing flextensional actuators by applying topology optimization techniques. The problem is posed as the design of a flexible structure coupled to the piezoceramic that maximizes the output displacement and generative force in some

313

J. Holnicki-Szulc and J. Rodellar (eds.), Smart Structures, 313–320.
© 1999 *Kluwer Academic Publishers. Printed in the Netherlands.*

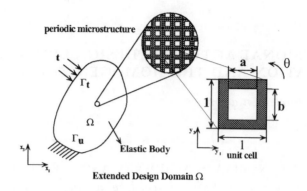

Figure 1. Microstructure used for optimization.

specified direction. Only static and low-frequency applications (inertia effects are negligible) are considered. The topology optimization method applied is based on the homogenization design method developed by Bendsøe and Kikuchi [3]. FEM is applied to the structural analysis in the optimization procedure. Even though two-dimensional (plane strain) topologies of flextensional actuators are presented to illustrate the implementation, the method can be extended to three-dimensional topologies.

2. Topology Optimization Procedure

2.1. HOMOGENIZATION DESIGN METHOD

The topology optimization problem consists of finding the optimal distribution of material properties in a extended fixed domain that maximizes some structure cost function. Therefore, the domain does not change during the optimization process which makes easy the calculation of the derivatives of any function defined over the domain Ω.

To allow the appearance of intermediate (or composite) materials rather than only void or full material in the final solution, a microstructure proposed by Bendsøe and Kikuchi [3] is defined in each point of the domain. This microstructure consists of a unit cell with a rectangular hole inside (see Fig. 1), and provides enough relaxation for the problem. The design variables are the dimensions a, b, and the orientation θ of the microstructure hole, as shown in Fig. 1. In this sense, the new problem consists of finding the optimal material distribution in a perforated domain with infinite microscale voids. In each point of the domain there is a composite material defined by the microstructure. The effective elasticity properties of this composite material relative to the principal coordinate axes of the material, are obtained applying the homogenization method [3].

2.2. FORMULATION OF OPTIMIZATION PROBLEM

The design of flextensional actuators requires two different objective functions: *mean transduction* and *mean compliance*. Mean transduction is related to the electromechanical conversion between two regions, Γ_{d_1} (electrical one) and Γ_{t_2} (mechanical one) (see Fig. 2), of the design domain. The larger this function, the larger the displacement generated in a certain direction and in region Γ_{t_2} due to an input electrical charge in region Γ_{d_1}. The concept of mean transduction is obtained by extending the reciprocal theorem in elasticity to the piezoelectric medium [4]. The mean transduction is calculated considering two load cases described in case 1 of Fig. 2. Therefore, the maximization of the output displacement is obtained by maximizing the mean transduction which is given by the expression [4]:

$$L_2(\boldsymbol{u}_1, \phi_1) = L_t(\boldsymbol{t}_2, \boldsymbol{u}_1) = L_d(d_1, \phi_2) = B(\phi_2, \boldsymbol{u}_1) - C(\phi_1, \phi_2) \quad (1)$$

where the following operators were defined to make the notation compact:

$$A(\boldsymbol{u}, \boldsymbol{v}) = \int_\Omega \varepsilon(\boldsymbol{u})^t \boldsymbol{c}^E \varepsilon(\boldsymbol{v}) d\Omega \qquad B(\phi, \boldsymbol{v}) = \int_\Omega (\boldsymbol{\nabla}\phi)^t \boldsymbol{e}^t \varepsilon(\boldsymbol{v}) d\Omega$$

$$C(\phi, \varphi) = \int_\Omega (\boldsymbol{\nabla}\phi)^t \boldsymbol{\epsilon}^S \boldsymbol{\nabla}\varphi d\Omega \qquad L_t(\boldsymbol{t}_i, \boldsymbol{v}_i) = \int_{\Gamma_{t_i}} \boldsymbol{t}_i \cdot \boldsymbol{v}_i d\Gamma$$

$$L_d(d_i, \varphi_i) = \int_{\Gamma_{d_i}} d_i \varphi_i d\Gamma \qquad\qquad\qquad\qquad\qquad (2)$$

and Ω is the fixed domain considered for design (it can also contain non-piezoelectric materials), $\boldsymbol{\nabla}$ is the gradient operator, $\varepsilon(.)$ is the strain operator, $\boldsymbol{t}_i$ is a traction vector, d_i is the surface electrical charge, and $\boldsymbol{c}^E$, $\boldsymbol{e}$, and $\boldsymbol{\epsilon}^S$ are the elastic, piezoelectric, and dielectric properties, respectively, of the medium. $\boldsymbol{u}$ and $\boldsymbol{v}$ are displacements, and ϕ and φ are electric potentials.

If only the electromechanical function is considered, a structure with no stiffness at all, may be obtained. Therefore, a structural function must also be considered to provide sufficient stiffness in the coupling structure between the regions Γ_{t_2} and Γ_{d_1}. The coupling structure stiffness is also related to the generative force. The larger the stiffness, the larger the generative force. This stiffness can be obtained by minimizing the mean compliance ($L_3(\boldsymbol{u}_3, \phi_3)$) in the contact region actuator/body, which is given by the expression [4]:

$$L_3(\boldsymbol{u}_3, \phi_3) = \int_{\Gamma_{t_2}} \boldsymbol{t}_3 \cdot \boldsymbol{u}_3 d\Gamma = A(\boldsymbol{u}_3, \boldsymbol{u}_3) + B(\phi_3, \boldsymbol{u}_3) \quad (3)$$

To combine both optimization problems the following objective function is proposed:

$$\mathcal{F}(\boldsymbol{x}) = w * ln\left(L_2(\boldsymbol{u}_1, \phi_1)\right) - (1 - w) ln(L_3(\boldsymbol{u}_3, \phi_3)) \quad (4)$$

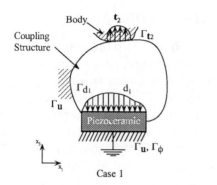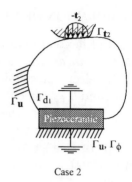

Figure 2. Load cases for calculation of mean transduction (case 1) and mean compliance (case 2).

where $0 \leq w \leq 1$ is a weight coefficient. This objective function allows us to control the contributions of the mean transduction (Eq. 1) and the mean compliance (Eq. 3) in the design. Therefore, the new optimization problem is stated as:

$$Maximize: \quad \mathcal{F}(\boldsymbol{x})$$
$$a, b, \text{ and } \theta$$
$$subject\ to: \quad \text{Equilibrium conditions}$$
$$0 \leq a \leq a_{sup} < 1$$
$$0 \leq b \leq b_{sup} < 1$$
$$\Theta(a, b) = \int_S (1 - ab)dS - \Theta_S \leq 0$$

S is the design domain Ω without including the piezoceramic, Θ is the volume of this design domain, and Θ_S is an upper bound volume constraint to control the maximum amount of material used to build the coupling structure. The index i assumes value 1 or 3 because the problem is considered in the plane 1-3. The piezoceramic is polarized in the #3 direction. Since the domain is discretized in finite elements, the above definitions must be substituted by their equivalent discretized ones using FEM [4]. In addition, the variables a, b, and θ, which theoretically are a continuous function of $\boldsymbol{x}$, became sets of continuous design variables a_n, b_n, and θ_n defined for the n finite element subdomain in the numerical problem. The upper bounds a_{sup} and b_{sup} specified for a and b, respectively, are necessary to avoid the singularity of the stiffness matrix in the finite element formulation (due to the presence of a material with "null" stiffness in the domain). In this work, the upper bounds a_{sup} and b_{sup} were chosen to be 0.995. Numerically, regions with $a = b = 0.995$ have practically no structural significance and can be considered void regions.

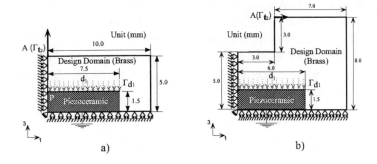

Figure 3. Design domains considered.

2.3. NUMERICAL IMPLEMENTATION

The optimization problem is solved using Sequential Linear Programming (SLP) which consists of the sequential solution of linearized problems defined by writing a Taylor series expansion for the objective and constraint functions around the current design points a_n and b_n in each iteration step. θ_n is obtained by considering the local principal stresses direction in each finite element after each optimization step. The sensitivities of the objective function necessary for the linearization of the problem are derived in Silva [4]. In each iteration, moving limits are defined for the design variables. After optimization, a new set of design variables a_n and b_n is obtained and updated in the design domain.

3. Results

Two examples will be presented to illustrate the design of flextensional actuators using the proposed method. The design domains used for the examples below are described in Fig. 3a and b. They consist of a domain of piezoceramic that remains unchanged during the optimization and a domain of brass where the optimization is conducted. The mechanical and electrical boundary conditions for both domains are described in the same figures. Only one quarter of the domain is considered since the applications considered have two symmetry axes. Electrical degrees of freedom are considered only in the ceramic domain. For each example, the correspondent interpretation of the topology, obtained using Image Based technique described in Kikuchi et al.[5], is also presented.

Table 1 describes the piezoelectric material properties used in the simulations. The Young's modulus and Poisson's ratio of the brass are equal to 106 GPa and 0.3, respectively. Two-dimensional elements under plane strain assumption are used in the finite element analysis. For all these examples, the total volume constraint of the material Θ_S is considered to be 30% of the volume of the whole domain Ω without piezoceramic (domain S). The initial value of the microscopic design variables a_n and b_n is 0.9, and that of θ is 0.0 in all elements. The amount of electrical charge applied to the piezoceramic electrode is 4 $\mu C/m^2$. Any value can be applied since the problem is linear.

TABLE 1. Material Properties of PZT5

c_{11}^E $(10^{10}$ N/m$^2)$	12.1	e_{13} (C/m^2)	-5.4
c_{12}^E $(10^{10}$ N/m$^2)$	7.54	e_{33} (C/m^2)	15.8
c_{13}^E $(10^{10}$ N/m$^2)$	7.52	e_{15} (C/m^2)	12.3
c_{33}^E $(10^{10}$ N/m$^2)$	11.1	$\epsilon_{11}^S/\epsilon_0$	1650
c_{44}^E $(10^{10}$ N/m$^2)$	2.30	$\epsilon_{33}^S/\epsilon_0$	1700
c_{66}^E $(10^{10}$ N/m$^2)$	2.10		

EXAMPLE 1: THE "MOONIE" TRANSDUCER

The objective of this example is to obtain the design of the "moonie" transducer, a well-known flextensional actuator [1]. This design was developed using simple analytical models and experimental techniques, and seems to be close to an optimum solution. The initial domain considered for this example is described in Fig. 3a (800 finite elements). The optimization problem is defined as the maximization of the deflection at point A in the direction of the dummy load when electrical charges d_1 are applied to the piezoceramic at electrode Γ_{d_1} (see Fig. 3a), while the mean compliance at point A is to be minimized since the actuator is supposed to have contact with a body at point A. The coefficient w was considered equal to 0.6, 0.7, and 0.9. The coupling structures obtained using the proposed method corresponds to the expected "arc" of the moonie transducer as shown in Figs. 4a, b, and c (the figures must be reflected to both symmetry axes). The decrease of the structure stiffness in the vertical direction as the coefficient w increases is reflected in the vertical displacement ratio between points A and P which is equal to 2.77, 6.20, and 9.74 times for Figs. 4a, b, and c, respectively. The generative force in the vertical direction is expected to decrease as w increases since the stiffness in this direction is reduced. Figure 5 shows the image of the moonie actuator obtained by reflecting Fig. 4c to both symmetry axes, and the corresponding deformed shape obtained using FEM.

EXAMPLE 2: FLEXTENSIONAL CLAMP

This example illustrates the design of a flextensional clamp, that is, a clamp actuated by a piezoceramic. The initial domain for this problem is shown

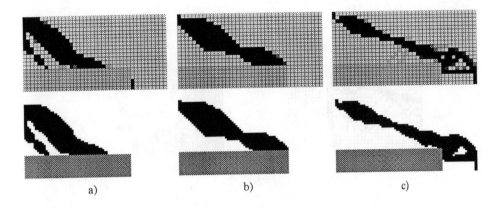

Figure 4. Optimal topologies obtained for the first example and correspondent image interpretations. a) $w = 0.6$; b) $w = 0.7$; c) $w = 0.9$.

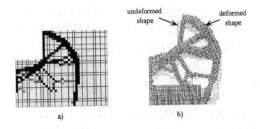

Figure 5. Flextensional actuator type "moonie".

in Fig. 3b (1136 finite elements). The function of the clamp is to (1) hold a workpiece even though no electrical charge is applied to the piezoceramic (2) deform along the direction of the dummy force at point A to release the workpiece when electrical charges are applied to electrode Γ_{d_1}. The coefficient w is equal to 0.5. The topology of flextensional clamp obtained is shown in Fig. 6a (the figure must be reflected to the vertical symmetry axis). Figure 6b shows the deformed shape obtained using FEM. The FEM model was generated using Image Based Technique.

4. Conclusions

A method for designing flextensional actuators for static and low-frequency applications has been proposed. This method is based upon topology optimization using the homogenization design method. The method consists of designing a flexible structure connected to a piezoceramic (or stack of piezoceramics) that amplifies and converts the output piezoceramic dis-

320

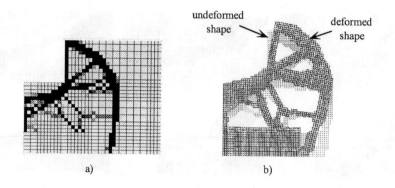

Figure 6. a) Optimal topology obtained for second example; b) deformed configuration using FEM.

placement. The examples show that new types of flextensional actuators that produce large output displacements in different directions can be designed using the proposed method. The complex topologies obtained can be manufactured using rapid prototyping techniques.

Acknowledgments

The first author thanks CNPq - Conselho Nacional de Desenvolvimento Científico e Tecnológico (Brazilian National Council for Scientific and Technologic Development) and University of São Paulo (São Paulo – Brazil) for supporting him in his graduate studies. The second author thanks the support from TOYOTA Central R.&D. Labs., Inc.

References

1. Dogan, A., Yoshikawa, S., Uchino, K., and Newnham, R. E. (1994) The Effect of Geometry on the Characteristics of the Moonie Transducer and Reliability Issue *Proceedings of IEEE 1994 Ultrasonic Symposium* 935–939.
2. Dogan, A., Uchino, K., and Newnham, R. E. (1997) Composite Piezoelectric Transducer with Truncated Conical Endcaps "Cymbal" *IEEE Transactions on Ultrasonics, Ferroelectrics and Frequency Control* **44** No. 3 May 597–605.
3. Bendsøe, M. P. and Kikuchi, N. (1988) Generating Optimal Topologies in Structural Design Using a Homogenization Method *Computer Methods in Applied Mechanics and Engineering* **71** 197–224.
4. Silva, E. C. N. (1998) *"Design of Piezocomposite Materials and Piezoelectric Transducers Using Topology Optimization"*, Ph.D. Dissertation, The University of Michigan-Ann Arbor.
5. Kikuchi, N., Hollister, S., and Yoo, J. (1997) A Concept of Image-Based Integrated CAE for Production Engineering *Proceedings of International Symposium on Optimization and Innovative Design* Japan 75–90.

ACTIVE CONTROL OF NONLINEAR 2-DEGREES-OF-FREEDOM VEHICLE SUSPENSION UNDER STOCHASTIC EXCITATIONS

L. SOCHA
Institute of Transport, Silesian Technical University,
Katowice, Poland

1. Introduction

The optimal suspension design is one of the basic problems in ride comfort analysis. In recent years considerable interest has been concentrated on the use of active vehicle suspension which can improve the comfort and safety during high-speed driving of the vehicle in comparison to the passive suspension system. Usually, both the active control and the vehicle suspension models were considered as linear ones. However, over the last ten years, there has been an increasing effort devoted to nonlinear models of vehicle suspensions under stochastic excitations, see for instance [7], [8]. The solution of the active optimal control was approximately obtained by using statistical linearization. The linearization coefficients were determined from the mean - square criterion. Since in the vibration analysis of stochastic systems in mechanical and structural engineering, a few linearization approaches were proposed, the objective of this paper is to compare them in application to the determination of active control. The detailed analysis was presented for three criteria of statistical linearization, namely the mean - square error of the displacement, equality of the second order moments of nonlinear and linearized elements, and the mean-square error of the potential energies.

2. Vehicle model

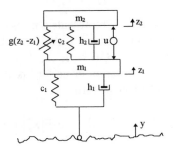

Figure 1. 2 -DOF vehicle model.

J. Holnicki-Szulc and J. Rodellar (eds.), Smart Structures, 321–327.

Consider the linear 2 - DOF vehicle model shown in Figure 1 with one nonlinear suspension spring between masses m_1 and m_2. u(t) denotes the active suspension force (acting independently of the forces in the passive elements).

The equations of motion of the system after simple transformation can be written as

$$dx_1 = x_3 dt,$$

$$dx_2 = x_4 dt,$$

$$dx_3 = \frac{1}{m_1}\left[-c_1 x_1 - h_1 x_3 + c_2 x_2 + h_2 x_4 + g(x_2) - u + m_1\left(a_1 a_2 x_5 + (a_1 + a_2)x_6\right)\right]dt - q d\xi,$$

$$dx_4 = \frac{1}{m_2}\left[-c_2 x_2 - h_2 x_4 - g(x_2) + u\right]dt + \frac{1}{m_1}\left[c_1 x_1 + h_1 x_3 - c_2 x_2 - h_2 x_4 - g(x_2) + u\right]dt,$$

$$dx_5 = x_6 dt,$$

$$dx_6 = \left[-a_1 a_2 x_5 - (a_1 + a_2)x_6\right]dt + q d\xi, \tag{1}$$

where the state variables are defined by

$$x_1 = z_1 - y_1, \quad x_2 = z_2 - z_1, \quad x_3 = \dot{z}_1 - \dot{y}, \quad x_4 = \dot{z}_2 - \dot{z}_1, \quad x_5 = y, \quad x_6 = \dot{y}, \tag{2}$$

and c_1 and c_2 are stiffness constant parameters; h_1 and h_2 are damping constant parameters; ξ is a standard Wiener process, a_1, a_2 and q are constant parameters of linear filter defined by

$$a_1 = a_1^{\bullet} v, \quad a_2 = a_2^{\bullet} v, \quad q = q^{\bullet}\sqrt{a_1 a_2 v}, \tag{3}$$

where $a_1^{\bullet}, a_2^{\bullet}$ and $q^{\bullet}$ are constant parameters of random road profile and v is the constant speed of vehicle.

3. Performance index and optimal active control

The objective of the use of the active control u is to minimize the joint performance index I defined by stationary characteristics of system (1)

$$I = \rho_1 I_1 + \rho_2 I_2 + \rho_3 I_3 + \rho_4 I_4, \tag{4}$$

with the partial performance indexes: I_1 represents a measure of ride comfort, I_2 limits the space required for the suspension, I_3 avoids loosing contact between the wheel and the road, I_4 limits the control force; ρ_i (i = 1, ..., 4) are weight coefficients. This criterion is a modified version of a criterion given for linear model in [3]. In new state variables, the performance index I has the form

$$I = \frac{\rho_1}{m_2^2} E\left[\left(c_2 x_2 + h_2 x_4 + g(x_2) - u\right)^2\right] + \rho_2 E\left[x_2^2\right] + \rho_3 E\left[x_1^2\right] + \rho_4 E\left[u^2\right]. \tag{5}$$

Here the stationary moments are considered. If the nonlinear stiffness $g(x_2)$ can be expressed by a linearized form

$$g(x_2) = \alpha k x_2 \tag{6}$$

where α is a constant parameter and k is a linearization coefficient, then the optimal control problem can be transformed to the standard one

$$dx = [A(k)x + Bu]dt + Gd\xi, \tag{7}$$

$$I = E[x^T Q(k)x + 2x^T N(k)u + ru^2] \rightarrow \min, \tag{8}$$

where $x = [x_1, \ldots, x_6]^T$ is the state vector $A(k)$ and $Q(k)$ are matrices dependent on the linearization coefficient k , B, G, $N(k)$ are vectors and r is a scalar defined by the parameters of Equations (1). We note that if $g(0) = 0$ then k is a nonlinear function of the stationary second order moment i.e. $k = k\left(E[x_2^2]\right)$.

In this paper we will compare three methods of statistical linearization for Gaussian excitations corresponding to the following criteria

a) Mean - square error of the displacement [5]

$$E\left[(k_a x - g(x))^2\right] \rightarrow \min. \tag{9}$$

b) Equality of the second order moments of nonlinear and linearized elements [5]

$$E\left[(g(x))^2\right] = E\left[(k_b x)^2\right]. \tag{10}$$

c) Mean - square error of the potential energies [1]

$$E\left[\left(\int_0^x (k_c x - g(x))dx\right)^2\right] \rightarrow \min. \tag{11}$$

As an example of the nonlinear function $g(x_2)$ we consider

$$g(x_2) = \alpha x_2^3. \tag{12}$$

One can show [1], [5] that the corresponding linearization coefficients have the form

$$k_a = 3E[x_2^2], \quad k_b = \sqrt{15}E[x_2^2], \quad k_c = 2.5E[x_2^2]. \tag{13}$$

To determinate the optimal control for nonlinear system with nonlinear criterion, we apply the idea proposed in the literature, for instance in [7], [9], where the statistical linearization and stochastic optimal control method for linear system with mean - square criterion are used. These two standard approaches are the basic steps in an iterative procedure which is given in the next section.

4. Iterative procedure

The following procedure is a modified version of a standard one given in [9]
Step 1°. First put k = 0 in (6) and solve the algebraic Riccati equation

$$P\left(A - \frac{1}{r}BN^T\right) + \left(A - \frac{1}{r}BN^T\right)^T P - \frac{1}{r}PBB^TP + \left(Q - \frac{1}{r}NN^T\right) = 0. \qquad (14)$$

The solution is a symmetric positive - definite matrix P. The optimal control has the form

$$u(t) = -Cx(t) = -\frac{1}{r}\left(N^T + B^TP\right)x(t). \qquad (15)$$

The covariance matrix of the state vector $V(t) = E\left[x(t)x^T(t)\right]$ for $t = \infty$ is given by solution of the matrix algebraic Lyapunov equation

$$(A - BC)V + V(A - BC)^T + GG^T = 0. \qquad (16)$$

Step 2° Substitute $E\left[x_2^2\right]$ obtained in step 1 into the linearization coefficient k.
Step 3° Calculate P, U and V using the linearized method obtained in the last step.
Step 4° Repeat step 3° until P and V converge.
Step 5° Calculate the optimal value of criterion I_{opt} using the solution of the Riccati equation obtained in step 4°

$$I_{opt} = tr\left(PGG^T\right), \qquad (17)$$

where tr denotes trace.
Fast converge algorithms for solving the Lyapunov and Riccati equations one can find, for instance in [2], [6]. In the work reported here, the Smith and Kleinman algorithms were used for the Lyapunov end Riccati equation, respectively.

5. Numerical results

To illustrate the obtained results, a comparison of criterion I_{opt} defined by (17) versus parameter α has been shown. In this comparison, three criteria of linearization, namely the mean-square error of the displacement, equality of the second order moments of nonlinear and linearized elements, and the mean-square error of the potential energies were considered. The numerical results denoted by lines with stars, circles and squares, respectively, are presented in Figure 2. The parameters selected for calculations and further simulations are: $m_1 = 100$, $m_2 = 500$, $c_1 = 100$, $c_2 = 50$, $h_1 = 1$, $h_2 = 5$, $a_1^* = 0.025$ $a_2^* = 0.075$ $q^* = \sqrt{0.0067}$, $v = 20$ $\rho_1 = 1$, $\rho_2 = 1000$, $\rho_3 = 10000$, $\rho_4 = 1$.

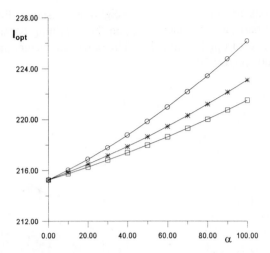

Figure 2. Comparison of optimization criteria obtained by application of different statistical linearization techniques versus parameter α.

The numerical results obtained by approximate methods were verified by simulations, where the optimal control determined by an approximate method was applied to the original nonlinear system with nonlinear criterion. The comparison of the relative errors Δ_{opt} is presented in Figure 3, where

$$\Delta_{opt} = \Delta_{opt}(\alpha) = \frac{\left|I_{opt}(\alpha) - I_{sim}\right|}{I_{sim}}. \tag{18}$$

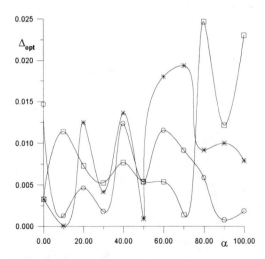

Figure 3. Comparison of relative errors of optimal criteria obtained by application of different statistical linearization techniques versus parameter α.

326

Further comparisons are given for other weighting constants. Figure 4 shows the dependence of the performance index I_{opt} upon the weighting constant (of suspension force) ρ_4 as it changes from 10^{-1} to $5\cdot10^4$. The other parameters are the same except $\rho_2 = 100$, $\rho_3 = 1000$ and $\alpha = 20$.

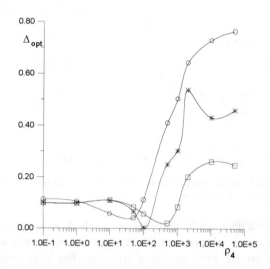

Figure 4. Comparison of relative errors of optimal criteria obtained by application of different statistical linearization techniques versus parameter ρ_4.

Figure 5 shows the dependence of the performance index I_{opt} upon the speed of the vehicle as it changes from 10^0 to 10^2. The other parameters are the same except $\rho_2 = 100$, $\rho_3 = 1000$ and $\alpha = 20$.

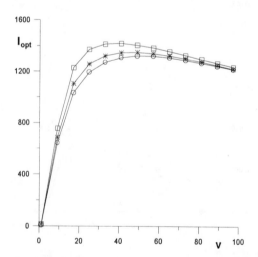

Figure 5. Comparison of relative errors of optimal criteria obtained by application of different statistical linearization techniques versus parameter v.

6. Conclusions

In this paper the problem of optimal active control for nonlinear 2-degrees-of-freedom vehicle suspension under stochastic excitations with nonlinear criterion has been studied. To determine the optimal control, a modified version of standard iterative procedure was used, where statistical linearization method and optimal control method for a linear system with mean - square criterion are used. This procedure was tested for three statistical linearization methods. Numerical studies show that for a given criterion of minimization (5), there are no significant differences between the considered linearization methods. This observation has already been made for one-dimensional simple examples of nonlinear systems with mean-square quadratic criterion by Heess [4]. We note that the smallest value of criterion I_{opt} defined by (17) for optimal control and the linearization method corresponding to the criterion of the mean - square error of the potential energies was obtained. However, in this case the relative error is the greatest.

Acknowledgments

This study is supported in part by Committee of Scientific Research in Poland under Grant No 7 TO7A 074 08 and by Silesian Technical University under Grant No BW - 410/RM10/98. Computer programming assistance given by M. Sc. K. Gawlikowicz is gratefully acknowledged.

References

1. Elishakoff, I. and Zhang, R. (1992) Comparison of the New Energy - Based   Versions of Stochastic Linearization Technique, in N. Beilomo and F. Casciati (eds), *Nonlinear Stochastic Mechanics*, Springer, Berlin, pp. 201 - 212.
2. Gajic, Z. and Qureshi, M. (1996) *Lyapunov Matrix Equation in System Stability and Control*, Academic Press, San Diego.
3. Hać, A. (1985) Suspension Optimization of 2 - DOF Vehicle Model Using a Stochastic Optimal Control Technique, *J. Sound and Vibr.*, **100**, pp. 343 - 357.
4. Heess, G. (1970) Application of state space statistical linearization to optimal stochastic control of non-linear systems, *Int. J. Control* **11**, pp. 697 - 701.
5. Kazakov, I. (1956) Approximate Probabilistic Analysis of the Accuracy of Operation of Essentially Nonlinear Systems, *Avtomat. i Telemekhan.* **17**,   pp. 423 - 450.
6. Kwakernak, H. and. Sivan, R. (1972) *Linear Optimal Control Systems*, Wiley - Interscience, New York.
7. Narayanan, S. and Raju, G. V. (1992) Active Control of Non-stationary Response of Vehicle with Nonlinear Suspension, *Vehicle System Dynamics* **21**, pp. 73 - 87.
8. Oueslati, F. and Sankar, S. (1994) A Class of Semi - Active Suspension Schemes for Vehicle Vibration Control, *J. Sound and Vibration.* **172**, pp. 391 - 411.
9. Yoshida, K. (1984) A method of optimal control of non-linear stochastic systems with non-quadratic criteria, *Int. J. Control* **39**, pp. 279 - 291.

ACTIVE STRUCTURAL CONTROL AGAINST WIND

T.T. SOONG
Department of Civil, Structural and Environmental Engineering
State University of New York at Buffalo
Buffalo, New York 14260, USA

H. GUPTA
Spars International/Aker Maritime
11757 Katy Freeway
Houston, TX 77079, USA

1. Introduction

In the past two decades, tremendous progress has been made in structural control research (Soong, 1990; Housner et al., 1994). In general, structural control as applied to wind excited motion of structures can be divided into two distinct approaches: structural, where mechanical properties of the structure are altered for control purposes; and aerodynamic, where aerodynamic modifications are made to the structural geometry. Aerodynamic control approach for mitigation of wind induced motion provides a distinct advantage over the traditional structural approach. Aerodynamic control devices are usually energy efficient since the energy in the flow is used to produce the desired control forces. Several passive aerodynamic devices have been developed in the past (Zdravkovich, 1981; Ogawa et al., 1987; Shiraishi et al., 1987; Kubo et al., 1992). Soong and Skinner (1981) performed experiments on an active device, called aerodynamic appendage, for controlling along wind motion of tall buildings. More recently Mukai et al. (1994) used a modified appendage mechanism to accommodate the changes in the angle of attack of the wind flow with respect to the building.

This paper presents an active aerodynamic control method to mitigate the bidirectional wind induced vibrations of tall buildings. The proposed control device, hereafter called Aerodynamic Flap System (or, AFS for short), consists of two flaps or appendages placed symmetrically at the leading separation edges of a structure. The flaps rotate about the leading edges of the structure like the flapping of a bird's wings. AFS is an active system driven by a feedback control algorithm which guides its operation based on information obtained from the vibration sensors. The area of the flaps and angular amplitude of rotation are the principal design parameters of AFS.

J. Holnicki-Szulc and J. Rodellar (eds.), Smart Structures, 329–336.

AFS is based on the idea of controlling the free shear layers and thus the vortices emerging from the separation edges of the structure by applying time varying geometrical perturbations at these edges. It produces an extra drag force through the exposed area (of the flap), a feature that can be exploited for control of along-wind motion. Experiments performed by Soong and Skinner (1981) fundamentally support the action of AFS in the along-wind direction. AFS also acts as a vortex generator, with the capacity of producing vortices at a desired frequency and phase. This feature can be exploited to control the across-wind motion.

2. Mathematical Modeling

A generic bluff body model is considered first to develop the mathematical description of the control system dynamics. Consider the bluff body in Fig. 1. For simplicity, the stiffness and damping are assumed to be the same in both directions. It is assumed that the applied aerodynamic force acts at the elastic center of the body, thus eliminating any torsional motion. It is also assumed that there is no mechanical or aerodynamic coupling in the two directions of motion and hence the equations can be developed separately in the two directions.

The motion in the along-wind direction primarily results from fluctuations in the upstream flow and, in terms of drag coefficient, the net force acting on the along-wind direction can be written as

$$F(t) = \frac{1}{2}\rho A C_D U^2 + \rho A C_D U u(t) \tag{1}$$

where δ is air density, A is projected area of the bluff body, C_D is drag coefficient and U is the mean wind speed. The wind speed fluctuation, $u(t)$, is usually assumed to be a stationary Gaussian process.

The major contribution of AFS to the along-wind dynamics comes from the exposed surface area of the flaps. Thus, for the along-wind direction, AFS acts as a force producing device. The force generated in the along-wind direction due to AFS can be written as

$$F_c(t) = 2\left[\frac{1}{2}\rho A_p(t) C_D (U + u(t))^2 S\right] \tag{2}$$

The factor two appears because of two symmetrically attached flaps. The exposed surface area $A_p(t)$ is a function of time since flaps rotate about the separation edge and it can be expressed as

$$A_p(t) = L(d\sin\theta_o + v_f(t)) \tag{3}$$

where L is length of the flap, d is width of the flap, θ_o is mean flap angle, and $v_f(t)$ is tip displacement of the flap $(= \Delta\theta(t)d)$. In Eq. 2, S is the control switching function which guides the flap operation. Thus, the dynamic equation of motion in the along-wind direction can be expressed simply as

$$\ddot{x} + 2\xi\omega\dot{x} + \omega^2 x = \frac{1}{m}\left[\frac{1}{2}\rho AC_D U^2 + \rho AC_D U u(t) + \rho A_p(t)SC_D(U + u(t))^2\right] \quad (4)$$

The approach used in this study for modeling of across-wind vortex induced vibrations follows closely the approach of Goswami et al., (1993). The motion of the flaps changes the vortex shedding pattern and subsequently the wake. The wake, an oscillator by itself as suggested by Billah (1989), responds only at the natural vortex shedding frequency of the bluff body in the absence of the flaps. Following Billah (1989) and Goswami et al. (1993), the equation of motion in the across-wind direction can be written as

$$\ddot{y} + 2\xi\omega\dot{y} + \omega^2 y = \frac{\rho U^2 D}{2m}\left[Y_1(R)\frac{\dot{y}}{U} + Y_2(R)\frac{y^2\dot{y}}{D^2 U} + B(R)v(t)\frac{y}{D^2}S\right] \quad (5)$$

where ξ is damping ratio, ω is natural frequency, m is mass per unit length, D is a dimension parameter, R is the reduced frequency, $Y_1(R)$ and $Y_2(R)$ are parameters of across-wind motion, and $B(R)$ is the non-dimensional control influence coefficient. The term involving $v(t)$ is the influence of AFS on across-wind dynamics of the bluff body. It is noted that the control influence on Eq. 5 is that of parametric excitation.

The bidirectional coupling through the control term introduces a two-way interaction in the dynamics of the bluff body. The switching function in the along-wind direction directly influences the operational time of the AFS for across-wind direction. The oscillation of flaps for across-wind direction changes the surface area of the appendage exposed to the wind thus changes the forces AFS can generate in the along-wind direction. The bidirectional coupling introduces interesting features to the AFS. On the one hand, it opens the possibility of controlling the two independent vibratory directions by using a single controller. On the other, it necessitates a trade-off, between the degree of control that can be achieved in the two directions.

3. Control Algorithms for AFS

In this study, a control approach suggested by Fujino et al. (1993) is adopted, giving

$$v(t) = a_v \cos(2\omega t + \pi/2) \quad (6)$$

for optimal energy dissipation. When amplitude and frequency of y varies slowly with time, Eq. 6 can be better implemented in a feedback form as

$$v(t) = \frac{2a_v}{\omega}\frac{y\dot{y}}{\left[y^2 + \frac{\dot{y}^2}{\omega^2}\right]} \quad (7)$$

where a_v is the control gain, which is stroke of the flap and is equal to the length of the flap multiplied by the angular amplitude. The flaps are initially positioned at an angle of 45° and the rotations take place around this angle. The motion of the individual flaps is at the frequency of ω, the natural frequency of motion. The two flaps attached to the two separation edges shed vortices alternately and thus create two vortices in one oscillation of the body. The feedback form of the individual flap motion can be determined as,

$$v_f(t) = \frac{a_v y}{\sqrt{y^2 + \frac{\dot{y}^2}{\omega^2}}} \tag{8}$$

The force generated by AFS in the along-wind direction can be utilized for control purposes through a switching function. The switching function adopted in this study for along-wind direction is a simple on-off type function, i.e.,

$$S = 1; \dot{x} \leq 0 \quad \text{and} \quad S = 0; \dot{x} > 0 \tag{9}$$

which can also be written as

$$S = \frac{1}{2}[1 - sgn(\dot{x})] \tag{10}$$

where sgn is the signum function.

However, the power of AFS in producing bidirectional control can be exploited if the switching function can be modified to create trade-off. For this purpose, an Exponential Switching Function (ESF) is introduced. The form of the function is

$$S = exp[-\lambda \dot{x}(1 + sgn(\dot{x}))] \tag{11}$$

It can be seen that λ is a parameter of ESF. Moreover, ESF collapses to Eq. 10 when $\lambda \to \infty$ and becomes one when $\lambda = 0$. The above behavior of ESF makes it useful for trade-off between the along-wind control and across-wind control. When the value of λ is zero, all the control effort is dedicated to the across-wind direction and, when the value of λ approaches ∞, the control effort is dedicated to the along-wind direction.

4. Dynamics of Tall Buildings with AFS

The generic bluff body equations can be extended to the dynamic modeling of tall buildings. The fundamental idea is that a small segment (dh) along the height of the building can be represented by bluff body equations and three dimensional effects are then incorporated into the prototype response through integrations over the entire structure.

The extension of Eq. 4 to tall buildings is fairly straightforward and quasi-steady assumptions can be used to advantage. Let $x(h,t) = \chi(t)\phi(h)$, then the mode generalized equation for a tall building in the first mode, $\phi(h)$, can be written as

$$M\left[\ddot{\chi} + 2\xi_m\omega\dot{\chi} + \omega^2\chi\right] = F_e(t) + F_c(t) \tag{12}$$

where χ is the generalized coordinate and M is the mode generalized mass. $F_e(t)$ is the mode generalized wind excitation force and $F_c(t)$ is the mode generalized control force in the along-wind direction with

$$F_e(t) = \frac{\rho D C_D U_H^2 H}{2(2a+b+1)} + \frac{\rho C_D D u(t) U_H H}{(a+b+1)} \tag{13}$$

where a is the exponent for wind speed (0.34), b is the mode shape exponent, H is building height, and U_H is wind speed at building height; and

$$F_c(t) = \rho C_D U_H^2 (1 - 2I_u) A_p(t) S \tag{14}$$

where I_u is the turbulence intensity at building height.

Following similar procedures, the equation of motion of a tall building in the across-wind direction can be obtained as

$$M\left[\ddot{\zeta} + 2\xi_m\omega\dot{\zeta} + \omega^2\zeta\right] = \Psi_1\dot{\zeta} + \Psi_2\zeta^2\dot{\zeta} + \Psi_3 Sv(t)\zeta \tag{15}$$

where

$$\Psi_1 = \frac{\rho D^2 Y_1 H f(1 - I_u)}{20}\left(\frac{U_h}{fD}\right)\left(\frac{h}{H}\right)^{2b} \tag{16}$$

$$\Psi_2 = \frac{\rho Y_2 H f(1 - I_u)}{20}\left(\frac{U_h}{fD}\right)\left(\frac{h}{H}\right)^{4b} \tag{17}$$

$$\Psi_3 = \frac{1}{2D}\rho U_H^2 B L_c \tag{18}$$

5. Design of AFS for Tall Buildings

The design of AFS for tall buildings essentially implies estimating the parameters of AFS based on the peak uncontrolled response. The procedure involves estimating the uncontrolled response and then calculating the AFS parameters for the desired reduction in response. To facilitate this procedure, approximate design formulae can be derived in terms of damping added to the structure due to AFS.

Let the switching function be defined by Eq. 11 (ESF). The fluctuating part in the controlled equation in the along-wind direction can be recast as

$$\ddot{\chi} + [2\xi_m\omega\dot{\chi} - 2C_c A_p(t)exp(-\lambda\dot{\chi}(1 + sgn(\dot{\chi})))] + \omega^2\chi = C_x u(t) \tag{19}$$

Using the technique of equivalent linearization, Eq. 19 can be approximated by a linear equation with additional linear damping ξ_a, which can be simplified to

$$\xi_a = \frac{C_c A_p}{2\sqrt{2\pi}\omega\sigma_{\dot{z}}}[1 - exp(-2\lambda\sigma_{\dot{z}})] \tag{20}$$

where $\sigma_{\ddot{z}}$ is the response due to additional damping, i.e.,

$$\sigma_{\ddot{z}}^2 = \frac{C_x^2 S_{uu}(f)}{16\pi(\xi_m + \xi_a)f} \tag{21}$$

where $S_{uu}(f)$ is spectral density of $u(t)$. Once the additional damping is found, the controlled response can be evaluate by Eq. 19, but with total damping $(\xi_m + \xi_a)$.

The additional damping in the across-wind direction can be similarly found. Under a set of simplifying assumptions, one can show that

$$\xi_a = \frac{\rho L_c B(2b+1)(d\Delta\theta)}{64\pi^2 \rho_b H D} \left(\frac{U_H}{fD}\right)^2 (1 + exp(-2\lambda\sigma_{\ddot{z}})) \tag{22}$$

The most important issue in the design of an AFS is the estimation of peak uncontrolled response over the life of a structure. The design process is further complicated by the fact that bidirectional control is desired from the AFS. Three primary conditions can be visualized on the worst case uncontrolled response from AFS operational point of view: (1) Along-wind motion dominates response; (2) Across-wind motion dominates response; and (3) both along-wind and across-wind responses are equally important. These three conditions will usually occur independently at different times over the life of the structure and AFS is expected to deliver the desired performance in all these cases. The first step in the design process is thus to determine under what circumstances these conditions occur.

The next step in design is to evaluate the AFS parameters for the three situations individually and a final design is selected from the three cases. The design parameters of AFS are: physical area of appendage, two times the area of an individual flap $(A_p^n = dL_c)$; the mean angle of flaps, θ_o; the flap oscillation amplitude, $a_v = d\Delta\theta$; and λ, the trade-off parameter.

6. Design Example

Step 1: Estimation of Building Parameters: A tall building is presented as an example for AFS design. The building is 200m (H) high with a square cross section of 40m (D). Other parameter values are: $\rho_b = 150$ kg/m^3, $a = 0.34$, $b = 1.5$, $f = 0.2$ Hz, $\xi_m = 1\%$ of critical, and design wind speed at the 10m height is 40m/s. The aerodynamic constants for the building are: $St = 0.15$, $C_D = 2.6$, $Y_1 = 26$, $Y_2 = 5382$ and $B = 250$. The Davenport spectrum (Davenport, 1961) is assumed for statistical description of turbulence.

Step 2:: Estimation of AFS Parameters for Peak Along-Wind Case: Only along-wind response is of importance in this case. Assuming a desired response reduction of 60%, the additional damping ratio can be computed from the controlled response as $\xi_a = 5.2\%$. The required area can then be found as $A_p^n = 160 \ m^2$. In the above, the binary switching function (BSF) is chosen since it provides optimal control in this case.

Step 3:   Estimation of AFS Parameters for Peak Across-Wind Case: The resonance occurs at the top of the building with $U_H = fD/St = 54\ m/s$. This corresponds to $U_{10} = 54(10/200)^{0.34} = 19.5\ m/s$. Only across-wind response is
considered in this case. The desired response reduction is again assumed to be
60%. The additional damping required can now be computed from the controlled
response as $\xi_a = 0.8\%$. The AFS parameters can be computed as $L_c d\Delta\theta = 53.8$.
The $\Delta\theta$ is chosen as 15° and this gives $A_p^n = L_c d = 200\ m^2$.

Step 4: Estimation of AFS Parameters for the Worst Bidirectional Response:
The along-wind and across-wind responses are equal at $U_{10} = 21.12\ m/s$. The
desired response reduction is again assumed to be 60%. The net area of flaps,
A_p^n, will be first computed from along-wind direction using BSF. The additional
damping is computed from the controlled along-wind response as $\xi_a = 5.2\%$. The
required area can then be found as $A_p^n = 128\ m^2$. The oscillation amplitude is
now computed from across-wind response. The additional damping required for
across-wind direction can be evaluated as $\xi_a = 0.026\%$. Oscillation amplitude is
calculated for BSF as $L_c d\Delta\theta = 3.43$. $A_p^n = 128\ m^2$ gives $\Delta\theta$ as 1.5^o. So choose
$A_p^n = 128\ m^2$ and $\Delta\theta = 1.5^o$.

Step 5: Selection of AFS Parameters: The AFS parameters can now be selected.
The parameters for Step 3 are largest in value and are thus chosen as final design
parameters of AFS. Hence $A_p^n = 200\ m^2$ and $\Delta\theta = 15^o$.

7.   Conclusions

The development of a new active aerodynamic control device (AFS) as shown in
Fig. 2 is presented in this paper. The device is simple to implement and has the
capability of controlling the along-wind as well as the across-wind vibrations. The
primary forcing in the along-wind direction is provided by the incident turbulence.
The device produces control forces based on the exposed surface area of the flaps. A
switching function is presented to effectively use the force produced by the device
for controlling the along-wind motion. Across-wind vibrations are modeled as
vortex shedding resonant vibrations. The effect of the control device is to produce
vortices at the desired phase and frequency. The influence of the device on the
across-wind vibrations is derived as a parametric stiffness term. An energy optimal
control algorithm is proposed for the across-wind direction. A bidirectional control
algorithm is also derived to control the two vibratory directions simultaneously.

Several practical features of AFS make it especially attractive for tall building
applications. AFS can be used for control of both along-wind and across-wind
vibrations, making it a general purpose device. AFS is not energy intensive and
does not need large actuating forces. This feature makes AFS a considerably
economical alternative. AFS is particularly attractive for retrofit as it requires
almost no structural changes to the building and is an external device. Finally,
AFS can be appropriately designed to add an aesthetic and architectural appeal
to the building.

336

References

1. Billah, K.Y.R. (1989), *A Study of Vortex-Induced Vibration*, Ph.D. Dissertation, Princeton University, Princeton, New Jersey.
2. Davenport, A.G. (1961), The Application of Statistical Concepts to Wind Loading on Structures, *Proc. Inst. of Civil Engrg.*, **19**, 449-472.
3. Fujino, Y., Warnitchai, P. and Pacheo, B.M. (1993), Active Stiffness Control of Cable Vibration, *Journal of Appl. Mech.*, **60**, 948-953.
4. Goswami, I., Scanlan, R.H. and Jones, N.P. (1993), Vortex Induced Vibrations of Circular Cylinders I: Experimental Data, *Journal of Eng. Mech.*, **119**, 2270-2287.
5. Goswami, I., Scanlan, R.H. and Jones, N.P. (1993), Vortex Induced Vibrations of Circular Cylinders II: New Model, *Journal of Eng. Mech.*, **119**, 2289-2302.
6. Housner, G.W., Soong, T.T. and Masri, S.F. (1994), Second Generation of Active Structural Control in Civil Engineering, Keynote Paper, *Proc. of First World Conference on Structural Control*, Pasadena, CA.
7. Kubo, Y, Modi, V.J., Yasuda, H. and Kato, K (1992), On the Suppression of Aerodynamic Instabilities through the Moving Surface Boundary Layer Control, *Proc. Eighth Int. Conf. on Wind Engrg.* Part 1, London, Canada.
8. Mukai, Y., Tachibana, E. and Inoue, Y. (1994), Experimental Study of Active Fin System for Wind Induced Structural Vibrations, *Proc. of First World Conf. on Struct. Control*, Pasadena, CA.
9. Ogawa, K., Sakai, Y. and Sakai, F. (1987), Aerodynamic Device for Suppressing Wind Induced Vibrations of Rectangular Sections, *Proc. Seventh Int. Wind Engrg. Conf.*, Aachen, Germany, **2**.
10. Shiraishi, N., Matsumoto, M., Shirato, H., and Ishizaki, H. (1987), On the Aerodynamic Stability Effects for Bluff Rectangular Cylinders with their Corner-cut, *Proc. Seventh Int. Wind Engrg., Conf.*, Aachen, Germany, **2**.
11. Soong, T.T. and Skinner, G.T. (1981), Experimental Study of Active Structural Control, *Journal of Eng. Mech.*, **107**, 1057-1068.
12. Soong, T.T. (1990), *Active Structural Control: Theory and Practice*, Longman Scientific and Technical, London and Wiley, New York.
13. Zdravkovich, M.M. (1981), Review and Classification of Various Aerodynamic and Hydrodynamic Means for Suppression of Vortex Shedding, *Journal of Wind Engrg. and Ind. Aerodynamics*, **7**, 145-189.

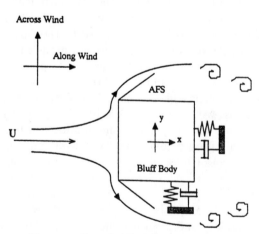

Figure 1. Generic Bluff Body

Figure 2. Schematic of Aerodynamic Flap System

LOW-CYCLE FATIGUE BEHAVIOR OF COPPER ZINC ALUMINUM SHAPE MEMORY ALLOYS

A. SUBRAMANIAM*, N. RAJAPAKSE*, D. POLYZOIS* and B.YUE[†]
* University of Manitoba, [†] Manitoba Hydro
Winnipeg, Canada R3T 5V6

1. Introduction

Fatigue and vibration failures continue to plague civil engineering structures. Transmission lines and supporting towers are exposed to wind and develop sustained motions. These sustained motions cause costly damage to conductors and other mechanical and structural subsystems. Wind induced vibrations of tower members have caused the fatigue failure of connecting members [1]. Many structural components in engineered structures are subjected to static and dynamic loading. Static loads, in the form of dead loads, are relatively harmless to a structure as they are anticipated and can be taken into consideration in the design. Dynamic loads, in the form of wind induced vibrations or in the form of earthquakes, are the more critical type of loads experienced by structures. The variability of naturally occurring dynamic loads is critical because their severity is unpredictable. In the construction industry, steel is the material of choice in building structures such as transmission towers. These structures have stood and continue to stand and serve their function rather well. However, over time, with continued exposure to dynamic loading (particularly wind), fatigue sets in and a member of the structure fails. Failure of the single member eventually leads to the failure of the entire structure. Fatigue failures of structures are expensive and dangerous. In terms of service life, steel subjected to cyclic loading fails in fatigue after a certain number of cycles. Shape Memory Alloys (SMA) have been reported to have a very high damping capacity and favorable fatigue properties. Recent studies have examined the possibility of using SMA for seismic energy dissipation. In civil engineering applications passive damping devices are more commonly found when compared to active devices. Employing SMA in structures could possibly reduce fatigue and vibration related problems in structures thereby increasing the service life of these structures. In order to use Copper Zinc Aluminum (CuZnAl) SMA in structural applications, a thorough study of its fatigue properties is required. In this study the low-cycle fatigue (LCF) behavior of CuZnAl in the austenitic phase at room temperature and CuZnAl in the martensitic phase at room temperature are examined.

337

J. Holnicki-Szulc and J. Rodellar (eds.), Smart Structures, 337–344.
© 1999 Kluwer Academic Publishers. Printed in the Netherlands.

2. Experimental

CuZnAl in the austenitic phase at room temperature and CuZnAl in the martensitic phase at room temperature were studied in this work. The CuZnAl alloys were obtained from the Memry Corporation, Connecticut. Austenitic CuZnAl alloy was supplied in the form of 6 mm (0.236 in.) diameter straightened rods. Each rod was 1.5 ft. in length. Memry Corporation had subjected this batch of material to the following heat treatment: 500°C for 30 minutes, followed by 800 °C for 15 minutes and water quenched thereafter. The transformation temperatures and the chemical composition of the austenitic CuZnAl are given in Table 1.

The martensitic CuZnAl alloy was received as a 6 mm (0.236 in.) diameter coiled rod. The supplier had not carried the heat treatment to completion on this material. As a result, the final heat treatment of the martensitic CuZnAl was done at the Metallurgical Laboratories of the University of Manitoba. The 6 mm coiled rod was cut into pieces, 11 inches in length. Each piece was swaged to a diameter of 5.84 mm (0.230 in.). The swaging process successfully straightened the rods with a small loss in diameter. The martensitic CuZnAl specimens were heat treated as follows: 800 °C for 30 minutes followed by water quenching. Following the heat treatment, the specimens were annealed at 100°C for 24 hours. The transformation temperatures along with chemical composition of the martensitic CuZnAl are given in Table 1.

The LCF test specimens were designed according to ASTM E466 [2]. Given the small diameter of the CuZnAl alloy rods, specimens were made so as to have the maximum test section diameter while still conforming to the ASTM Standard requirements for small sized specimens. The test specimens were round in cross-section with a test section diameter of 3.175mm and a gauge length of 12mm. The LCF test specimens were polished in a five-stage process. A lathe was employed in the polishing process to ensure uniformity. The samples were longitudinally polished using 320, 400 and finally 600 grit paper in that order. This was followed by polishing using 6-micron diamond paste. A final polish was done using 1-micron diamond paste.

Fully reversed, total strain controlled LCF tests were conducted in an air atmosphere at room temperature (~22°C) using a servohydraulic Instron testing machine (model 8502) with a digital control system (model 8500 plus). A sinusoidal waveform was employed with the strain varying between ε_{max} and ε_{min}. The test machine was equipped with a 250 kN load cell. Strain measurements were obtained using an Instron extensometer with a range of ±10%. The LCF tests were conducted using Instron LCF software that relied on the extensometer for feedback control of strain during the tests. Before the start of each LCF test, the LCF test machine measured the modulus of elasticity of each test specimen by applying a cyclic load of 500N at a low frequency. The LCF software monitored the test and collected the test data. The data recorded for each test included the tensile peak stress, σ_{max}, the compressive peak stress, σ_{min}, the stress range, $\Delta\sigma = \sigma_{max} + |\sigma_{min}|$, the tensile peak total strain, ε_{max}, the compressive peak total strain, ε_{min}, the total strain range, $\Delta\varepsilon_t = \varepsilon_{max} + |\varepsilon_{min}|$, and the number of cycles to failure, N_f. All LCF tests were carried out at a frequency of 0.5 Hz.

3. Results and Discussion

The low-cycle fatigue test program was carried out to investigate the fatigue behavior of both austenitic and martensitic CuZnAl in the plastic range. The tests were conducted with the total cyclic strain amplitude, ε_a, as the independent variable. The strain amplitudes selected, between 0.3% and 4%, were strains which would produce failure between 1 and 50000 cycles. However, attempts to test specimens at total strain amplitudes greater than 1% resulted in specimen buckling. This was attributed to the slenderness of the specimens. Therefore, five samples each of austenitic and martenstic CuZnAl were tested at total strain amplitudes between 0.3 and 1.0% as shown in Table 2.

The total axial strain amplitude can be expressed as

$$\varepsilon_a = \frac{\Delta\varepsilon_t}{2} \tag{1}$$

where $\Delta\varepsilon_t$ = total strain range.
The total axial strain amplitude can be further broken down into its separate elastic and plastic components:

$$\frac{\Delta\varepsilon_t}{2} = \frac{\Delta\varepsilon_e}{2} + \frac{\Delta\varepsilon_p}{2} \tag{2}$$

where: $\Delta\varepsilon_e$ = elastic-strain range
  $\Delta\varepsilon_p$ = plastic-strain range
Using Hooke's law, the elastic-strain range, $\Delta\varepsilon_e$, can be calculated by the relationship

$$\Delta\varepsilon_e = \frac{\Delta\sigma}{E} \tag{3}$$

Using the result of Eqs. (3), the total axial strain amplitude can be expressed as

$$\frac{\Delta\varepsilon_t}{2} = \frac{\Delta\sigma}{2E} + \frac{\Delta\varepsilon_p}{2} \tag{4}$$

where E is the modulus of elasticity. By rearranging Eqs. (4), the plastic strain amplitude can be calculated as follows:

$$\frac{\Delta\varepsilon_p}{2} = \frac{\Delta\varepsilon_t}{2} - \frac{\Delta\sigma}{2E} \tag{5}$$

Modulus of elasticity values used in the calculation of the plastic strain amplitude were measured before the start of each LCF test by the LCF test machine. The moduli of elasticity values were 120,000 MPa and 90,000 MPa for the austenitic and martensitic CuZnAl alloys respectively. The low-cycle fatigue results for the austenitic CuZnAl are shown in Table 2. Sample ALCF3 was not tested to failure. It was tested up to 48001 cycles which was very much greater than the LCF region ($N_f < 10^4$) [3]. The calculated

LCF parameters for the austenitic and martensitic CuZnAl alloys are given in Table 2. The average stress amplitude for sample ALCF6 obtained from the LCF software program appears to be incorrect. For this reason, sample ALCF6 is left out of computations involving the stress amplitude. However, sample ALCF6 follows the strain-life trend set by the other samples, as shown in Figure 6. For this reason it was decided to include sample ALCF6 in the strain-life analysis.

The cyclic stress response curves for austenitic and martensitic CuZnAl are shown in Figures 1 and 2 respectively. The average stress amplitude, $\Delta\sigma/2$, of each cycle is plotted against the number of cycles to failure. At all strain levels tested, the austenitic CuZnAl shows cyclic hardening within the first 10 cycles. Following the hardening during the first few cycles, the alloy shows cyclic stability up to fracture. The martensitic CuZnAl samples showed gradual cyclic hardening up to the point of fracture. The degree of hardening at any particular strain amplitude, $H_{\Delta\varepsilon t/2}$, was calculated using the following relationship [4]:

$$H_{\Delta\varepsilon_t/2} = \frac{(\Delta\sigma/2)_m - (\Delta\sigma/2)_1}{(\Delta\sigma/2)_1} \times 100 \tag{6}$$

For the austenitic CuZnAl alloy, $(\Delta\sigma/2)_1$ is taken as the average stress amplitude of the first cycle and $(\Delta\sigma/2)_m$ as the average stress amplitude level attained after initial hardening. In the case of martensitic CuZnAl, $(\Delta\sigma/2)_m$ is the average stress amplitude level just before fracture and $(\Delta\sigma/2)_1$ is taken as the average stress amplitude of the first cycle. The calculated degrees of hardening for both materials are included in Table 2. The austenitic alloy shows a moderate cyclic hardening of 9.5 - 18.6% with the degree of hardening increasing with increased total strain amplitude. The extent of cyclic hardening in the martensitic alloy was large and varied, from 33% to 176%. The cyclic hardening of 176% was observed in sample MLCF2 which was tested at a total strain amplitude of 0.5%. However, sample MLCF1 which was also tested at a total strain amplitude of 0.5%, hardened by 50%. Excluding the cyclic hardening of sample MLCF2, it is observed that, for martensitic CuZnAl cyclic hardening decreases with increasing total strain amplitude, which is in contrast to the cyclic hardening behavior exhibited by austenitic CuZnAl.

The cyclic stress-strain data of both alloys are analyzed in terms of the power law relationship

$$(\Delta\sigma/2)_{0.5N_f} = K'(\Delta\varepsilon_p/2)^{n'}_{0.5N_f} \tag{7}$$

where $(\Delta\sigma/2)_{0.5Nf}$ and $(\Delta\varepsilon_p/2)_{0.5Nf}$ are the cyclic stress and plastic strain amplitudes at half-life, K' is the cyclic strength coefficient, and n' is the cyclic work hardening exponent. The stress amplitude, $(\Delta\sigma/2)_{0.5Nf}$, and plastic strain amplitude, $(\Delta\varepsilon_p/2)_{0.5Nf}$, are plotted logarithmically in Figure 3. Lines are drawn through the data points using regression analysis to determine the values of the cyclic stress-strain constants. The values of K' and n' derived from the analysis are tabulated in Table 3. The cyclic strength coefficient and cyclic work-hardening exponent of the austenitic CuZnAl alloy is shown to be twice that of the martensitic alloy. For both materials, the cyclic strain hardening exponent falls in the range, 0.10 - 0.25, reported for most metals [5].

The data was further analyzed using the Coffin-Manson law. The Coffin-Manson law is an empirical relation showing that the plastic strain amplitude ($\Delta\varepsilon_p/2$) - number of reversals ($2N_f$) data can be linearized on log-log coordinates. The Coffin-Manson law can be expressed as:

$$\frac{\Delta\varepsilon_p}{2} = \varepsilon_f' \left(2N_f\right)^c \tag{8}$$

where $\Delta\varepsilon_p/2$ is the plastic strain amplitude, $2N_f$ is the number of reversals to failure, ε_f' is the fatigue ductility coefficient and c is the fatigue ductility exponent.

Coffin-Manson plots for austenitic and martensitic CuZnAl are shown on Figure 4. The values of ε_f' and c are also shown on Figure 4. The value of ε_f' for austenitic CuZnAl is higher than ε_f obtained from the tension tests. The fatigue ductility exponent, c, for austenitic CuZnAl is -0.59 and falls within the range of values reported for normal metals, i.e. -0.5 < c < -0.7 (Bannantine et al., 1990). For martensitic CuZnAl, the fatigue ductility coefficient and fatigue ductility exponent are determined to be 5.29 and -0.93 respectively. The value of ε_f' is much higher than ε_f and the value of c, -0.93, is lower than the value reported for most metals.

The Basquin relationship is an empirical relationship showing that the stress amplitude at half-life ($\Delta\sigma/2)_{0.5Nf}$ - number of reversals to failure ($2N_f$) can be linearized on log-log coordinates. The Basquin relationship can be expressed as:

$$\left(\frac{\Delta\sigma}{2}\right)_{0.5N_f} = \sigma_f' \left(2N_f\right)^b \tag{9}$$

where σ_f' is the fatigue strength coefficient and b is the fatigue strength exponent (Basquin's exponent).

The Basquin relationship plots for austenitic and martensitic CuZnAl are shown in Figure 5. The line through the data in both plots is obtained by linear regression analysis. Values of σ_f' and b determined using Eqs. (9) for both alloys are shown on Figure 5. The fatigue strength coefficient, σ_f', of austenitic CuZnAl is almost double the value of σ_f measured during monotonic tensile tests while the value of b is in the higher end of the range reported for most metals, i.e. -0.05 < b < -0.12. In the case of martensitic CuZnAl, the value of σ_f' is much closer to the value of σ_f and the value of b is also in the higher end of the range reported for most metals. Values of K' and n' were also derived from the following relationships:

$$K' = \frac{\sigma_f'}{\left(\varepsilon_f'\right)^{b/c}} \tag{10}$$

TABLE 1. Alloy compositions and M_s values.

Alloy	Nominal Composition, wt%					M_s, °C
	Cu	Zn	Al.	Zr	Ag	
Austenitic CuZnAl	69.5	26.5	3.8	.09	-	-7.3
Martensitic CuZnAl	70	25.7	4.05	.09	.06	42

TABLE 2: Low-cycle fatigue data

Sample	$\Delta\varepsilon_t/2$, %	$(\Delta\sigma/2)_{0.5Nf}$, MPa	$\Delta\varepsilon_p/2$, %	$H_{\Delta\varepsilon t/2}$, %	$2N_f$
Austenitic CuZnAl					
ALCF3	0.3	299.70	0.0335	9.5	96002
ALCF2	0.5	441.25	0.1465	10.3	7326
ALCF5	0.5	436.77	0.149	15.1	16982
ALCF6	0.75	296.31	0.4145	-	2198
ALCF4	1.0	532.01	0.577	18.6	850
Martensitic CuZnAl					
MLCF7	0.3	301.01	0.0125	89	89114
MLCF1	0.5	365.25	0.135	50.2	5426
MLCF2	0.5	441.11	0.0915	176	13008
MLCF8	1.0	519.52	0.497	46	2284
MLCF9	1.0	464.82	0.499	33	1558

TABLE 3: Cyclic stress-strain constants of CuZnAl alloys

Material	Cyclic stress-strain constants [Eqs. (7)]		Cyclic stress-strain constants [Eqs. (10) & (11)]	
	K', MPa	n'	K', MPa	n'
Austenitic CuZnAl	1580	0.203	1488	0.197
Martensitic CuZnAl	961	0.128	915	0.118

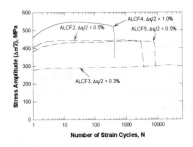

Figure 1. Cyclic stress response curves of austenitic CuZnAl showing variation of Δσ/2 with the number of strain cycles

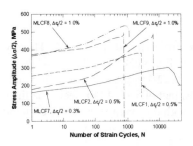

Figure 2. Cyclic stress response curves of martensitic CuZnAl showing variation of Δσ/2 with the number of strain cycles

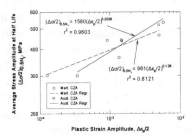

Figure 3. Variation of the average stress amplitude at half-life with corresponding plastic strain amplitude

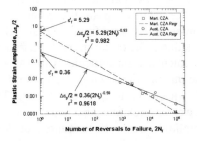

Figure 4. Coffin-Manson plots for austenitic and martensitic CuZnAl

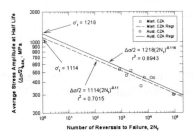

Figure 5. Variation of low-cycle fatigue with Δσ/2 shown on log-log basis

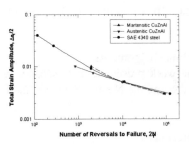

Figure 6. Comparing the behavior of CuZnAl to SAE 4340 steel in low-cycle fatigue

$$n' = \frac{b}{c} \tag{11}$$

and are included in Table 3. These values are in good agreement with those derived using Eqs. (10) and (11).

$\Delta\varepsilon_t/2$-$2N_f$ plots for austenitic and martensitic CuZnAl are shown in Figure 6 along with that of SAE 4340 steel. The lines through the data points is the best-fit line drawn through the average values of each total strain amplitude value.

4. Conclusions

Fully reversed, total strain controlled low-cycle fatigue experiments were conducted on austenitic and martensitic CuZnAl alloys. The cyclic stress response of the austenitic CuZnAl was characterized by cyclic hardening within the first 10 cycles followed by cyclic stability up to the point of fracture. For the martensitic CuZnAl, cyclic hardening occurred from the first cycle with the process accelerating towards the point of failure. The cyclic strength coefficient, K', and the cyclic work hardening exponent, n', calculated for austenitic CuZnAl alloy was almost double the values calculated for the martensitic CuZnAl. Both alloys exhibited linear Coffin-Manson plots. However, ε_f' is not equal to ε_f for either alloy. The constants of the stress amplitude-fatigue life power law relationship, namely σ_f' and b, are similar for both alloys. Total strain amplitude - reversals to failure plot shows a similarity in the behavior of austenitic and martensitic CuZnAl compared to that of steel. Further studies are needed to establish the complete fatigue behavior of CuZnAl SMA. The material is relatively inexpensive when compared NiTi SMA and has the potential for use in civil engineering applications if larger diameter rods can be manufactured. The use of fiber reinforced plastics with SMA reinforcements may also have potential application in civil engineering and merit future experimental investigations.

Acknowledgements

The work presented in this paper was supported by research grants from Manitoba Hydro and the Natural Sciences and Engineering Research Council (Grant A-6507).

References

1. Goel, A. P., "Fatigue Problems in Power Transmission Lines, Canadian Civil Engineer, February 1994.
2. American Society for Testing and Materials. 1990. "ASTM E 466 Conducting Constant Amplitude Axial Fatigue Tests of Metallic Materials," Philadelphia.
3. ASM Handbook, Vol. 8 - Mechanical Testing, 1992.
4. Prasad, N.E., Malakondaiah, G., Kutumbarao, V.V., and Rama Rao, P. 1996. "In-plane Anisotropy in Low-Cycle Fatigue Properties of and Bilinearity in Coffin-Manson Plots for Quaternary Al-Li-Cu-Mg 8090 Alloy Plate." Material Science and Technology, July 1996, Vol. 12. pp. 563 - 577
5. Bannantine, J. A., Comer, J. J., and Handrock, J. L. 1990. "Fundamentals of Metal Fatigue Analysis," Prentice Hall, Englewood Cliffs, New Jersey.

ON TWO PROBLEMS OF EXACT CONTROLLABILITY FOR ELASTIC ANISOTROPIC SOLIDS

J.J. TELEGA

Institute of Fundamental Technological Research
Polish Academy of Sciences
ul. Świętokrzyska 21, 00-049 Warsaw, Poland,
e-mail: jtelega@ippt.gov.pl

AND

W.R. BIELSKI

Institute of Geophysics, Polish Academy of Sciences
ul. Księcia Janusza 64, 01-452 Warsaw, Poland
e-mail: wbielski@igf.edu.pl

1. Introduction

In the study of exact controllability and stabilization problems of two- and three-dimensional bodies (elastic, viscoelastic) and structures like plates and shells it is usually assumed that those solids are made of isotropic and homogeneous materials (see Komornik [3], Lagnese [4], Lagnese and Lions [5], Lasiecka et al. [6], Lions [8]).

In the paper by Telega and Bielski [11] the problem of exact controllability of 3D anisotropic homogeneous elastic solids was solved. The control function of Dirichlet type act on a part of the boundary. The aim of the present contribution is twofold. First, in Section 2 the problem of exact controllability of a thin vibrating plate has been solved. The plate is made of an anisotropic and homogeneous linear elastic material. It is assumed that the boundary conditions are given by (3), though different controls can also be investigated (see [5]). Second, in Section 3 the problem of exact controllability for a 3D linear elastic solids is combined with homogenization. More precisely, it is assumed that the composite material the body is made of exhibits a microperiodic structure (see also Lions [8], vol.2).

J. Holnicki-Szulc and J. Rodellar (eds.), Smart Structures, 345–354.

2. Exact controllability of anisotropic Kirchhoff plate

Let $\Omega \subset \mathbb{R}^2$ be a bounded sufficiently regular domain. This domain represents the midplane of the undeformed thin plate. By h and w we denote the thickness of the plate and transverse displacement, respectively. The equation of motion is assumed in the following form:

$$\ddot{w} - \gamma \Delta \ddot{w} + \frac{\partial^2}{\partial x_\alpha \partial x_\beta} \left(D_{\alpha\beta\lambda\mu} \frac{\partial^2 w}{\partial x_\lambda \partial x_\mu} \right) = 0, \quad \text{in} \quad Q. \tag{1}$$

Here $\gamma = h^2/12$, $D_{\alpha\beta\lambda\mu} = \frac{1}{\varrho h} C_{\alpha\beta\lambda\mu}$, and ϱ stands for the density. Moreover, $\ddot{w} = \frac{\partial^2 w}{\partial t^2}$, t denotes time and Greek indices take values 1, 2.
Initial conditions are given by

$$w(0) = w^0, \qquad \dot{w}(0) = w^1 \quad \text{in} \quad \Omega. \tag{2}$$

Here we use the following convenient notation: $w(t) = \{w(x,t)|x \in \Omega\}$.
Boundary conditions are assumed in the form

$$\begin{cases} w = 0 & \text{on} \quad \Sigma = \Gamma \times (0,T), \\ \frac{\partial w}{\partial \boldsymbol{n}} = \begin{cases} v \text{ on } \Sigma_0 \subset \Sigma, & \Sigma_0 \text{ suitably chosen,} \\ 0 \text{ on } \Sigma \backslash \Sigma_0, \end{cases} \end{cases} \tag{3}$$

where $\Gamma = \partial\Omega$ and $\boldsymbol{n}$ denotes the outward unit normal vector to Γ. Other boundary conditions and controls are also plausible. For instance, assume that $\Gamma = \Gamma_0 \cup \Gamma_1$. We can act on $\Sigma_0 = \Gamma_0 \times (0,T)$ by

$$\begin{cases} w = 0 & \text{on} \quad \Sigma_0, \\ \frac{\partial w}{\partial \boldsymbol{n}} = v_0 & \text{on} \quad \Sigma_0 \text{ (or on part of } \Sigma_0, \frac{\partial w}{\partial \boldsymbol{n}} = 0 \text{ outside).} \end{cases}$$

Further, we can act on $\Sigma_1 = \Gamma_1 \times (0,T)$ by

$$\begin{cases} D_{\alpha\beta\lambda\mu} \frac{\partial^2 w}{\partial x_\lambda \partial x_\mu} n_\alpha n_\beta = v_1, \\ \frac{\partial M_\tau}{\partial s} + \frac{\partial M_{\alpha\beta}}{\partial x_\beta} n_\alpha - \gamma \frac{\partial \ddot{w}}{\partial \boldsymbol{n}} = v_2 \text{ on } \Sigma_1 \text{ (or on part } \Sigma_1, v_2 = 0 \text{ on the remaining part).} \end{cases}$$

Here $M_\tau = M_{\alpha\beta} n_\alpha \tau_\beta$, $M_{\alpha\beta} = D_{\alpha\beta\lambda\mu} w_{,\lambda\mu}$, $w_{,\alpha\beta} = \frac{\partial^2 w}{\partial x_\alpha \partial x_\beta}$, whilst τ is the tangent unit vector and s denotes a curvilinear abscissa on Γ measured positively in the direction of τ. The elasticity tensor $\boldsymbol{D} = (D_{\alpha\beta\lambda\mu})$ possesses the usual symmetry and coercivity properties. Here $\mathbb{E}_s^2$ stands for the space of symmetric 2×2 matrices. Our purpose is to drive the plate described by (1), (2) and (3) to rest at a given time T; i.e., $w(T) = \dot{w}(T) = 0$.

2.1. BASIC INEQUALITIES

First, we consider the equation

$$\ddot{\theta} - \gamma\Delta\ddot{\theta} + \frac{\partial^2}{\partial x_\alpha \partial x_\beta}(D_{\alpha\beta\lambda\mu}\frac{\partial^2\theta}{\partial x_\lambda \partial x_\mu}) = 0 \quad \text{in} \quad Q, \tag{4}$$

subject to initial conditions

$$\theta(0) = \theta^0, \qquad \dot{\theta}(0) = \theta^1 \text{ in } Q \tag{5}$$

and to *homogeneous* Dirichlet boundary conditions

$$\theta = 0, \qquad \frac{\partial\theta}{\partial n} = 0 \text{ on } \Sigma. \tag{6}$$

Let us pass to the variational formulation of (4) and (6). Multiplying (4) by $\bar{\theta}$ satisfying $\bar{\theta} = \frac{\partial\bar{\theta}}{\partial n} = 0$ on Σ and performing integration by parts we get

$$(\ddot{\theta}, \bar{\theta}) + \gamma(\nabla\ddot{\theta}, \nabla\bar{\theta}) + d(\theta, \bar{\theta}) = 0, \tag{7}$$

where

$$(\phi, \psi) = \int_\Omega \phi(x)\psi(x)\,dx, \qquad d(\phi, \psi) = \int_\Omega D_{\alpha\beta\lambda\mu}\kappa_{\alpha\beta}(\phi)\kappa_{\lambda\mu}(\psi)\,dx,$$
$$\kappa_{\alpha\beta}(\phi) = \phi_{,\alpha\beta} = \frac{\partial^2\phi}{\partial x_\alpha \partial x_\beta}. \tag{8}$$

Let $E_\gamma(t)$ stand for the total energy of the plate obeying (4)-(6). We have (see [5] in the case of isotropy)

$$E_\gamma(t) = \frac{1}{2}[\|\dot{\theta}(t)\|_{L^2}^2 + \gamma\|\nabla\dot{\theta}(t)\|_{L^2}^2 + d(\theta(t), \theta(t))]. \tag{9}$$

It can be shown that (the system is conservative) $\frac{d}{dt}E_\gamma(t) \equiv 0$. Thus we have

$$E_\gamma(t) = E_{0\gamma} = \frac{1}{2}[\|\theta^1\|_{L^2}^2 + \gamma\|\nabla\theta^1\|_{L^2}^2 + d(\theta^0, \theta^0)]. \tag{10}$$

Consequently we conclude that if $\theta^0 \in H_0^2(\Omega)$ and $\theta^1 \in H_0^1(\Omega)$ then the problem (4)-(6) admits a unique solution, say $\theta = \theta_\gamma$, which satisfies (see [5])

$$\theta_\gamma \in C([0, T]; H_0^2(\Omega)) \cap C^1([0, T]; H_0^1(\Omega)). \tag{11}$$

Direct inequality. Multiplying (4) by $h_\alpha\frac{\partial\theta}{\partial x_\alpha}$, where $h_\alpha \in C^1(\bar{\Omega})$ satisfies $h_\alpha = n_\alpha$ on Γ, we write

$$\int_Q (\ddot{\theta} - \gamma\Delta\ddot{\theta} + \frac{\partial^2}{\partial x_\alpha \partial x_\beta}D_{\alpha\beta\lambda\mu}\frac{\partial^2\theta}{\partial x_\lambda \partial x_\mu})h_\delta\frac{\partial\theta}{\partial x_\delta}\,dx dt = 0.$$

Finally, we obtain

$$\int_\Sigma D_{\alpha\beta\lambda\mu}\kappa_{\alpha\beta}(\theta)\kappa_{\lambda\mu}(\theta)\,d\Sigma \le c(T)E_{0\gamma}, \tag{12}$$

which is valid for $T > 0$ arbitrarily small.

Inverse inequality. Let $x^0 = (x_\alpha^0) \in R^2$. We set

$$m_\alpha(x) = x_\alpha - x_\alpha^0, \qquad \Gamma(x^0) = \{x \in \Gamma | \boldsymbol{m}(x) \cdot \boldsymbol{n}(x) > 0\}, \tag{13}$$

$$X = [(\dot\theta, \boldsymbol{m} \cdot \nabla\theta) - \gamma(\Delta\dot\theta, m \cdot \nabla\theta)]|_0^T. \tag{14}$$

Multiplying (4) by $\boldsymbol{m} \cdot \nabla\theta = m_\alpha \frac{\partial\theta}{\partial x_\alpha}$ we find

$$0 = \int_0^T [(\ddot\theta(t), \boldsymbol{m} \cdot \nabla\theta(t)) - \gamma(\Delta\ddot\theta, \boldsymbol{m} \cdot \nabla\theta(t) +$$

$$(\frac{\partial^2}{\partial x_\alpha \partial x_\beta} D_{\alpha\beta\lambda\mu}\theta_{,\lambda\mu}(t), \boldsymbol{m} \cdot \nabla\theta(t))]\,dt =$$

$$= X - \int_Q \frac{m_\alpha}{2}\frac{\partial\dot\theta^2}{\partial x_\alpha}\,dxdt + \gamma \int_Q (\Delta\dot\theta)m_\alpha\frac{\partial\dot\theta}{\partial x_\alpha}\,dxdt + \int_0^T d(\theta, \boldsymbol{m} \cdot \nabla\theta)\,dt -$$

$$\int_\Sigma D_{\alpha\beta\lambda\mu}\theta_{,\lambda\mu}n_\alpha\frac{\partial}{\partial x_\beta}(m_\omega\frac{\partial\theta}{\partial x_\omega})\,d\Sigma. \tag{15}$$

On Γ we have $\frac{\partial\theta}{\partial x_\beta} = n_\beta\frac{\partial\theta}{\partial \boldsymbol{n}} = 0$, since $\frac{\partial\theta}{\partial \boldsymbol{n}} = 0$. Then

$$d(\theta, \theta) = \int_\Gamma D_{\alpha\beta\lambda\mu}\theta_{,\alpha\beta}\theta_{,\lambda\mu}\,d\Gamma = \int_\Gamma D_{\alpha\beta\lambda\mu}n_\alpha n_\beta n_\lambda n_\mu (\frac{\partial^2\theta}{\partial\boldsymbol{n}\partial\boldsymbol{n}})^2\,d\Gamma. \tag{16}$$

We also have

$$\frac{1}{2}\int_Q [\dot\theta^2 + \gamma|\nabla\dot\theta|^2 + d(\theta, \theta)]\,dxdt = \int_0^T E_{0\gamma}\,dt = TE_{0\gamma}. \tag{17}$$

After lengthy calculation we finally obtain

$$[T - 2\lambda_1(\gamma)]E_{0\gamma} \le \max_{x\in\bar\Omega}|\boldsymbol{m}(x)||n|\frac{1}{2}\int_{\Sigma(x^0)} D_{\alpha\beta\lambda\mu}\kappa_{\alpha\beta}(\theta)\kappa_{\lambda\mu}(\theta)\,d\Sigma =$$

$$\frac{R(x^0)}{2}\int_{\Sigma(x^0)} D_{\alpha\beta\lambda\mu}\kappa_{\alpha\beta}(\theta)\kappa_{\lambda\mu}(\theta)\,d\Sigma, \tag{18}$$

where, see [5]

$$\lambda_1^2 = \frac{1}{\lambda_0^2} + \gamma c^2,$$

that is

$$\lambda_1 = \lambda_1(\gamma) = (\frac{1}{\lambda_0^2} + \gamma c^2)^{1/2}. \tag{19}$$

We observe that (18) makes sense for $T > 2\lambda_1(\gamma)$, since T must be positive.

To apply HUM we set

$$\|\{\theta^0, \theta^1\}\|_F^2 = \int_{\Sigma(x^0)} D_{\alpha\beta\lambda\mu}\kappa_{\alpha\beta}(\theta)\kappa_{\lambda\mu}(\theta)\, d\Sigma. \tag{20}$$

From (12) and (18) it follows that for $T > 2\lambda_1(\gamma)$, γ fixed, $\|\{\theta^0, \theta^1\}\|_F^2$ is a norm on $F = H_0^2(\Omega) \times H_0^1(\Omega)$ equivalent to the $H_0^2(\Omega) \times H_0^1(\Omega)$ - norm.

Theorem 1. Let T be given and such that $T > 2/\lambda_0$, where λ_0 is the constant appearing in (19). Then for γ sufficiently small and given $\{w^0, w^1\} \in H_0^1(\Omega) \times L^2(\Omega)$ there exists $v \in L(\Sigma(x^0))$ such that the solution w of (1), (2) and (3) satisfies

$$w(T) = \dot{w}(T) = 0. \tag{21}$$

Proof. We take γ such that $T > 2\lambda_1(\gamma)$. For a prescribed $\{\theta^0, \theta^1\} \in H_0^2(\Omega) \times H_0^1(\Omega)$, θ is the solution to (4)–(6). Next, we define η as the solution to the adjoint state:

$$\ddot{\eta} - \gamma\Delta\ddot{\eta} + \frac{\partial^2}{\partial x_\alpha \partial x_\beta}\left(D_{\alpha\beta\lambda\mu}\frac{\partial\eta}{\partial x_\lambda \partial x_\mu}\right) = 0 \quad \text{in } Q,$$

$$\eta(T) = \dot{\eta}(T) = 0 \quad \text{in } \Omega,$$

$$\eta = 0 \text{ on } \Sigma, \tag{22}$$

$$\frac{\partial\eta}{\partial n} = \begin{cases} D_{\alpha\beta\lambda\mu}n_\alpha n_\beta \kappa_{\lambda\mu}(\theta) & \text{on } \Sigma(x^0), \\ 0 & \text{on } \Sigma \setminus \Sigma(x^0). \end{cases}$$

Multiplying $(22)_1$ by θ, integrating by parts and taking into account (5), (6) and $(22)_{2,3,4}$ we get

$$(\dot{\eta}(0) - \gamma\Delta\dot{\eta}(0), \theta^0) + (-\eta(0) + \gamma\Delta\eta(0), \theta^1) = \int_{\Sigma(x^0)} [D_{\alpha\beta\lambda\mu}n_\alpha n_\beta \kappa_{\lambda\mu}(\theta)]^2\, d\Sigma. \tag{23}$$

For $C > 0$ being the coercivity constant, we have

$$D_{\alpha\beta\lambda\mu}n_\alpha n_\beta n_\lambda n_\mu \geq C|n|^2|n|^2 = C. \tag{24}$$

On Γ we have $\kappa_{\alpha\beta}(\theta) = n_\alpha n_\beta \frac{\partial}{\partial n}\left(\frac{\partial\theta}{\partial n}\right)$, provided that $\theta = 0$; therefore

$$(\dot{\eta}(0) - \gamma\Delta\dot{\eta}(0), \theta^0) + (-\eta(0) + \gamma\Delta\eta(0), \theta^1) \geq C \int_{\Sigma(x^0)} D_{\alpha\beta\lambda\mu}\kappa_{\alpha\beta}(\theta)\kappa_{\lambda\mu}(\theta)\, d\Sigma. \tag{25}$$

We define now the operator Λ_γ by

$$\Lambda_\gamma\{\theta^0, \theta^1\} = \{\dot{\eta}(0) - \gamma\Delta\dot{\eta}(0), -\eta(0) + \gamma\Delta\eta(0)\} =$$

$$\{(1 - \gamma\Delta)\dot{\eta}(0), -(1 - \gamma\Delta)\eta(0)\}. \tag{26}$$

By using (23) and (25) we write

$$\langle \Lambda_\gamma \{\theta^0, \theta^1\}, \{\theta^0, \theta^1\}\rangle = \int_{\Sigma(x^0)} [D_{\alpha\beta\lambda\mu} n_\alpha n_\beta \kappa_{\lambda\mu}(\theta)]^2 \, d\Sigma \geq$$
$$C \int_{\Sigma(x^0)} D_{\alpha\beta\lambda\mu} \kappa_{\alpha\beta}(\theta) \kappa_{\lambda\mu}(\theta) \, d\Sigma. \tag{27}$$

Consequently, Λ_γ defines an isomorphism from $F = H_0^2(\Omega) \times H_0^1(\Omega)$ onto $F' = H^{-2}(\Omega) \times H^{-1}(\Omega)$. Next, for a given $\{w^0, w^1\} \in H_0^1(\Omega) \times L^2(\Omega)$ we solve the equation:

$$\Lambda_\gamma \{\theta_\gamma^0, \theta_\gamma^1\} = \{(1 - \gamma\Delta)w^1, -(1 - \gamma\Delta)w^0\}. \tag{28}$$

By θ_γ we denote the corresponding solution to (4)–(6) and take

$$v = D_{\alpha\beta\lambda\mu} n_\alpha n_\beta \kappa_{\lambda\mu}(\theta_\gamma) \qquad \text{on } \Sigma(x^0), \tag{29}$$

(in fact $v = v_\gamma$). This is the control driving the plate to rest at time $t = T$.
□

Remark 1. (i) $\lambda_1(\gamma) \to 1/\lambda_0$ as $\gamma \to 0$. (ii) $J(v_\gamma) = \inf\{J(v)|v \in \mathsf{U}_{ad}\}$, where

$$J(v) = \int_{\Sigma(x^0)} v^2 \, d\Sigma, \tag{30}$$

and the set of admissible controls is defined by

$$\mathsf{U}_{ad} = \{v \in L^2(\Sigma(x^0)) | \{w^0, w^1\} \in H_0^1(\Omega) \times L^2(\Omega)$$
$$\text{is given, } w(T; v) = \dot{w}(T; v) = 0\}. \tag{31}$$

3. Exact controllability and homogenization: three-dimensional elasticity

Let now Ω be a bounded domain in $\mathbb{R}^3$, with sufficiently regular boundary $\Gamma = \partial\Omega$. The elastic moduli are εY-periodic, where Y is a so called basic cell (see [1]) and ε is a small parameter. We set

$$a_{ijkl}^\varepsilon = a_{ijkl}\left(\frac{x}{\varepsilon}\right), \quad x \in \Omega. \tag{32}$$

The moduli $a_{ijkl}(y)$ $(y = x/\varepsilon)$ are Y-periodic. Here the Latin indices take values 1, 2, 3. We assume that

$$a_{ijkl} = a_{jikl} = a_{klij} \in L^\infty(Y). \tag{33}$$

$$\exists C > 0 \text{ such that } \forall \epsilon \in \mathbb{E}_s^3 \ a_{ijkl}\epsilon_{ij}\epsilon_{kl} \geq C|\epsilon|^2, \tag{34}$$

where $\mathbb{E}_s^3$ stands for the space of symmetric 3×3 matrices. The equation of dynamic elasticity are specified by

$$\ddot{\boldsymbol{u}}^\varepsilon - \text{div}(\boldsymbol{a}^\varepsilon \boldsymbol{e}(\boldsymbol{u}^\varepsilon)) = \boldsymbol{v}, \qquad \text{in } Q = \Omega \times (0, T), \tag{35}$$

$$u^\varepsilon(0) = u^0, \quad \dot{u}^\varepsilon(0) = u^1 \quad \text{in } \Omega; \quad u^\varepsilon = 0 \quad \text{on } \Sigma = \Gamma \times (0,T), \quad (36)$$

where $e_{ij}(u) = (\frac{\partial u_i}{\partial x_j} + \frac{\partial u_j}{\partial x_i})/2$. In (35), v denotes the control function. Let $T > 0$ given. Since $v(\cdot, t)$ is defined on Ω, therefore T may be arbitrarily small. Let us suppose that

$$u^0 \in H_0^1(\Omega)^3, \quad u^1 \in L^2(\Omega)^3. \quad (37)$$

We set

$$\mathcal{V}(u^0, u^1) = \{v \in L^2(Q)^3 | u^\varepsilon(T; v) = \dot{u}^\varepsilon(T; v) = 0\}. \quad (38)$$

It can be shown that $\mathcal{V}(u^0, u^1) \neq \emptyset$, see [8, vol. 1, p. 62]. Denote by v^ε the minimizer of the following problem:

$$(P_\varepsilon) \qquad \min\{\frac{1}{2}\int_Q |v|^2 \, dxdt \mid v \in \mathcal{V}(u^0, u^1)\}. \quad (39)$$

Consider the homogeneous system

$$\ddot{\Phi} - \text{div}(a^\varepsilon e(\Phi)) = 0 \quad \text{in } Q,$$
$$\Phi(0) = \Phi^0, \quad \dot{\Phi}(0) = \Phi^1 \quad \text{in } Q; \quad \Phi = 0 \text{ on } \Sigma. \quad (40)$$

Direct inequality. There exists a positive constant, independent of ε, such that, cf. [8, vol. 2]

$$\int_0^T \|\dot{\Phi}(t)\|_{L^2}^2 \, dt \leq cT(\|\Phi^1\|_{L^2}^2 + \|e(\Phi^0)\|_{L^2}^2). \quad (41)$$

Inverse inequality. This inequality states that there exists a constant $m > 0$, independent of ε, such that

$$\int_0^T \|\dot{\Phi}(t)\|_{L^2}^2 \, dt \geq m(\|\Phi^1\|_{L^2}^2 + \|e(\Phi^0)\|_{L^2}^2). \quad (42)$$

Indeed, (40) yields

$$\left| \begin{array}{l} (\ddot{\Phi}, \hat{\Phi}) + a^\varepsilon(\Phi, \hat{\Phi}) = 0 \quad \forall \hat{\Phi} \in H_0^1(\Omega)^3, \\ \Phi(t) \in H_0^1(\Omega)^3, \end{array} \right. \quad (43)$$

where

$$a^\varepsilon(\Phi, \hat{\Phi}) = \int_\Omega a^\varepsilon_{ijkl}(x) e_{ij}(\Phi) e_{kl}(\hat{\Phi}) \, dx. \quad (44)$$

From (43), taking $\hat{\Phi} = q(t)\Phi$ with $q(t) = t^2(T-t)^2$ and exploiting Korn's inequality, we get the desidered inequality (42).

Estimation of Φ

Lemma 1. There exists a constant $m_1 > 0$, independent of ε, such that (see [9])

$$m_1^{-1}[\|\Phi^0\|_{L^2}^2 + \|\Phi^1\|_{H^{-1}}^2] \leq \int_0^T \|\Phi(t)\|_{L^2}^2 \, dt \leq m_1[\|\Phi^0\|_{L^2}^2 + \|\Phi^1\|_{H^{-1}}^2]. \tag{45}$$

Proof. Let $\chi^\varepsilon \in H_0^1(\Omega)^3$ be a solution to

$$-\operatorname{div}(a^\varepsilon(x)e(\chi^\varepsilon)) = -\Phi^1 \qquad \text{in } \Omega. \tag{46}$$

We introduce θ^ε by

$$\theta^\varepsilon(t) = \int_0^t \Phi(\tau) \, d\tau + \chi^\varepsilon. \tag{47}$$

By using the direct and inverse inequalities to $\dot{\theta}^\varepsilon(t)$, as well as the inequality (34) and Korn's inequality we establish the formula (45). $\square$

Let us pass now to application of HUM. Consider the following problem, cf. (40)

$$\ddot{\Phi}^\varepsilon - \operatorname{div}(a^\varepsilon(x)e(\Phi)) = 0 \qquad \text{in } \mathbf{Q},$$
$$\Phi^\varepsilon(0) = \Phi^0, \quad \dot{\Phi}^\varepsilon(0) = \Phi^1 \quad \text{in } \Omega, \quad \Phi^\varepsilon = 0 \text{ on } \Sigma. \tag{48}$$

Next, we solve the problem:

$$\ddot{\Psi}^\varepsilon - \operatorname{div}(a^\varepsilon(x)e(\Psi^\varepsilon)) = -\Phi^\varepsilon \quad \text{in } Q,$$
$$\Psi^\varepsilon(T) = \dot{\Psi}^\varepsilon(T) = 0 \quad \text{in } \mathbf{Q}, \quad \Psi^\varepsilon = 0 \quad \text{on } \Omega. \tag{49}$$

We define the operator Λ_ε by

$$\Lambda_\varepsilon\{\Phi^0, \Phi^1\} = \{\dot{\Psi}^\varepsilon(0), -\Psi^\varepsilon(0)\}. \tag{50}$$

Multiplying $(49)_1$ by Φ^ε, after simple calculation we get

$$\langle \Lambda_\varepsilon\{\Phi^0, \Phi^1\}, \{\Phi^0, \Phi^1\}\rangle = \int_0^T \|\Phi^\varepsilon(t)\|_{L^2}^2 \, dt. \tag{51}$$

Let the space F be given by $F = L^2(\Omega)^3 \times H^{-1}(\Omega)^3$. Then the natural norm is equivalent to

$$\|\{\Phi^0, \Phi^1\}\|_F = \left(\int_0^T \|\Phi^\varepsilon(t)\|^2 \, dt\right)^{1/2}. \tag{52}$$

By (51) we conclude that Λ_ε is an isomorphism from F onto $F' = L^2(\Omega)^3 \times H_0^1(\Omega)^3$. Consequently

$$\Lambda_\varepsilon\{\Phi^0, \Phi^1\} = \{u^1, -u^0\}, \tag{53}$$

admits a unique solution.

We are now in a position to formulate the main result of this section.

Theorem 2. Under the assumptions (33), (34) we have

$$v^\varepsilon \to v \quad \text{in} \quad L^2(Q) \text{ strongly}, \tag{54}$$

where v is the control which renders a minimum value to

$$(P) \quad \min\{\int_Q |z|^2 \, dx dt \mid z \in \mathcal{U}(u^0, u^1)\}, \tag{55}$$

where $\mathcal{U}(u^0, u^1) = \{z \in L^2(Q)^3 \mid u(T; z) = \dot{u}(T; z) = 0\}$, and the solution $u = u(v)$ of

$$\ddot{u} - \operatorname{div}(a^h e(u)) = v \quad \text{in } Q,$$
$$u(0) = u^0, \quad \dot{u}(0) = u^1 \quad \text{in } \Omega, \quad u = 0 \quad \text{on } \Sigma, \tag{56}$$

satisfies $u(T; v) = \dot{u}(T; v) = 0$ in Ω. The homogenized elastic moduli a_{ijkl}^h are given by (see [10])

$$a_{ijkl}^h = \frac{1}{|Y|} \int_Y a_{ijnm}(y)(\delta_{mk}\delta_{nl} + e_{mn}^y(w^{(kl)})) \, dy, \tag{57}$$

where $w^{(kl)} \in H_{per}(Y)$ are solutions to

$$\forall w \in H_{per}(Y), \quad \int_Y a_{ijmn}(y) e_{mn}^y(w^{(kl)}) e_{ji}^y(w) \, dy = - \int_Y a_{ijkl} e_{ij}^y(w) \, dy. \tag{58}$$

Here $e_{ij}^y(w) = (\frac{\partial w_i}{\partial y_j} + \frac{\partial w_j}{\partial y_i})/2$, and

$$H_{per}(Y) = \{w \in H^1(Y)^3 \mid w \text{ takes equal values on the opposite sides of } Y\}. \tag{59}$$

Sketch of the proof. By (51) and (53) we obtain

$$\langle \Lambda_\varepsilon\{\Phi_\varepsilon^0, \Phi_\varepsilon^1\}, \{\Phi_\varepsilon^0, \Phi_\varepsilon^1\}\rangle = \int_0^T \|\Phi^\varepsilon\|_{L^2}^2 \, dt = \int_\Omega (u^1 \cdot \Phi_\varepsilon^0 - u^0 \cdot \Phi_\varepsilon^1) \, dx. \tag{60}$$

The optimal control v^ε is given by: $v^\varepsilon = -\Phi^\varepsilon$. By using Lemma 1 and (60) we conclude that $\{\Phi^\varepsilon\}_{\varepsilon>0}$ is bounded in $L^2(0, T; L^2(\Omega)^3)$. Thus, up to a subsequence, we have $\Phi^\varepsilon \rightharpoonup \Phi$ in $L^2(0, T; L^2(\Omega)^3)$ weak. It can be shown that

$$\ddot{\Phi} - \operatorname{div}(a^h e(\Phi)) = 0 \quad \text{in } Q,$$
$$\Phi(0) = \Phi^0, \quad \dot{\Phi}(0) = \Phi^1 \text{ in } \Omega; \quad \Phi = 0 \text{ on } \Sigma,$$

354

and

$$\int_0^T \|\mathbf{\Phi}^\varepsilon\|_{l^2}^2 \, dt \to \int_\Omega (\mathbf{u}^1 \cdot \mathbf{\Phi}^0 - \mathbf{u}^0 \cdot \mathbf{\Phi}^1) \, dx.$$

Hence we conclude that

$$\int_\Omega (\mathbf{u}^1 \cdot \mathbf{\Phi}^0 - \mathbf{u}^0 \cdot \mathbf{\Phi}^1) \, dx = \int_0^T \|\mathbf{\Phi}\|_{L^2}^2 \, dt,$$

and (54) follows. □

Remark 2. (i) Suppose that the density is εY-periodic, i.e., $\varrho^\varepsilon(x) = \varrho(\frac{x}{\varepsilon})$, $x \in \Omega$. Then (35) is to be replaced by

$$\varrho^\varepsilon \ddot{\mathbf{u}}^\varepsilon - \operatorname{div}(\mathbf{a}^\varepsilon e(\mathbf{u})) = \mathbf{v} \quad \text{in } Q.$$

The homogenized equation is given by

$$\varrho^h \ddot{\mathbf{u}} - \operatorname{div}(\mathbf{a}^h e(\mathbf{u})) = \mathbf{v} \quad \text{in } Q,$$

where $\varrho^h = \frac{1}{|Y|} \int_Y \varrho(y) \, dy$. Theorem 2 remains valid.

(ii) More realistic cases of control, where $\mathbf{v}$ acts only on $\omega \subset \Omega$, can also be studied (see [7], [12]).

Acknowledgment. The authors were supported by the State Committee for Scientific Research (Poland) through the grant No 7 T07 A016 12.

References

1. Bensoussan, A., Lions J.-L. and Papanicolaou, G. (1978) *Asymptotic Analysis for Periodic Structures*, North Holland, Amsterdam.
2. Ciarlet, P. G. and Rabier, P. (1980) *Les Equations de von Kármán*, Springer-Verlag, Berlin.
3. Komornik, V. (1994) *Exact Controllability and Stabilization: The Multiplier Method*, John Wiley and Sons, Chichester, Masson, Paris.
4. Lagnese, J. F. (1989) *Boundary Stabilization of Plates*, SIAM Studies in Applied Mathematics, vol.10, Philadelphia.
5. Lagnese, J. F. and Lions, J.-L. (1988) *Modelling Analysis and Control of Plates*, Masson, Paris.
6. Lasiecka I., Triggiani, R. and Valente, V. (1996) Uniform stabilization of spherical shells by boundary dissipation, *Adv. in Diff. Eqs.*, **1**, pp. 635–674.
7. Lebeau, G. and Zuazua, E. (1998) Null-controllability of a system of linear thermoelasticity, *Arch. Rat. Mech. Analysis*, **141**, pp. 297–329.
8. Lions, J.-L. (1988) *Contrôlabilité Exacte. Perturbations et Stabilisation des Systèmes Distribués, t.1: Contrôlabilite Exacte, t.2: Perturbations*, Masson, Paris.
9. Lions, J.-L. (1988a) Contrôlabilite exacte et homogénéisation, *Asymptotic Anal.*, **1**, pp. 3–11.
10. Sanchez-Palencia, E. (1980) *Non-Homogeneous Media and Vibration Theory*, Springer, Berlin.
11. Telega, J. J. and Bielski, W. (1996) Exact controllability of anisotropic elastic bodies, in: *Modelling and Optimization of Distributed Parameter Systems: Applications to Engineering*, ed. by K. Malanowski, Z. Nohorski and M. Peszyńska, pp. 254–262, Chapman and Hall, London.
12. Zuazua, E. (1994) Approximate controllability for linear parabolic equations with rapidly oscillating coefficients, *Control and Cybernetics*, **27**, pp. 793–801.

NONLINEAR PIEZOELECTRIC COMPOSITES: DETERMINISTIC AND STOCHASTIC HOMOGENIZATION

J.J. TELEGA, B. GAMBIN, A. GALKA
Institute of Fundamental Technological Research
00-049 Warsaw, Swietokrzyska 21, Poland

1. Introduction

The aim of the present contribution is to perform *nonlinear* homogenization of piezoelectric composites with periodic or random microstructure. As it was argued by Tiersten [10], for stronger electric fields one has to take into account higher order terms in the electric field **E**. The form of the electric enthalpy $H(\mathbf{e}, \mathbf{E})$ proposed by this author, being of the third order in **E**, cannot be *concave* in **E**; here **e** denotes the strain tensor. It should be remembered that $H(\mathbf{e}, \mathbf{E})$ is here understood as a partial *concave* conjugate of the internal energy function $U(\mathbf{e}, \mathbf{D})$ with respect to **D**, **D** being the electric displacement vector. By using the Γ-convergence theory we obtain general form of the macroscopic potential (the effective internal energy function). The homogenization process smears out the microscopic inhomogeneities, cf. [1,2,5,7,8,11,12]. To perform this procedure the duality argument will play an essential role. The dual approach is also useful in finding bounds on effective properties. The homogenized potential U_h and its dual U_h^* are also formulated for a statistically homogeneous ergodic medium.

2. Basic relations

Let $V \subset \mathbb{R}^3$ be a bounded, sufficiently regular domain such that its closure $\overline{V}$ stands for a considered piezoelectric composite in its natural state. By $\gamma = \partial V$ we denote the boundary of V. If $\mathbf{u} = (u_i)$ is a displacement field, then $e_{ij}(\mathbf{u}) = u_{(i,j)}$ is the strain tensor; i,j=1,2,3. By $\mathbf{D} = (D_i)$, $\mathbf{E} = (E_i)$ and $\boldsymbol{\sigma} = (\sigma_{ij})$ we denote the electric displacement vector, the electric field and the stress tensor, respectively. As usual, we set $E_i(\varphi) = -\varphi_{,i}$, where φ is the electric potential. Let $\epsilon > 0$ be a small parameter and $\epsilon = \frac{l}{L}$. Here l, L are typical length scales associated with microinhomogeneities and the

J. Holnicki-Szulc and J. Rodellar (eds.), Smart Structures, 355–364.
© 1999 *Kluwer Academic Publishers. Printed in the Netherlands.*

region V, respectively. The internal energy is $U = U(y, \mathbf{e}, \mathbf{D})$, $y \in Y$. Here Y is a so-called basic cell, cf. [1,7,9]. We set

$$U_\epsilon(x, \mathbf{e}, \mathbf{D}) = U(\frac{x}{\epsilon}, \mathbf{e}, \mathbf{D}), \tag{1}$$

where $x \in V$, $\mathbf{e} \in \mathbb{E}_s^3$ and $\mathbf{D} \in \mathbb{R}^3$; $\mathbb{E}_s^3$ stands for the space of symmetric 3×3 matrices. We observe that the case of quadratic internal energy has been studied in [7]. In the general case the constitutive equations are given by

$$\sigma = \frac{\partial U}{\partial \mathbf{e}}, \quad \mathbf{E} = \frac{\partial U}{\partial \mathbf{D}}. \tag{2}$$

We make the following assumption:
(A) The function $U : (y, \epsilon, \rho) \in \mathbb{R}^3 \times \mathbb{E}_s^3 \times \mathbb{R}^3 \to U(y, \epsilon, \rho) \in \mathbb{R}$ is measurable and Y-periodic in y, convex in $(\mathbf{e}, \mathbf{D})$ and such that

$$\exists c_1 \geq c_0 > 0, \ c_0(|\epsilon|^p + |\rho|^q) \leq U(y, \epsilon, \rho) \leq c_1(|\epsilon|^p + |\rho|^q) \tag{3}$$

for each (y, ϵ, ρ). Here $p > 1$ and $q > 1$. As usual, we set $\frac{1}{p} + \frac{1}{p'} = 1$, $\frac{1}{q} + \frac{1}{q'} = 1$.
As a particular case of the internal energy one can consider the following one:

$$U(\frac{x}{\epsilon}, \mathbf{e}, \mathbf{D}) = U_1(\frac{x}{\epsilon}, \mathbf{e}, \mathbf{D}) + U_2(\frac{x}{\epsilon}, \mathbf{e}, \mathbf{D}), \tag{4}$$

where $U_1(\frac{x}{\epsilon}, \mathbf{e}, \mathbf{D})$ is a positive definite quadratic form in $\mathbf{e}$ and $\mathbf{D}$, typical for linear piezocomposites, cf. [4,7]. On the other hand, the function $U_2(\frac{x}{\epsilon}, \mathbf{e}, \mathbf{D})$ collects non-quadratic, higher order terms. A simple example is provided by

$$U_2(y, \mathbf{e}, \mathbf{D}) = \tilde{U}_2(y, \mathbf{D}) = \frac{1}{4} b_{ijkl}(y) D_i D_j D_k D_l, \quad y = \frac{x}{\epsilon}, \tag{5}$$

where $b_{ijkl} \in L^\infty(Y)$ is a completely symmetric tensor. Further restrictions on material functions are imposed by the requirement of $\tilde{U}_2(y, \mathbf{D})$ being convex in $\mathbf{D}$. The present contribution is confined to small deformations and the internal energy $U(y, \epsilon, \rho)$ convex in ϵ and ρ. Finite deformations are properly described by a nonconvex internal energy. Nonconvex homogenization is out of scope of the present contribution.
Let us pass to the formulation of the minimum principle. The following boundary conditions are assumed:

$$\mathbf{u} = 0 \quad \text{on } \gamma_0, \quad \sigma_{ij} n_j = \Sigma_i \quad \text{on } \gamma_1, \tag{6}$$

$$\varphi = \varphi_0 \quad \text{on } \gamma_2, \quad D_i n_i = 0 \quad \text{on } \gamma_3, \tag{7}$$

where Σ_i are the surface tractions, $\gamma = \overline{\gamma}_0 \cup \overline{\gamma}_1$, $\gamma_0 \cap \gamma_1 = \emptyset$; $\gamma = \overline{\gamma}_2 \cup \overline{\gamma}_3$, $\gamma_2 \cap \gamma_3 = \emptyset$, and $\mathbf{n} = (n_i)$ is the outward unit normal vector to γ; obviously $\emptyset$ denotes the empty set. For fixed $\epsilon > 0$ we set

$$F_\epsilon(\mathbf{u}, \mathbf{D}) = \int_V U_\epsilon(x, \mathbf{e}(\mathbf{u}), \mathbf{D})dx - L(\mathbf{u}, \mathbf{D}), \tag{8}$$

where

$$L(\mathbf{u}, \mathbf{D}) = \int_V b_i u_i dx + \int_{\gamma_1} \Sigma_i u_i d\gamma - \int_{\gamma_2} \varphi_0 D_i n_i d\gamma, \tag{9}$$

and

$$\mathbf{u} \in \mathbf{W}(V, \gamma_0) = \left\{ \mathbf{v} = (v_i) \mid v_i \in W^{1,p}(V), \ \mathbf{v} = \mathbf{0} \text{ on } \gamma_0 \right\},$$

$\mathbf{D} \in \mathbf{W}(div; V, \gamma_3) =$
$\{\mathbf{D} = (D_i) \mid D_i \in L^q(V), \ div\mathbf{D} \in L^q(V), \ div\mathbf{D} = 0 \text{ in } V, \ \mathbf{D} \cdot \mathbf{n} = 0 \text{ on } \gamma_3\}$.

The equlibrium problem of physically nonlinear piezocomposites with the ϵ-Y periodic microstructure means evaluating

$(\mathcal{P}_\epsilon)$ $F_\epsilon(\mathbf{u}^\epsilon, \mathbf{D}^\epsilon) = inf\{F_\epsilon(\mathbf{u}, \mathbf{D}) \mid \mathbf{u} \in \mathbf{W}(V, \gamma_0), \ \mathbf{D} \in \mathbf{W}(div; V, \gamma_3)\}$.

The assumption (A) implies the existence of unique $(\mathbf{u}^\epsilon, \mathbf{D}^\epsilon) \in \mathbf{W}(V, \gamma_0) \times \mathbf{W}(div; V, \gamma_3)$ solving the problem $(\mathcal{P}_\epsilon)$.

Γ-*convergence of the sequence of functionals* $\{F_\epsilon\}_{\epsilon>0}$. We proceed to find the limit functional

$$\Gamma\left[(s - L^p(V)^3) \times (w - L^q(V)^3)\right] - \lim_{\epsilon \to 0} F_\epsilon = F_h, \tag{10}$$

where $L^p(V)^3 = [L^p(V)]^3$ and $s - L^p(V)^3(w - L^q(V)^3)$ stands for the strong topology of $L^p(V)^3$ (the weak topology of $L^q(V)^3$). The loading functional L may be assumed to be continous in the topology $\tau = (s - L^p(V)^3) \times (w - L^q(V)^3)$. To this end it is sufficient to assume that $\mathbf{b} \in L^{p'}(V)^3$, $\Sigma \in L^{p'}(\gamma_1)^3$ and $\varphi_0 \in W^{1-\frac{1}{q'},q'}(\gamma_2)$. As a particular case, one can impose φ_0 being continous on γ_2 and vanishing on $\partial\gamma_2$. Then the functional L plays the role of a perturbation functional. Consequently it suffices to study the $\Gamma(\tau)$- limit of the following sequence of functionals $\{J_\epsilon\}_{\epsilon>0}$ given by

$$J_\epsilon(\mathbf{u}, \mathbf{D}) = \int_V U_\epsilon(x, \mathbf{e}(\mathbf{u}), \mathbf{D})dx. \tag{11}$$

A detailed presentation of the theory of Γ-convergence is provided by Dal Maso [5]. Periodic homogenization is summarized in the form of the following theorem.

Theorem 1. Let the assumption (A) be satisfied. The sequence of functionals $\{J_\epsilon\}_{\epsilon>0}$ is $\Gamma(\tau)$-convergent to the functional

$$J_h(\mathbf{u}, \mathbf{D}) = \int_V U_h(\mathbf{e}(\mathbf{u}), \mathbf{D})dx, \tag{12}$$

where $\mathbf{u} \in W^{1,p}(V)^3$, $\mathbf{D} \in L^q(V)^3$ and

$$U_h(\boldsymbol{\epsilon}, \boldsymbol{\rho}) = \inf \left\{ \frac{1}{|Y|} \int_Y U(y, \mathbf{e}^y(\mathbf{v}(y)) + \boldsymbol{\epsilon}, \mathbf{d}(y) + \boldsymbol{\rho})dy \mid \right.$$

$$\left. \mathbf{v} \in W_{per}^{1,p}(Y)^3, \mathbf{d} \in \Delta_{per}(Y) \right\}. \tag{13}$$

Here $\boldsymbol{\epsilon} \in \mathbb{E}_s^3$, $\boldsymbol{\rho} \in \mathbb{R}^3$, $e_{ij}^y(\mathbf{v}) = \frac{1}{2}\left(\frac{\partial v_i}{\partial y_j} + \frac{\partial v_j}{\partial y_i}\right)$ and

$$W_{per}^{1,p}(Y)^3 = \{\mathbf{v} \in W^{1,p}(Y)^3 | \mathbf{v} \text{ is } Y-\text{periodic}\}, \tag{14}$$

$$\Delta_{per}(Y) = \left\{\mathbf{d} \in L^q(Y)^3 \mid div_y\mathbf{d} = 0 \text{ in } Y, \langle\mathbf{d}\rangle = 0, \ \mathbf{d} \text{ is antiperiodic,} \right\} \tag{15}$$

$$\langle\mathbf{d}\rangle = \frac{1}{|Y|} \int_Y \mathbf{d}(y)dy. \qquad \square$$

Remark. 1. A function $\mathbf{v} \in W_{per}^{1,p}(Y)^3$ is Y-periodic if the traces of $\mathbf{v}$ on the opposite faces of Y are equal. Similarly, if $\mathbf{d} \in \Delta_{per}(Y)$, then the traces $\mathbf{d} \cdot \mathbf{N}$ are opposite on the opposite faces of Y. Here $\mathbf{N}$ stands for the outward unit normal vector to ∂Y.

Properties of U_h

(i) The function U_h is convex.

(ii) $\exists c_1 \geq c_0' > 0$ such that

$$c_0'(|\boldsymbol{\epsilon}|^p + |\boldsymbol{\rho}|^q) \leq U_h(\boldsymbol{\epsilon}, \boldsymbol{\rho}) \leq c_1(|\boldsymbol{\epsilon}|^p + |\boldsymbol{\rho}|^q),$$

for each $\boldsymbol{\epsilon} \in \mathbb{E}_s^3$, $\boldsymbol{\rho} \in \mathbb{R}^3$, .The constant c_1 is the same as in (3). The *dual macroscopic* potential U_h^* is given by

$$U_h^*(\boldsymbol{\epsilon}^*, \boldsymbol{\rho}^*) = \inf \{\langle U^*(y, \mathbf{t}(y) + \boldsymbol{\epsilon}^*, \mathbf{E}^y(\xi) + \boldsymbol{\rho}^*)\rangle \mid$$

$$\mathbf{t} \in \mathcal{S}_{per}(Y), \xi \in W_{per}^{1,q'}(Y)\}, \tag{16}$$

where U^* is the Fenchel conjugate of $U(y, \cdot, \cdot)$, $\boldsymbol{\epsilon}^* \in \mathbb{E}_s^3$, $\boldsymbol{\rho}^* \in \mathbb{R}^3$ and

$$\mathcal{S}_{per}(Y) = \left\{\mathbf{t} \in L^{p'}(Y, \mathbb{E}_s^3)| \ div_y\mathbf{t} = 0 \text{ in } Y, \langle\mathbf{t}\rangle = 0, \ \mathbf{t} \cdot \mathbf{N} \text{ is antiperiodic}\right\}. \tag{17}$$

Remark 2. The coercivity condition (3) can be weakened and incorporated into the scheme of nonuniform homogenization, cf. [9].

3. Stochastic homogenization

A random medium is modelled by a probability space $(\Omega, \mathcal{A}, P)$, where Ω is the set of all possible realizations, $\mathcal{A}$ is a σ-algebra on Ω, and the probability P is a positive measure on $(\Omega, \mathcal{A})$ such that $P(\Omega) = 1$. We shall always assume that $\mathcal{A}$ is P-complete. A real random variable X is measurable map from Ω into $\mathbb{R}$. $\mathcal{E}$ denotes the expectation (ensemble-average) symbol. We recall the definition of statistically homogeneous ergodic (S.H.E.) medium [5]. The statistical homogeneity means that two geometric points of the space are statistically indistinguishable. In other words, the statistical properties of the medium are invariant under the action of translations. Therefore, we assume the existence of a measure preserving flow $\tau_x, x \in \mathbb{R}^3$, which is a set of bijective representations of Ω into itself [6],

$$\tau_x : \Omega \to \Omega \quad (\omega \to \tau_x\omega),$$

satisfying the group property:

$$\forall x, y, \quad \tau_x \cdot \tau_y = \tau_{x+y}, \quad \tau_0 = \text{Identity},$$

the invariance property:

$$\forall x, \ \forall A \ \in \mathcal{A}, \tau_x A = \{\omega : \tau_{-x} \in A\} \in \mathcal{A} \text{ and } P(\tau_x A) = P(A),$$

and the assumption of measurability of the map $(\omega, x) \to \tau_x\omega$.

The existence of measure preserving flow enables us to connect uniquely with each random variable, X, the statistically homogeneous random process, $\tilde{X}$, defined by:

$$\forall \ (\omega, x) \in \Omega \times \mathbb{R}^3 \quad \tilde{X}(\omega, x) = X(\tau_{-x}\omega).$$

The above relationship means that the value of $\tilde{X}$ at point x is equal to its value at the origin after translation of the medium by $-x$. We recall that "almost surely"(a.s.) means "for every ω in a subset of Ω of probability one". We say that the measure preserving flow τ is ergodic if constants are the only real random variables such that for each x, $X \circ \tau_x = X$ a.s. We say that a medium is S.H.E. if it is modelled by a probability space on which acts ergodic measure preserving flow. An important consequence of Birkhoff's theorem is that, almost surely, we have the following weak convergence in $L^p_{loc}(\mathbb{R}^3)$ [5]:

$$\lim_{\epsilon \to 0_+} \tilde{X}(\omega, \cdot//\epsilon) = \mathcal{E}X.$$

Theorem 2. Consider a S.H.E. medium described by the probability space $(\Omega, \mathcal{A}, \mathcal{P})$ and measurable ergodic flow $\tau_x, x \in \mathbb{R}^3$. Let U be a map from $\Omega \times \mathbb{E}_s^3 \times \mathbb{R}^3$ into $\mathbb{R}^+$ having the following properties: $U(\cdot, \mathbf{e}, \mathbf{D})$ is mesurable for all $\mathbf{e} \in \mathbb{E}_s^3$ and $\mathbf{D} \in \mathbb{R}^3$, $U(\omega, \cdot, \cdot)$ is convex almost surely and

$$\exists c_1 \geq c_0 > 0, \quad c_0(|\boldsymbol{\epsilon}|^p + |\boldsymbol{\rho}|^q) \leq U(\omega, \boldsymbol{\epsilon}, \boldsymbol{\rho}) \leq c_1(|\boldsymbol{\epsilon}|^p + |\boldsymbol{\rho}|^q) \qquad \text{a.s.}$$

Then, the map $\tilde{U}(\omega, \cdot, \cdot, \cdot)$ is defined by: $\tilde{U}(\omega, x, \mathbf{e}, \mathbf{D}) = \mathbf{U}(\tau_{-\mathbf{x}}\omega, \mathbf{e}, \mathbf{D})$.

The random medium modelled by this map is almost surely macroscopicaly homogeneous. Moreover, the homogenized potential U_h and its conjugate U_h^* are given by:

$$U_h(\mathbf{e}, \mathbf{D}) = \inf\{\mathcal{E}U(\omega, \mathbf{k}(\omega)+\mathbf{e}, \mathbf{d}(\omega)+\mathbf{D}) | \mathbf{k} \in \mathcal{K}_1, \mathbf{d} \in \mathcal{K}_2\}, \mathbf{e} \in \mathbb{E}_s^3, \mathbf{D} \in \mathbb{R}^3,$$

$$U_h^*(\boldsymbol{\sigma}, \mathbf{E}) = \inf\{\mathcal{E}U^*(\omega, \mathbf{t}(\omega)+\boldsymbol{\sigma}, \mathbf{F}(\omega)+\mathbf{E}) | \mathbf{t} \in \mathcal{S}_1, \mathbf{F} \in \mathcal{S}_2\}, \boldsymbol{\sigma} \in \mathbb{E}_s^3, \mathbf{E} \in \mathbb{R}^3,$$

where
$\mathcal{K}_1 = \left\{\mathbf{k} \in L^p(\Omega, \mathbb{E}_s^3) | \mathcal{E}\mathbf{k} = 0, \tilde{\mathbf{k}}(\omega, \cdot) \quad \text{is a.s. a kinem. adm. strain field}\right\}$,
$\mathcal{K}_2 = \{\mathbf{d} \in L^q(\Omega, \mathbb{R}^3) | \mathcal{E}\mathbf{d} = 0, \tilde{\mathbf{d}}(\omega, \cdot) \quad \text{is a.s. a divergence free field}\}$,
$\mathcal{S}_1 = \{\mathbf{t} \in L^{p'}(\Omega, \mathbb{E}_s^3) | \mathcal{E}\mathbf{t} = 0, \tilde{\mathbf{t}}(\omega, \cdot) \quad \text{is a.s. a divergence free field}\}$,
$\mathcal{S}_2 = \{\mathbf{F} \in L^{q'}(\Omega, \mathbb{R}^3) | \mathcal{E}\mathbf{F} = 0, \tilde{\mathbf{F}}(\omega, \cdot) \quad \text{is a.s. a rotation free field}\}$.

4. Specific case

Let the internal energy be specified by (4) and (5). Consider a layered medium with x_3-axis perpendicular to the layers.

We assume that $Y = (0, 1)$, moreover $\nabla \to (0, 0, \dfrac{d}{dy})$.

The nonzero local strain components appearing in (13) are:

$$e_{13}^y = \frac{1}{2}\frac{dv_1}{dy}, \quad e_{23}^y = \frac{1}{2}\frac{dv_2}{dy}, \quad e_{33}^y = \frac{dv_3}{dy},$$

whilest the local electric displacement vector is: $\mathbf{d} = (d_1(y), d_2(y), 0)$.
The elementary cell Y consists of two layers: one of them is linear and the second one is nonlinear. In the linear range the layers exhibit the symmetry of PZTA. We also assume that

$$b_{ijkl}(y) = b(y)\delta_{ij}\delta_{kl},$$

where

$$b(y) = \left\{ \begin{array}{ll} b & \textit{in the nonliner layer with the volume fraction } \varphi_1, \\ 0 & \textit{otherwise.} \end{array} \right.$$

Solving the minimization problem in (13) with respect to **v** we find

$$\hat{e}_{13}^y = \frac{h_{15}}{a_{44}}d_1(y), \quad \hat{e}_{23}^y = \frac{h_{15}}{a_{44}}d_2(y).$$

Here h_{15}, is the piezoelectric coefficient and a_{44} stands for the elastic moduli (in Voigt notation). Then (13) reduces to

$$U_h(\rho) = \inf \left\{ \int_0^1 \left[\frac{1}{4}b(y) \left((d_1(y) + \rho_1)^2 + (d_2(y) + \rho_2)^2 + (\rho_3)^2 \right)^2 + \right. \right. \quad (18)$$

$$\left. \left. \frac{1}{2}k \left((d_1(y) + \rho_1)^2 + (d_2(y) + \rho_2)^2 \right) \right] dy \mid d_1, d_2 \text{ are } Y - \text{periodic}, < \mathbf{d} >= \mathbf{0} \right\}$$

where $k = \kappa_{11} - \frac{h_{15}^2}{a_{44}}$, and κ_{11} is the dielectric coefficient. Let

$$\begin{cases} D_\alpha(y) = \rho_\alpha + d_\alpha(y) \\ \int_0^1 D_\alpha(y)dy = \rho_\alpha \end{cases} \quad \alpha = 1, 2. \quad (19)$$

The local Euler equations are:

$$\frac{d}{dy}\left[b(y)(D_1^2 + D_2^2 + (\rho_3)^2)D_1 + kD_1 \right] = 0 \quad \text{in } Y, \quad (20)$$

$$\frac{d}{dy}\left[b(y)(D_1^2 + D_2^2 + (\rho_3)^2)D_2 + kD_2 \right] = 0 \quad \text{in } Y. \quad (21)$$

Hence

$$D_1 \left[b(y)(D_1^2 + D_2^2 + (\rho_3)^2) + k \right] = c_1 \quad \text{in } Y, \quad (22)$$

$$D_2 \left[b(y)(D_1^2 + D_2^2 + (\rho_3)^2) + k \right] = c_2 \quad \text{in } Y. \quad (23)$$

The nondimensional form of (22), (23) is given by

$$\overline{D}_1 \left[\overline{b}(y)\overline{D}_1^2 \left(1 + (\frac{\rho_2}{\rho_1})^2 \right) + \overline{b}(y) + \overline{k} \right] = \overline{c}, \quad \overline{D}_2 = \frac{\rho_2}{\rho_1}\overline{D}_1, \quad (24)$$

where

$$E_i = \kappa_{11}\rho_3 \overline{E}_i, \quad D_i = \rho_3 \overline{D}_i, \quad h_{ijk} = \kappa_{11}\rho_3 \overline{h}_{ijk},$$

$$b = \frac{\kappa_{11}}{(\rho_3)^2}\overline{b}, \quad a_{ijkl} = \kappa_{11}(\rho_3)^2 \overline{a}_{ijkl}, \quad k = \kappa_{11}\overline{k}, \quad c_1 = \kappa_{11}\rho_3 \overline{c}$$

Hence

$$\overline{D}_1 = \overline{c}\frac{1}{\overline{b}(y) + \overline{k}}G\left(\frac{[\overline{b}(y)(1 + (\frac{\rho_2}{\rho_1})^2)]^{1//2}|\overline{c}|}{(\overline{b}(y) + \overline{k})^{3//2}} \right). \quad (25)$$

Integration over the cell Y yields

$$\bar{\rho}_1 = \bar{c} \left\{ \varphi_1 \frac{1}{\bar{b} + \bar{k}} G \left(\frac{[\bar{b}(1 + (\frac{\rho_2}{\rho_1})^2)]^{1//2} |\bar{c}|}{(\bar{b} + \bar{k})^{3//2}} \right) + (1 - \varphi_1) \frac{1}{\bar{k}} G(0) \right\}, \qquad (26)$$

where $G(x)$

$$G(x)x = y, \quad y^3 + y = x.$$

For x in the interval $[-5, 5]$ the function G can be approximated by:

$$G_a(x) = \frac{2}{2 + |x|}$$

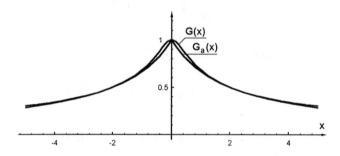

Figure1. Plot of the functions $G(x)$ and $G_a(x)$

To perform numerical calculation we assume: $\rho_1 \neq 0$, $\rho_2 = 0$, $\rho_3 = 1, \epsilon_{13} = 0$ and approximate the function G by G_a. For PZT7a we have $k = 0.246 - (2.3)^2//47.2$ $[\frac{m}{nF}]$. The macroscopic constitutive relation for the electric field $\mathbf{E} = \rho^*$ has now the following form

$$\rho_1^* = \frac{\partial U_h(\rho)}{\partial \rho_1} = c_1(\rho_1)$$

where $c_1(\rho_1)$ is calculated from the relation (28)
In Figs. 2 and 3 $\bar{\rho}_1^*$ versus $\bar{\rho}_1$ for various volume fractions are depicted. It is assumed that $\bar{b} = 0.01$ (Fig. 2) and $\bar{b} = 0.1$ (Fig. 3)

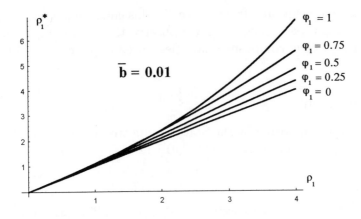

Figure 2.

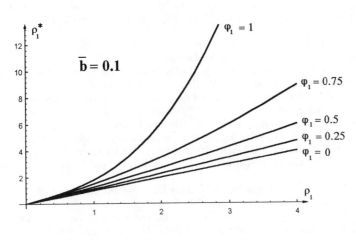

Figure 3.

5. Final remarks

To model the behaviour of piezoelectric composites with periodic or stochastic structure and subjected to stronger electric fields, nonlinear homogenization has been used. Our considerations are confined to small deformations. Such a case has its practical value, cf. [6]. Th. 1 justifies also the homogenization results obtained by the first author in [3] for linear piezocomposites. The primal and dual effective potential cannot be explicitly found, except in particular cases. Hence the need for bounding the effective potential U_h from below and from above. To this end, the nonlinear bounding techniques can be applied [12]. We observe that bounding techniques for linear piezoelectric composites have been used in [3], cf. also [4,6,11]

Acknowledgement

The authors were supported by the State Committee for Scientific Research (Poland) through the grant No 7 T07 A016 12.

References

1. Dal Maso, G., (1993) An Introduction to Γ-convergence, Birkhäuser, Boston.
2. Dal Maso, D., Modica, L., (1986) Nonlinear stochastic homogenization, *Ann. di Mat. Pur. ed Applic.*, 4, **144**, pp. 374-389.
3. Galka, A., Gambin, B., Telega, J.J., (1998) Variational bounds on the effective moduli of piezoelectric composites, *Arch. Mech.*, **50**, No 4.
4. Gibiansky, L.V., Torquato, S., (1997) On the use of homogenization theory to design optimal piezocomposites for hydrophone applications, *J. Mech. Phys. Solids*, **45**, pp. 689-708.
5. Sab, K., (1994) Homogenization of nonlinear random media by duality method. Application to plasticity, *Asymptotic Analysis*, **9** , pp. 311-336.
6. Sigmund, O., Torquato, S., Aksay, I.A., (1997) On the design of 1-3 piezocomposites using topology optimization, DCAM, Report No 542, April 1997.
7. Telega, J.J., (1991) *Piezoelectricity and homogenization. Application to biomechanics*, in: Continuum Models and Discrete Systems, vol 2, ed. by G.A. Maugin, pp. 220-229, Longman, Scientific and Technical, Harlow, Essex.
8. Telega, J.J., Gambin, B., (1996) *Effective properties of an elastic body damaged by random distribution of microcracks*, Proceedings of 8 th International Symposium, June 11-16 1995, Varna, Bulgaria, ed. K. Z. Markov, World Scientific, Singapore, pp. 300-307
9. Telega, J.J., Galka, A., Gambin, B., (1998) Effective properties of physically nonlinear piezoelectric composites, *Arch. Mech.*, **50**, pp. 321-340.
10. Tiersten, H.F., (1993) *Equations for the extension and flexure of relatively thin electroelastic plates undergoing large electric fields*, in: Mechanics of Electromagnetic Materials and Structures, ed. by J.S. Lee, G.A. Maugin, Y. Shindo, pp. 21-34, AMD-vol.161, ASME, New York.
11. Torquato, S., (1991) Random heterogeneous media: Microstructure and improved bounds on effective properties, *Appl. Mech. Rev.* **44**, pp. 37-77
12. Willis, J.R., (1986) Variational estimates for the overall response of an inhomogeneous nonlinear dielectric, *Homogenization and Effective Properties of Materials and Media*, ed. J.L. Ericksen, D. Kinderlehrer, R.V. Kohn and J.-L. Lions, pp. 245-263. Springer-Verlag, New York.

SIMULATION EXAMINATION OF ANNULAR PLATES EXCITED BY PIEZOELECTRIC ACTUATORS

A. TYLIKOWSKI
Warsaw University of Technology
Institute of Machine Design Fundamentals
Narbutta 84, 02-524 Warsaw, Poland

1. Introduction

The structural vibration due to the excitation of a piezoelectric actuator has been modelled by Jie Pan et al. [1] for beams, by Dimitriadis et al. [1] for rectangular plates, and by Van Niekerk et al. [3] for circular plates. Van Niekerk, Tongue and Packard presented comprehensive static model for a circular actuator and a coupled circular plate. Their static results were used to predict the dynamic behaviour of the coupled system, particularly to reduce acoustic transmissions. Piezoelectric transducers can be modeled as two-dimensional devices. The approach allows the distributed transducer shape to be included into the control design process for two-dimensional structures as an additional design parameter. The essence of the approach involves replacing the piezoactuators by forces and moments distributed along the piezoelement edges (e.g. [4]).

The purpose of this work is to study a dynamic model of the response of thin rotating annular plates to excitation by an actuator using annular piezoceramics glued to the plate surface. The elastic plate is clamped in the radial direction at the inner radius and free at its outer edge. Between the coordinates a and b, an annular piezoelectric actuator is attached. In this model the free stress boundary conditions characterizing the finite size piezoceramic actuators are also taken into account. The model is able to predict the response of the plate driven by the piezoelectric actuators glued to lower and upper plate surfaces. The actuators are driven by a pair of electrical fields with the same amplitute and in opposite phase. The actuators were used to excite steady-state harmonic vibrations in the plate. Annular piezoelectric actuators enable, obviously, obtaining only the axisymmetrical vibration modes. They are very useful in generating the required eigenforms selectively. It is yet to be emphasized that the use of annular piezoactuators can appear to be insufficient if such selective approach toward the vibration control is not desired, i.e. when all vibration modes need to be affected.

J. Holnicki-Szulc and J. Rodellar (eds.), Smart Structures, 365–372.
© 1999 *Kluwer Academic Publishers. Printed in the Netherlands.*

2. Two-Dimensional Distributed Actuators

Consider the rectangular plate with the rectangular piezoactuator. The equivalent external forces concentrated on the piezoelement edges are given as follows:

$$q(x,y) = M_o\big\{\big[H(x-x_1) - H(x-x_2)\big]\big[\delta'(y-y_1) - \delta'(y-y_2)\big]+$$

$$+\big[H(y-y_1) - H(y-y_2)\big]\big[\delta'(x-x_1) - \delta'(x-x_2)\big]\big\}. \tag{1}$$

For the two annular piezoactuators driving out-of-phase the radial moment and the force distributed along the outer and inner edges, are equivalent to the following transverse forces:

$$q(r) = C_o\Lambda(t)\Big[\delta'(r-a) - \delta'(r-b) + \frac{2}{r}(\delta(r-a) - \delta(r-b))\Big] \tag{2}$$

where C_o is a given constant, Λ is the piezoelectric strain.

3. Dynamic Equation of Axisymmetrical Plate Motion

Figure 1 shows a Kirchhoff thin annuler plate with identical piezoceramic actuators mounted on opposite sides of the plate. The annular actuators are driven by a pair of electrical fields with the same amplitude and in opposite phase. The actuators are perfectly bonded to the plate, which implies the strain continuity at the bonding interface. As the axisymmetrical motion is considered, the plate is divided into three annular sections as shown in Figure 1, and the motion of the plate in each section is analysed separately.

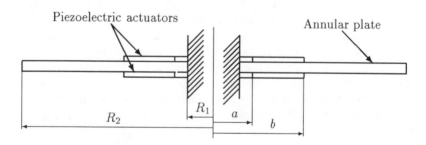

Figure 1. Geometry of plate with actuators

Consider a finite element of the radial length dr in the second section. The radial stresses in piezolayers are assumed to be uniformly distributed in the direction perpendicular to the plate due to the small thickness. For the radial actuator motion, the dynamics equation is as follows:

$$(r\frac{\partial \sigma_r}{\partial r} + \sigma_r - \sigma_t)t_a - \tau r = \rho_a t_a r \frac{\partial^2 u}{\partial t^2} \tag{3}$$

where σ_r - radial stress, σ_t - circumferential stress, τ - shear stress on the interface surface, u - radial displacement, t_a - thickness of actuator, ρ_a - density of actuator, r - radial coordinate, t - time. Using the Hooke's law and expressing the strains in actuators by the radial displacement

$$\epsilon_r = \frac{\partial u}{\partial r}, \qquad \epsilon_t = \frac{u}{r}, \tag{4}$$

$$\sigma_r = \frac{E_a}{1 - \nu_a^2}(\frac{\partial u}{\partial r} + \nu_a \frac{u}{r}), \tag{5}$$

$$\sigma_t = \frac{E_a}{1 - \nu_a^2}(\frac{u}{r} + \nu_a \frac{\partial u}{\partial r}), \tag{6}$$

we eliminate the normal stresses, and the dynamic actuator equation in displacement has the following form:

$$\frac{E_a t_a}{1 - \nu_a^2}\left\{r\frac{\partial}{\partial r}\left[\frac{1}{r}\frac{\partial}{\partial r}(ru)\right]\right\} - \tau r = \rho_a t_a r u_{,tt}, \tag{7}$$

where E_a and ν_a are modulus and Poisson's ratio of the actuator. Equation of the transverse plate motion w is as follows:

$$\frac{\partial(Tr)}{\partial r} = \rho_p t_p r \frac{\partial^2 w}{\partial t^2} \tag{8}$$

where T - shear force, ρ_p - plate density, t_p - plate thickness. The balance of moments has the form

$$\frac{\partial(M_r r)}{\partial r} - M_t - Tr + \tau t_p r = 0, \tag{9}$$

where M_r - radial moment, M_t - circumferential moment. Using Hooke's law we express the moments by the transverse plate displacement in the following form:

$$M_r = -D_p \left(\frac{\partial^2 w}{\partial r^2} + \frac{\nu_p}{r} \frac{\partial w}{\partial r} \right), \tag{10}$$

$$M_t = -D_p \left(\frac{1}{r} \frac{\partial w}{\partial r} + \nu_p \frac{\partial^2 w}{\partial r^2} \right), \tag{11}$$

where D_p is the plate cylindrical stiffness. Eliminating the shear force we rewrite equation (8) in the transverse displacement

$$D_p \nabla^2 w - \frac{1}{r} \frac{\partial(\tau t_p r)}{\partial r} + \rho_p t_p \frac{\partial^2 w}{\partial t^2} = 0 \tag{12}$$

where ∇^2 operator has the form $\nabla^2 = \frac{1}{r} \frac{\partial}{\partial r} \left\{ r \frac{\partial}{\partial r} \frac{1}{r} \left[\frac{\partial}{\partial r} \left(r \frac{\partial w}{\partial r} \right) \right] \right\}$.

Comparing the radial and circumferential strains on the interlayer surface we have

$$\epsilon_{ar} = \frac{\partial u}{\partial r} = \epsilon_{pr} = -\frac{t_p}{2} \frac{\partial^2 w}{\partial r^2}, \tag{13}$$

$$\epsilon_{at} = \frac{u}{r} = \epsilon_{pt} = -\frac{t_p}{2} \frac{1}{r} \frac{\partial w}{\partial r}, \tag{14}$$

the plate transverse displacement is related to the radial actuator displacement by

$$u = -\frac{t_p}{2} \frac{\partial w}{\partial r}, \tag{15}$$

Differentiating Equation (7) with respect to r, adding to Equation (12), the terms including τ vanish and the following equation is obtained for section 2:

$$(D_p + D_a)\nabla^4 w + \rho_p t_p \frac{\partial^2 w}{\partial t^2} - \frac{\rho_a t_a t_p^2}{2} \nabla^2 \left(\frac{\partial^2 w}{\partial t^2} \right) = 0, \qquad a < r < b, \tag{16}$$

where the actuator stiffness is denoted by $D_a = E_a t_a t_p^2 / 2(1 - \nu_a^2)$. The plate displacement equations for the plate section 1 and 3 have the classical form

$$D_p \nabla^4 w_1 + \rho_p t_p \frac{\partial^2 w_1}{\partial t^2} = 0, \qquad R_1 < r < a, \tag{17}$$

$$D_p \nabla^4 w_3 + \rho_p t_p \frac{\partial^2 w_3}{\partial t^2} = 0, \qquad b < r < R_2. \tag{18}$$

4. Boundary and Joint Conditions

The boundary conditions at $r = R_1$ and $r = R_2$ of the plate correspond the clamped and free edge, respectively.

$$w(R_1) = 0, \qquad \frac{\partial w}{\partial r}(R_1) = 0, \tag{19}$$

$$M_r(R_2) = -D_p \left(\frac{\partial^2 w}{\partial w^2} + \frac{\nu_p}{r} \frac{\partial w}{\partial r} \right)\Big|_{r=R_2} = 0, \ T(R_2) = -D_p \frac{\partial}{\partial r} \left(\frac{\partial^2 w}{\partial w^2} + \frac{1}{r} \frac{\partial w}{\partial r} \right)\Big|_{r=R_2} = 0. \tag{20}$$

At the joints between sections 1 and 2 and between sections 2 and 3, the continuity in plate deflection, slope and radial moment have to be satisfied

$$w_1(a) = w_2(a), \qquad \frac{\partial w_1}{\partial r}(a) = \frac{\partial w_2}{\partial r}(a), \tag{21}$$

$$w_2(b) = w_3(b), \qquad \frac{\partial w_2}{\partial r}(b) = \frac{\partial w_3}{\partial r}(b), \tag{22}$$

$$M_{r1}(a) = M_{r2}(a), \qquad M_{r2}(b) = M_{r3}(b). \tag{23}$$

The vanishing stress condition for the piezoelectric annular element at $r = a$ and $r = b$ constrains the deformations at these circles. Equating the radial stresses to zero we have

$$\sigma_{ra} = \frac{E_a}{1 - \nu_a^2} \left[\frac{\partial u}{\partial r} - \Lambda_a + \nu_a \left(\frac{u}{r} - \Lambda_a \right) \right] = 0, \tag{24}$$

where Λ_a is the piezoelectric strain assumed to be equal in the radial and circumferential directions. Substituting Equation (13), we have four joint conditions

$$\frac{\partial^2 w}{\partial r^2} + \frac{\nu_a}{r}\frac{\partial w}{\partial r} = -\frac{2}{t_p}\Lambda_a(1+\nu_a), \qquad r = a^-, \; a^+, \; b^-, \; b^+. \tag{25}$$

Equations (16), (17), (18), boundary conditions (19), (20) and joint conditions (21), (22), (23), (25) form a boundary value problem. Joint conditions (25) imply the equality of moments at $r = a$ and $r = b$. The dynamic plate response to the excitation by applied voltage U can be determined from the solution of this boundary value problem.

5. Analytical Solution

For a single frequency excitation Ω

$$\Lambda_a = d_a \frac{U}{t_a}\sin\Omega t \tag{26}$$

Equations (16), (17), and (18) become fourth-order linear ordinary homogeneous differential equations with r dependent coefficients. The solutions for w_1, w_2 and w_3 are respectively

$$w_1 = C_1 J_0(\kappa r) + C_2 Y_0(\kappa r) + C_3 I_0(\kappa r) + C_4 K_0(\kappa r), \tag{27}$$

$$w_2 = C_5 J_0(\alpha r) + C_6 Y_0(\alpha r) + C_7 I_0(\beta r) + C_8 K_0(\beta r), \tag{28}$$

$$w_3 = C_9 J_0(\kappa r) + C_{10} Y_0(\kappa r) + C_{11} I_0(\kappa r) + C_{12} K_0(\kappa r), \tag{29}$$

where the zeroth order Bessel functions of the first and the second kind are denoted by J_0, Y_0 and I_0, K_0, respectively. The wave number κ, α and β are calculated from the following formulae:

$$\kappa^4 = \rho_p t_p \Omega^2 / D_p, \tag{30}$$

$$\alpha^2 = \frac{1}{2}\left[\sqrt{\left(\frac{\rho_a t_a t_p^2}{2(D_p + D - a)}\Omega^2\right)^2 + \frac{4\rho_p t_p}{D_p + D_a}\Omega^2} + \frac{\rho_a t_a t_p^2}{2(D_p + D - a)}\Omega^2\right], \tag{31}$$

$$\beta^2 = \frac{1}{2}\left[\sqrt{\left(\frac{\rho_a t_a t_p^2}{2(D_p + D - a)}\Omega^2\right)^2 + \frac{4\rho_p t_p}{D_p + D_a}\Omega^2} - \frac{\rho_a t_a t_p^2}{2(D_p + D - a)}\Omega^2\right], \tag{32}$$

The twelve unknown constants in Eqs. (27) - (29) are determined by the simply supported boundary conditions at the plate edges and the joint conditions. For a given input piezoelectric strain Λ and excitation frequency Ω, the constants can be calculated from nonhomogeneous linear algebraic equations.

6. Results and Conclusions

Numerical calculations based on the formulas presented in the previous section are performed for a wide range of angular frequency and $\Lambda = 0.00114$. More precisely, $\Omega = 1\ 1/s$ for the static loading, in the first beam resonance $\Omega = 481\ 1/s$, and for higher frequencies. The parameters of the plate and piezoelectric elements used in calculations are listed in Table 1.

TABLE 1. Material parameters used in calculations

Material		Plate–Steel	Actuator–PZTG-1195
density	kg/m^3	7800	7275
modulus	N/m^2	21.6×10^{10}	63×10^9
thickness	m	0.002	0.0002
piezoelectric const.	m/V	–	1.9×10^{-10}
inner radius	m	0.02	0.03
outer radius	m	0.15	0.07

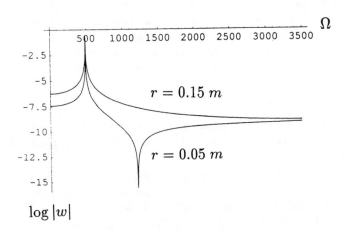

Figure 2. Plate transverse response $\log w$
(a) near field at $r = 0.05m$ (b) far field at the outer edge

Figure 2 shows the plate response (displacement) to a cyclic piezoelectric strains applied to both the piezoelectric actuators bonded to the plate. A comparison of

372

spatial distribution of deflection for a static case and in the first resonanse is given in Figure 3.

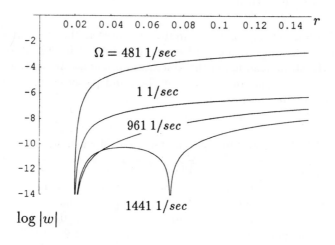

Figure 3. Spatial plate transverse response $\log w$

A dynamic model has been developed which is able to predict the response of the annular plate driven by the annular piezoactuators glued to the lower and upper surfaces. The actuators are used to excite steady-state harmonic vibrations in the plate. Spatial distributions of the plate deflection for the given characteristic values of the frequency and the frequency characteristics for fixed points are shown.

References

1. Dimitriadis, E., Fuller, C.R., and Rogers, C.A. (1991) Piezoelectric actuators for distributed vibration excitation of thin plates, *Journal of Applied Mechanics* **113**, 100-107.
2. Jie Pan, Hansen, C.H., and Snyder, S.D. (1992) A study of the response of a simply supported beam to excitation by a piezoelectric actuator, *Journal of Intelligent Material Systems and Structures* **3**, 3-16.
3. Van Niekerk, J.L., Tongue, B.H., and Packard, A.K. (1995) Active control of a circular plate to reduce transient noise transmission, *Journal of Sound and Vibration* **183**, 643-662.
4. Sullivan, J.M., Hubbard, J.E., and Burke, S.E. (1996) Modeling approach for two-dimensional distributed transducers of arbitrary spatial distribution, *Journal of Acoustical Society of America* **99**, 2965-2974.

NEW CONCEPT OF ACTIVE MULTIPLE-FREQUENCY VIBRATION SUPRESSION TECHNIQUE

M. VALÁŠEK (1), N. OLGAC (2)
*(1) Department of Mechanics, Faculty of Mechanical Engineering,
Czech Technical University in Prague,
(2) Department of Mechanical Engineering, University of Connecticut,
Storrs, CT-06269, USA*

1. Introduction

The paper deals with a new methodology for vibration suppression. It is based on an actively tuned vibration absorber which consists of an auxiliary mass m_a mounted on a spring (k_a), and a damper (c_a) elements with an additional control actuation force F_a. When this set of four components are tuned to show **peak response characteristics** at certain frequencies they can be used effectively as a vibration absorber against the excitations $F(t)$ at these frequencies (see Fig. 1). If the excitation frequencies agree with the sensitive frequencies of the absorber the primary system should show no response while the absorber gets excited and displays large amplitude motions.

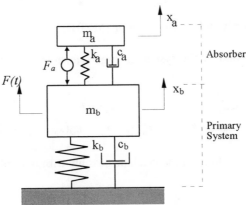

Figure 1. Delayed Resonator Principle

The most direct way of synthesizing the absorber section is to select a feedback control F_a such that the absorber subsection exhibits resonance behavior at

373

J. Holnicki-Szulc and J. Rodellar (eds.), Smart Structures, 373–382.
© 1999 *Kluwer Academic Publishers. Printed in the Netherlands.*

these frequencies. In other words this linear substructure possesses characteristic poles on the imaginary axis. Then the problem reduces to a pole placement operation. This paper presents an unconventional pole placement strategy. The state feedback control is suggested with **pure time delays**. The delay element introduces an augmentation of the system's order, thus the number of the characteristic poles increases. This is the key point of the proposed technique. It can be shown that, with the help of the delay element the number of the poles to be placed can be arbitrarily selected. The proposed technique brings an extension to the **Delayed Resonator** (DR) vibration absorption of [1]. In DR the time delayed feedback is utilized with a single frequency of resonance in contrast to multiple frequency possibilities here. It is also an extension to the **Dual Frequency Fixed Delayed Resonator** (DFFDR) vibration absorption of [2] where only double fixed frequencies are considered in contrast to multiple on-line tunable frequencies here.

Considering that the frequency content of *F(t)* may vary in time the absorber section should be tunable to track these variations. That is, the absorber's resonance frequencies should be real-time tunable. Practically, such a vibration absorber can absorb oscillatory energy at *n*-arbitrary frequencies which are time varying. This new setting can be effectively utilized as a tunable vibration suppresser at the respective *n* frequencies. In other words, when such an absorber is attached to a primary structure which is disturbed at these *n*-tonal (time varying frequencies) excitation the primary structure can be quieted at the point of attachment. This forms a very desirable and effective absorption scheme.

The highlight topics of the text are the **form of the time-delayed feedback** to achieve *n*-toppled resonance, the **stability** of the absorber structure and the combined setting of absorber and primary.

2. Multiple-Frequency Delayed Resonator (MFDR)

The equations of motion of the system in Fig. 1 together with the feedback are as follows

$$
\begin{aligned}
m_a \ddot{x}_a &= c_a(\dot{x}_b - \dot{x}_a) + k_a(x_b - x_a) + F_a \\
m_b \ddot{x}_b + c_b \dot{x}_b + k_b x_b &= -c_a(\dot{x}_b - \dot{x}_a) - k_a(x_b - x_a) - F_a + F
\end{aligned}
\tag{1}
$$

where F is the excitation force of the machine possibly with multiple discrete frequency components such as

$$
F = \sum_{k=1}^{K} f_k \sin(\omega_k t)
\tag{2}
$$

The control algorithm for the feedback forcing (F_a) is the key contribution of the proposed method. It is suggested to have only position feedback with multiple delayed terms with commensurate (i.e. integer multiple of a **given T**) delays as

$$F_a = \sum_{i=0}^{n} g_i (x_a(t - iT) - x_b(t - iT)) \tag{3}$$

or possibly only the acceleration with similar time delays

$$F_a = \sum_{i=0}^{n} g_i (\ddot{x}_a(t - iT) - \ddot{x}_b(t - iT)) \tag{4}$$

The number of feedback terms, n is twice the number of required eigenfrequencies K. The reason for this relation is simple: a complex pole can be placed by at least two real parameter (g_1 and g_2 gains for example). These gains of the delay terms are determined in such way that the absorber is in resonance with multiple exciting frequencies and these frequencies are on-line adjustable.

The characteristic equation of the absorber section alone (when it is installed to a perfect ground) is

$$CE(s) = m_a s^2 + c_a s + k_a - \sum_{i=0}^{n} g_i e^{-iTs} = 0 \tag{5}$$

This transcendental equation should have the required **resonant eigen-frequencies** , ω_k, plus other stable eigenvalues. To check the stability properties of this characteristic equation, however, is a difficult task due to the presence of infinite characteristic roots [3].

3. Synthesis of MFDR

Not only the MFDR should be stable itself, but also it should form a stable structure together with the primary system. The formation of such feedback, i.e. the synthesis, is the key problem here. An approach we follow is first the discretization and then the pole placement using full state feedback or output feedback. The discretization helps eliminate the transcendentality of the characteristic equation. Notice that the delay is simply a power of 'z' in the discrete domain. Thus, the characteristic roots become finite in number. The discretization is in fact a realistic model of the system in Fig. 1 because the control is provided by a microprocessor at discrete sampling times [4]. The discrete control scheme of the system (1), (3) is depicted on the Fig. 2.

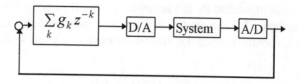

Figure 2 MFDR Control Scheme

First, the desired eigen-frequencies are placed for the absorber m_a itself. The system of the absorber

$$m_a \ddot{x}_a + c_a \dot{x}_a + k_a x_a = c_a \dot{x}_b + k_a x_b + F_a \tag{6}$$

is considered alone as attached directly to the frame

$$m_a \ddot{x}_a + c_a \dot{x}_a + k_a x_a = F_a \tag{7}$$

For the placement of desired eigen-frequencies this system is discretized with the sampling period equal T using the zero-order-hold method. The state variables

$$\begin{aligned} x_1 &= \dot{x}_a \\ x_2 &= x_a \end{aligned} \tag{8}$$

are introduced and the discretized description of (6) is obtained

$$\begin{bmatrix} x_1(k+1) \\ x_2(k+1) \end{bmatrix} = \mathbf{A}_d \begin{bmatrix} x_1(k) \\ x_2(k) \end{bmatrix} + \mathbf{B}_d F(k) \tag{9}$$

Then according to the number of terms in (4) corresponding additional state variables are introduced

$$\begin{aligned} x_3(k+1) &= x_2(k) \\ x_4(k+1) &= x_3(k) \\ &\cdots \\ x_{n-1}(k+1) &= x_{n-2}(k) \\ x_n(k+1) &= x_{n-1}(k) \end{aligned} \tag{10}$$

where $x_{j+2}(k) = x_a(k-j)$. These state equations (8) and (9) are rewritten in the form of an augmented system

$$\mathbf{x}(k+1) = \mathbf{A}\mathbf{x}(k) + \mathbf{B}F(k) \tag{11}$$

The gains g_i are determined as the gains $\mathbf{K} = [k_0, g_0, g_1, g_2, ..., g_{n+2}]$ for the pole placement of desired eigen-frequencies $\pm j\omega_k$ by the state feedback

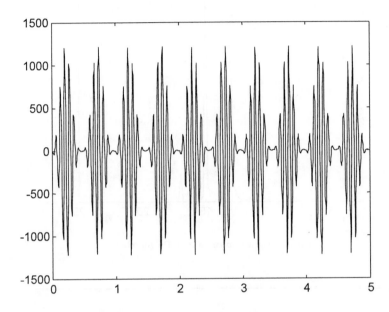

Figure 3 Time Behaviour of MFDR Absorber

$$F(k) = -\mathbf{K}\mathbf{x}(k) \tag{12}$$

The gain g_0 corresponds to the position feedback , k_0 corresponds to the velocity feedback and the other gains g_i correspond to the delayed position feedback. The controller with commensurate delays appears naturally in (12).

This strucutre represents a full state feedback. Therefore in this case the stability of the absorber is guaranteed.

An example of simulation with 4 eigenfrequencies [11, 13, 15, 17] Hz for the system $m_a = 1kg$, $c_a = 1Nm^{-1}s$, $k_a = 100Nm^{-1}$ with the delay $T = 0.01s$ is shown at the Fig. 3 as time behaviour and at the Fig. 4 as frequency spectrum.

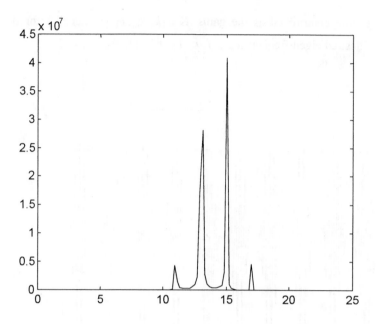

Figure4. Frequency Spectrum of MFDR Absorber

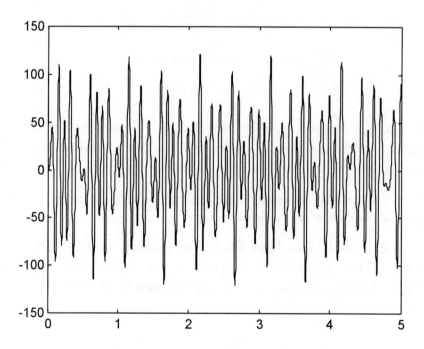

Figure 5. Time Behaviour of MFDR Absorber with Observer

The concern is that both states of the absorber (9), position and velocity, are necessary. They should either be measured or reconstructed from the position measurements p and acceleration a using the observer

$$v(k) = \frac{p(k) - p(k-1)}{T} + \frac{T}{2}a(k)$$

(13)

An example of simulation with 3 eigenfrequencies [7, 11, 13] Hz for the system $m_a = 1kg$, $c_a = 1Nm^{-1}s$, $k_a = 100Nm^{-1}$ with the delay $T = 0.01s$ is shown at the Fig. 4 as time behaviour. The sampling time of the observer is 10 times higher than the sampling time of the absorber itself.

This strucutre represents a full state feedback with an observer. Therefore in this case the stability of the absorber can be assured by the suitable choice of the observer. There exists also a solution for output (partial) state feedback where $k_2 = 0$.

4. Stability of the whole system

The stability of the absorber is not sufficient. The combined system (1) should also be stable and it can be tested relatively easily. The procedure of stability test is following. The combined system (1) is discretized with the sampling period T, including the delayed states (10) and state feedback (12). The finite number of eigenvalues of the resulting linear discrete system are tested for stability.

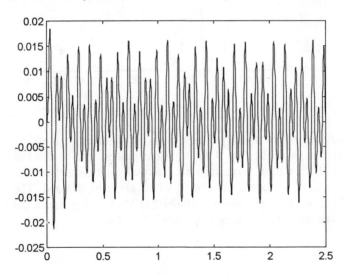

Figure 6. Time Behaviour of Excited Primary System without Absorber

The simplicity of this stability test is one of the main result of this study. This result is very important because the difficult problem of testing infinite number of eigenvalues is now transformed into the test of finite number of eigenvalues. It is worth emphasizing that, this is a realistic solution to the problem, not an approximation because of the discrete nature of the control used.

The simplicity of this test and its algebraic nature enable to provide a search for MFDR parameters. An example of simulation with 2 eigenfrequencies [11.5, 20] Hz for the primary system $m_b = 10kg$, $c_b = 100Nm^{-1}s$, $k_b = 35629Nm^{-1}$, absorber $m_a = 1kg$, $c_a = 1Nm^{-1}s$, $k_a = 17410Nm^{-1}$ with the delay $T = 0.025s$ is shown at the Fig. 6-9. The time behaviour of the primary system without the absorber is at the Fig. 6 and its frequency spectrum is at the Fig. 7.

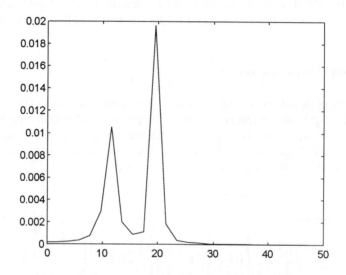

Figure 7. Frequency Spectrum of Excited Primary System without Absorber

The time behaviour of the primary system with the absorber is at the Fig. 8 and its frequency spectrum is at the Fig. 9 for transient process and at the Fig. 10 for the steady process. The exciting frequencies are removed, the steady state is determined by the resulting poles of the combined system which are least damped. The poles are - 0.6376 ±9.0542i, -0.2650 ±18.6946i, -0.0076 ±11.3721I. They prove the stability of the combined systems. The resulting amplitude is 0.004 m.

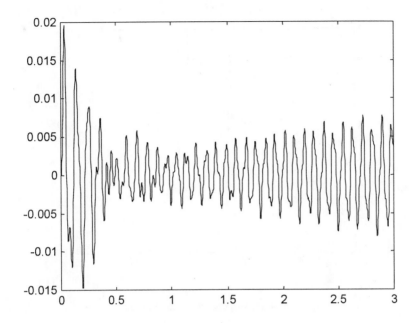

Figure 8. Time Behaviour of Excited Primary System with MFDR Absorber

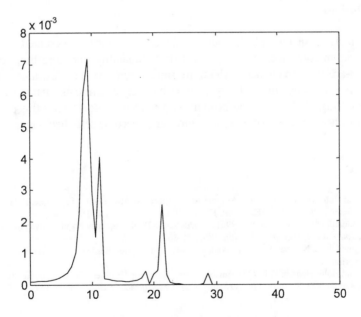

Figure 9. Transient Frequency Spectrum of MFDR

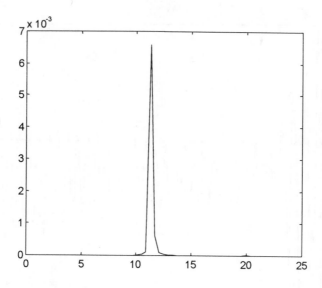

Figure 10. Steady Frequency Spectrum of MFDR

5. Conclusions

A new dimension in multiple-frequency delayed resonance is presented as an active vibration suppression technique. Its real-time tunability for absorbing oscillatory energy at n-arbitrary frequencies which are time varying is very desirable. The other important result is that the stability of the absorber structure is guaranteed and the combined setting of absorber and primary is provided. Additionally this stability test can be performed on-line enabling the controller to become self-checking.

References

1. Olgac, N. and Holm-Hansen, B. (1994) A Novel Active Vibration Absorption Technique: Delayed Resonator, *Journal of Sound and Vibration*, **176**, 93-104.
2. Olgac, N., Elmali, H., and Vijayan, S. (1996) Introduction to Dual Frequency Fixed Delayed Resonator (DFFDR), *Journal of Sound and Vibration*, **189**, 355-367.
3. Elmali, H. and Olgac, N. (1998) Discrete Domain Analysis of Delayed Resonator, *Jornal of Vibration and Control*, in print.
4. Olgac, N., and Holm-Hansen, B. (1995) Tunable Active Vibration Absorber: The Delayed Resonator, *Transaction of the ASME, Journal of Dynamic Systems, Measurement, and Control*, **117**, 513-519.

ADVANCED MONITORING SYSTEM FOR LARGE STRUCTURAL SYSTEMS

M.L. WANG AND D. SATPATHI
Department of Civil and Materials Engineering, Bridge Research Center
University of Illinois at Chicago, Chicago, IL-60607, USA

Abstract

The concept of a health monitoring system for civil structural systems is continuously evolving as a result of new requirements and the development of new sensor technologies. The area of structural health monitoring is still rather underdeveloped, and there is a lack of consensus on what needs to be monitored and how it can be implemented. The goal of a health monitoring system is to be able to provide advance notice of impending failures, and if possible provide objective data on preventive maintenance and retrofit requirements. In this article the authors have enumerated the general requirements of a successful health monitoring system for bridges, and have provided a real life example of a bridge monitoring system, that the authors are in the process of implementing.

1. Introduction

While a lot of research work has been done in the area of disaster prevention of large structural systems, very few methods have really been applied at field. The most frequently used technique for inspection is still based on visual methods, which is carried out at periodic intervals (~ 2 years for critical bridges).[1] NDT of bridges such as Golden Gate Bridge would have to incorporate techniques that can identify local hot spots created by corrosion, overstressing, fatigue related cracking etc. Condition monitoring of a number of bridges in the United States has been carried out in the last decade by a number of groups. [2-6] Most of these monitoring schemes have been based on localized strain monitoring and nondestructive evaluation of the presence of flaws in an existing structure. But in order for such a scheme to be implemented it is necessary to identify the problem areas in the structure. Global techniques based on mode shape comparison [7-9] on the other hand tend to suffer from the lack of sufficient sensitivity. It means that unless the damage is very large, it may be difficult to detect it using global techniques. A viable system for health monitoring has to incorporate both global techniques as well as local monitoring technologies.

A program based on the following guidelines is currently in the process of being implemented by the authors in the United States:

J. Holnicki-Szulc and J. Rodellar (eds.), Smart Structures, 383–390.

i) Monitor global structural response such as deformed shape, vibrational characteristics, operational stress and strain levels etc.

ii) Detect changes in the structural behavior or structural strength. This may include detection of change in stiffness, cracks in members, spalling of concrete decks and girders, corrosion in metallic structures, yielding and fracture of metallic components of a structure etc.

iii) Identify & monitor all the appropriate environmental quantities which interact with the structure such as traffic flow, traffic pattern and traffic loading; temperature cycles, wind directions and wind speeds, hydrological characteristics which might affect the soil structure or fluid structure interaction; and seismic events.

iv) Appropriate model that will analyze all the necessary information and make a decision on the safety and reliability of a structure.

2. A Health monitoring system

The first step in the development of a health monitoring system consists of implementing a sensing array. There are two major challenges to this problem. The first one involves how to get the maximum amount of information using the least number of sensors (optimization). This is of paramount importance, since it dictates the economics of the scheme. The other major hurdle has to deal with acquiring data from a large number of sensors and involves the physical process of wiring the sensors to their data acquisition system. In order to overcome this problem the possibility of adapting wireless technology for data transfer is being explored by a number of groups. [10-11]

While collecting the appropriate data represent one part of the challenge the other part of the challenge is posed by the problem of how to utilize this large amount of date to develop a suitable warning or collapse prevention system. While researchers unanimously agree that it is important to develop some type of a Finite Element base line model of the actual structure, there isn't a whole lot of agreement on what techniques to use to make the actual predictions. For global monitoring a number of techniques such as comparing the structural dynamic characteristics, the static influence line etc has been suggested in various literatures as a way of detecting a structural softening. They work only in a limited sense in terms of their damage detection capabilities. They have been able to detect damages of catastrophic magnitude in large bridges, but do not have the ability to detect the onset of a problem, since the damage has not reached a significant stage. Under the present circumstances it is probably appropriate to utilize a judicious combination of global and local monitoring system to put in place an effective health monitoring system. An ideal system should comprise of the following:

i) A good correlated finite element model of the structure to serve as the baseline. If a new structure is being considered as a candidate, appropriate measurements should be made to obtain both static and dynamic characteristics of the system and correlate the model with the actual test data.

ii) Based on the finite element model, the probable hot spots should be identified. A health monitoring system based on local hot spot monitoring using strain gauge technology, acoustic emission and ultrasonics, force sensors etc. should be employed for hot spot monitoring. At the same time a global monitoring system based on measurement of acceleration response or displacement response be put together along with sensors for monitoring the necessary environmental parameters.

iii) While the necessary models for global damage detection are not fully developed, a combination of current global damage detection techniques in conjunction with data from local structural monitoring sensors and environmental parameters collected from the appropriate sensors can be packaged together. Changes in the global damage detection parameters will have to be matched against data from localized sensors and environmental sensors. Lack of discrepancy between the different sensor groups will have to be resolved initially by using expert bridge inspection engineers and will form the basis of developing an expert system to resolve conflicts between the different set of sensors, and eventually make assessments of safety and reliability of the structure. In the event of a situation where the expert system is unable to resolve a situation it will recommend an immediate detailed manual inspection, during which process new hot spots may be discovered and will have to be monitored. Also any time a change in the structure is detected and identified, it will be necessary to incorporate the changes in the Finite Element model.

iv) If a scenario is deemed dangerous or has the potential of turning into a serious problem, appropriate retrofitting and repair should be undertaken.

3. A proposed health monitoring system

In the majority of situations, a health monitoring system is implemented as a forensic and investigative tool to identify the source of an unexpected behavior pattern experienced by a structure, and make recommendations for retrofitting. In the post retrofitting stage, the same system can be utilized for monitoring the efficiency of the retrofit. Therefore, depending on the nature of the problems, a health monitoring system has to be tailored towards a structure. For example, the health monitoring system configuration for a cable-stayed bridge could be different from that of a prestressed concrete box bridge. The authors are members of the Bridge Research Center at the University of Illinois, and are currently in the process of assisting the Illinois Department of Transportation (IDOT) in implementing a health monitoring system for a post-tensioned concrete segmental box bridge [Fig1. and Fig 2.]. In this case, the structure is believed to be suffering from insufficient shear capacity, and in consequence, the bridge is experiencing extensive shear cracking at the supports and at the shear keys [Fig 3.]. Temporary retrofit measures for increasing the shear capacity along the shear keys had been undertaken with only partial success. A proposed health monitoring plan for the structure is in the process of being implemented. The proposed scheme plans to identify the nature of the mechanism producing this large shear forces, predict the remaining useful service life of the structure, and assess the risk of operating the structure in it's current structural state. The first phase of the study, which is designed to investigate the actual performance of the structure, consists of the following tasks:

I.   Modal testing and analysis
II.   Fracture mechanics based investigation of crack growth and failure prediction.
III.   Investigation of corrosion.

IV.   Measurement of actual prestressing forces and revaluation of the bridge design
V.   Load testing and load rating.

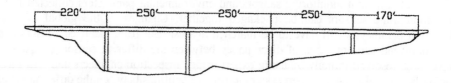

Figure 1. Longitudinal view of the post tensioned concrete box Kishwakee Bridge.

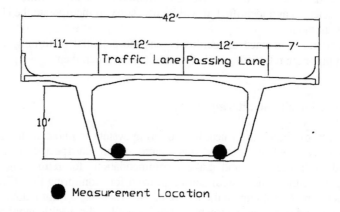

● Measurement Location

Figure 2. Cross-section of the box showing acceleration measurement locations.

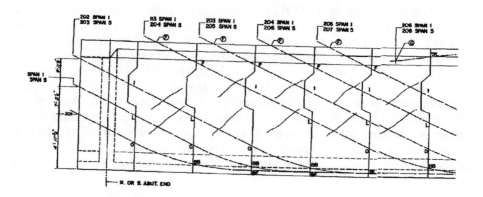

Figure 3. Typical shear cracks observed at Kishwakee Bridge.

3.1 MODAL TESTING AND ANALYSIS

Modal tests are routinely used in the automobile industry and the aerospace industry for identifying manufacturing defects and identifying potentially undesirable vibrational modes. In the case of the Kishwakee Bridge, it is apparent that the shear capacities of the sections are inadequate. It is however necessary to identify the sources of the shear forces. A part of the shear force is generated due to the beam effect, which in all likelihood had been accounted for in the design process. If the section is subjected to torsion, the shear stresses experienced by the sections will increase dramatically. The first objective of a modal test is thus to identify the presence of torsional modes. If these modal frequencies are in the vicinity of the ambient excitation frequencies produced by the traffic loading, we can conclude that torsional modes are responsible for the generation of shear stresses in excess of the design capacity. In order to accomplish this task we have proposed a three phase plan.

- i) Finite element modeling of the structure and numerical modal analysis.
- ii) Identification of the first ten modes of the structure (mode shape, modal frequencies, and modal damping).
- iii) Spectral characterization of traffic loading.

Preliminary tests have been carried out on the bridge. Four accelerometers with extended low frequency response (Dytran 3187B1 series, nominal sensitivity~1 V/g) were placed at the segment adjacent to the midspan segment of span 2 (between pier 1 and 2). The spectrum from the four sensors was recorded as a heavy vehicle passed on the deck overhead. The frequency averaged response from two such tests are shown in Fig 5 and Fig 6. It can be seen that the spectrums are not identical. This is because only those frequencies that are present in the input will be visible in the structural response. This introduces some amount of subjectivity in the data analysis. One way of overcoming this will be to carry out frequency domain averaging of the signal from any given location for a large number of occurrences. An incomplete modal analysis had been carried out by Construction Technologies Laboratory Incorporation (CTL) in 1987. Comparison of the modal frequencies as shown in Table 1 reveals that the first modal frequency has reduced by around 30 %. This implies that the structure has seen significant reduction in its stiffness since the last time it was tested.

TABLE 1 Comparison of modal frequencies

Mode number	Test set1 (BRC, 1998)	Test set 2 (BRC, 1998)	CTL test data (1987)	% change in data (approximate)	CTL Finite Element results
1	1.18	1.1	1.65-1.70	-32	1.67
2	1.87	1.77	2.10-2.15	-14	1.94
3	2.64	2.65	2.70-2.75	-3	2.29

388

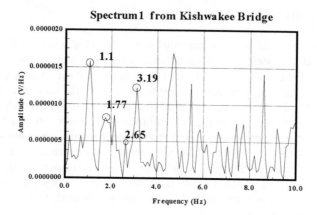

Figure 4. Frequency averaged spectrum response (1) from Kishwakee Bridge (ambient excitation due to normal traffic loading).

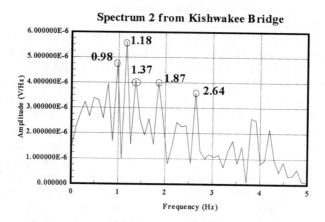

Figure 5. Frequency averaged spectrum response (2) from Kishwakee Bridge (ambient excitation due to normal traffic loading).

3.2. FRACTURE MECHANICS BASED PREDICTION OF CRACK GROWTH

The combined crack tip strain distribution in conjunction with data on the rate of crack opening and crack propagation will allow us to calculate certain critical fracture mechanics parameters. Based on these parameters, we can then model the crack growth, which can then be utilized to predict how much a certain crack will grow in a given time. We can also establish a loading criterion, which can lead to a unstable crack propagation, leading to collapse. Based on the frequency of occurrence of a given loading pattern we can then use a probabilistic model for calculating the chances of a complete collapse in a local segment.

3.3. INVESTIGATION OF CORROSION

Failure of prestressed wires or tendons occurs as a result of cross-sectional loss due to conventional electrochemical corrosion (or rusting) or stresses corrosion cracking. Both types of corrosion occur in the presence of water, and stress corrosion cracking can occur without the presence of chlorides or other contaminants. Water can enter the system as a result of several conditions:

- Exposure of a bare strand during shipping or storage.
- Improper protection of tendon
- Damage to tendons following placement in forms.
- Insufficient protection at end anchors.
- Improper grouting procedures.
- Cracks or porosity in the concrete.
- Insufficient concrete covers.

Procedures for investigating the condition of prestressed structures are significantly different from those used for conventionally reinforced structures. Generally, little external evidence of deterioration is apparent. Occasional cases of strands erupting through anchorage's or slab surfaces cannot be depended on to provide an indication of trouble. Because a prestressed structure can approach a state of critical structural inadequacy without cracking or deflection, it is incumbent on the engineer to develop an understanding of the condition of the post-tensioning system. A thorough inspection will answer several questions: a)Has moisture entered the system and if so, how? b) Is corrosion accelerating substances present? c) Is corrosion occurring? d) Are wires or strands broken?

3.4 MEASUREMENT OF ACTUAL PRESTRESSING FORCES

As a structure gets older the prestressing forces in the strands reduce (due to a number of reasons). The design codes provides guidelines for calculating prestress losses, but very often, the actual prestress losses may be up to 10% higher than predicted by the design equations. As prestress loss occurs, the load carrying capacity of the structure reduces, coupled with an increase in operating deflections due to a reduction in stiffness. It is therefore important to be able to measure the actual prestressing forces in the tendons in order to redesign and reevaluate the load carrying capacity of the bridge. Unfortunately, that is not possible on most bridges, because of the lack of suitable technology for measuring the total stress in a tendon. We are however proposing a novel magneto-elastic sensor, which can measure the strand forces without having to remove the polyethylene ducts. The sensor is designed on the principle that the hysteresis loop of a ferromagnetic material such as steel is dependent on its internal stress state.

4. Conclusions

In this article the authors have put forward the notion that a health monitoring system has to be devised on a case by case basis, and has to be tailored for a structure with the requirements and concerns of the organization responsible for the structure in mind. The authors have used a real life bridge as an example case study. Currently, the authors' group is in the process of designing and implementing the health monitoring program that has been discussed. The bridge is a post-tensioned segmental box bridge, with a total length of around 300 m and spans over the Kishwakee river on Interstate 39 in the state of Illinois in USA.

Acknowledgements

This research is currently being funded by two agencies, the National Science Foundation (grant no. 9622576), and the Office of the Vice Chancellor of Research (OVCR), University of Illinois at Chicago.

References

1. Silano, L. G. (1992), Bridge Inspection and Rehabilitation-A Practical Guide, Wiley InterScience, NY.
2. Shahawy, M. and Arockiasamy, M. (May 1996), "Field Instrumentation to study the time dependent behavior in sunshine skyway bridge. I", *ASCE Journal of Bridge Engineering*, pp.76-86.
3. Shahawy, M. and Arockiasamy, M. (May 1996), " Analytical and measured strains in Sunshine Skyway Bridge. II", *ASCE Journal of Bridge Engineering*, pp.87-97.
4. Carlyle, J. M. (1993) "Acoustic emission monitoring of the 1-10 Mississippi River bridge." *Report.* Prepared for U.S. Dept. of Transportation, Federal. Highway Administration, Washington, D.C., USA.
5. Maji, A. K., Satpathi, D. and Kratochvil, T. (1997), "Acoustic emission source location using lamb wave modes", *ASCE Journal of Engineering Mechanics*, v.123, no. 3, pp. 154-161.
6. Weischedel, H. R. and Hohle, H. W. (1995), "Quantitative nondestructive evaluation of stay cables of cable stayed bridges: methods and practical experience", in Chase, S. (ed.), *Nondestructive Evaluation of Aging Bridges and Highways, SPIE*, v. 2456, pp. 226-236.
7. Farrar, C.R., and K.M. Cone, (1995), "Vibration testing of the 1-40 bridge before and after the introduction of damage," *Proceedings of 13th International Modal Analysis Conference*, pp. 203-209.
8. Heo, G., Wang, M. L., and Satpathi, D., (in press) "A health monitoring system for large structural systems", *International Journal of Smart Materials and Structures*, USA.
9. Chen, J. C. and Garba, J. A. (1988), "On-Orbit Damage Assessment for Large Space Structures," *AIAA Journal*, 26 (9), pp.1119 - 1126.
10. Neuzil, P., Serry, F. M., Krenek, O. and Maclay, G. J. (1997), "An integrated circuit to operate a transponder with embeddable MEMS microsensors for structural health monitoring", *Proceedings of Structural Health Monitoring: Current Status and Perspectives,* Chang, F. K.(ed), pp.492-501, Technomic Publishing, Lancaster, USA.
11. Kiremidjian, A. S., Straser, E. G., Meng, T., Law, K., and Soon, H. (1997), "Structural damage monitoring for civil structures", *Proceedings of Structural Health Monitoring: Current Status and Perspectives,* Chang, F. K.(ed), pp.492-501, Technomic Publishing, Lancaster, USA.

SUBJECT INDEX